Bioremediation

Other Environmental Engineering Books from McGraw-Hill

American Water Works Association
WATER QUALITY AND TREATMENT

Chopey
ENVIRONMENTAL ENGINEERING FOR THE CHEMICAL PROCESS INDUSTRIES

Corbitt
STANDARD HANDBOOK OF ENVIRONMENTAL ENGINEERING

Freeman
HAZARDOUS WASTE MINIMIZATION

Freeman
STANDARD HANDBOOK OF HAZARDOUS WASTE TREATMENT AND DISPOSAL

Jain
ENVIRONMENTAL IMPACT ASSESSMENT

Levin and Gealt
BIOTREATMENT OF INDUSTRIAL AND HAZARDOUS WASTE

McKenna and Cunneo
PESTICIDE REGULATION HANDBOOK

Majumdar
REGULATORY REQUIREMENTS OF HAZARDOUS MATERIALS

Nanney
ENVIRONMENTAL RISKS IN REAL ESTATE TRANSACTIONS

Seldner and Cothrel
ENVIRONMENTAL DECISION MAKING FOR ENGINEERING AND BUSINESS MANAGERS

Waldo and Hines
CHEMICAL HAZARD COMMUNICATION GUIDEBOOK

Bioremediation

Katherine H. Baker
Environmental Microbiology Associates, Inc.
Harrisburg, Pennsylvania

Diane S. Herson
School of Life and Health Sciences
University of Delaware
Newark, Delaware

Environmental Microbiology Associates, Inc.
Harrisburg, Pennsylvania

McGraw-Hill, Inc.

New York San Francisco Washington, D.C. Auckland Bogotá
Caracas Lisbon London Madrid Mexico City Milan
Montreal New Delhi San Juan Singapore
Sydney Tokyo Toronto

Library of Congress Cataloging-in-Publication Data

Bioremediation / [edited by] Katherine H. Baker, Diane S. Herson.
    p.   cm.
    ISBN 0-07-003360-9
    1. Bioremediation.   I. Baker, Katherine H.   II. Herson, Diane S.
TD 192.5.B5567   1994
628.5′2—dc20   93-41457
                                                                    CIP

    2 3 4 5 6 7 8 9 0   DOC/DOC   9 0 9 8 7 6 5 4

ISBN 0-07-003360-9

The sponsoring editor for this book was Gail F. Nalven, the editing supervisor
was Joseph Bertuna, and the production supervisor was Suzanne W. Babeuf. It
was set in Palatino by McGraw-Hill's Professional Book Group composition
unit.

Printed and bound by R. R. Donnelley & Sons Company.

This book is dedicated to Blut and Steve for their support and love.

Contents

Contributors

Katherine Devine DEVO Enterprises, Inc., Washington, D.C. (CHAP. 10)

D. D. Hale Technology Applications Inc., Athens, Ga. (CHAP. 8)

Steven N. Liss School of Chemical Engineering, Ryerson Polytechnic University, Toronto, Ontario, Canada (CHAP. 9)

Carol D. Litchfield Chester Environmental, Monroeville, Pa. (CHAP. 7)

J. E. Rogers Environmental Research Laboratory, U.S. Environmental Protection Agency, Athens, Ga. (CHAP. 8)

Craig Rosenbaum-Wilkinson Lamont-Doherty Earth Observatory, Columbia University, NYC (CHAP. 3)

George J. Skladany ENVIROGEN, Inc., Lawrenceville, N.J. (CHAP. 4)

Marleen A. Troy OHM Remediation Services Corp., Princeton, N.J. (CHAP. 5)

D. A. Wabah Department of Biology, Towson State University, Towson, Md. (CHAP. 8)

Preface

Bioremediation involves a multidisciplinary approach encompassing many aspects of science, from macro-geological formations whose development may be plotted in eons to ephemeral microbiological events that may be virtually instantaneous. Because of this broad range, and because of the relative newness of the field, information vital to bioremediation is scattered throughout the corpus of scientific endeavor.

The purpose of *Bioremediation*, therefore, is twofold. First, it is intended as an introduction to the field, with fundamental information regarding both the background and the practice of bioremediation for the graduate student or professional coming to it from a grounding in a particular branch of science. Second, and perhaps even more important, this book should serve as an entry to the literature in the field; the chapter references are as significant as the chapter contents themselves.

In soliciting contributors, we have sought the participation of scientists who are all involved in the practical implementation of bioremediation. They are not academics only, but people with field experience.

In addition to the writers of the articles comprising this book, thanks are due to others whose talents and interests have been brought to bear on its shaping and polishing. Pamela Vercellone-Smith and Michael Waddington read the material and made invaluable suggestions. McGraw-Hill's sponsoring editor, Gail Nalven, believed in this project and encouraged us to carry it through. Our editing supervisor, Joe Bertuna, did a patient and thorough job of harmonizing disparate styles and clarifying the sometimes recondite material. Black Bear Productions assisted in the often tedious proofreading during the final stages of production. Desktop Design Associates designed and produced graphics illustrating concepts and disciplines which they had not been familiar with previously. R. E. Wright Associates graciously provided technical drawings. Finally, our families endured months of our distraction, frustration, and abstraction with as much patience and good humor as humanly possible.

Katherine H. Baker

Diane S. Herson

Bioremediation

1

Introduction and Overview of Bioremediation

Katherine H. Baker

*Environmental Microbiology
Associates, Inc
Harrisburg, Pa.*

Diane S. Herson

*School of Life and Health Sciences
University of Delaware
Newark, Del.*

One of the major problems facing the industrialized world today is the contamination of soils, ground water, sediments, surface water, and air with hazardous and toxic chemicals. The National Priority List currently contains over 1200 sites, with potential sites numbering over 32,000. In addition to these, it is estimated that a significant number of the over 7 million underground storage tanks in the United States are leaking (Kovalick, 1991). While regulatory steps have been implemented to reduce or eliminate the production and release to the environment of these chemicals, significant environmental contamination has occurred in the past and will probably continue to occur in the future.

The need to remediate these sites has led to the development of new technologies that emphasize the detoxification and destruction of the contaminants rather than the conventional approach of disposal. *Bioremediation,* the use of microorganisms or microbial processes to detoxify and degrade environmental contaminants, is among these new technologies. Although bioremediation is viewed as a new technology, microorganisms have been used routinely for the treatment and transformation of waste products for at least 100 years. The municipal waste water treatment industry is based on the exploitation of microorganisms in controlled and engineered systems. Both activated sludge and fixed-film treatment systems depend on the metabolic activities of microorganisms which degrade the wastes entering the treatment facility. Specialized waste treatment plants containing selected and acclimated populations of microorganisms are often used to treat industrial effluents (Eckenfelder, 1989).

What is new is the application of microbiological processes to the remediation of soils, ground water, and similar environmental media. These systems differ from traditional waste water treatment systems in terms of the types of chemicals which are being degraded and the matrix in which the degradation may be occurring. Sites are frequently contaminated with complex mixtures of organic compounds such as creosote, fuel oils, or combinations of industrial solvents. The concentrations of individual contaminants may vary significantly within the site, with some areas having extremely low concentrations of the materials present and other "hot spots" having concentrations over 1 million–fold higher. Often, inorganic wastes such as metals are also present. Superimposed on the heterogeneity of waste composition is the nature of the contaminated media. Unlike municipal waste treatment systems, such as activated sludge treatment facilities, in which conditions are optimized to ensure homogeneity and uniform mixing, bioremediation frequently must address multiphasic, heterogeneous environments such as soils in which the contaminant is present in association with the soil particles, dissolved in soil liquids, and in the soil atmosphere.

Because of these added complexities, successful bioremediation is dependent on an interdisciplinary approach involving such disciplines as engineering, microbiology, ecology, geology, and chemistry. Thus bioremediation often depends on assembling the right team of experts to evaluate and design the appropriate treatment system.

Bioremediation technologies can be broadly classified as ex situ or in situ. Table 1-1 summarizes the most commonly used bioremediation technologies. *Ex situ technologies* are those treatment modalities which

Table 1-1. Bioremediation Treatment Technologies

Bioaugmentation	Addition of bacterial cultures to a contaminated medium; frequently used in bioreactors and ex situ systems
Biofilters	Use of microbial stripping columns to treat air emissions
Biostimulation	Stimulation of indigenous microbial populations in soils and/or ground water; may be done in situ or ex situ
Bioreactors	Biodegradation in a container or reactor; may be used to treat liquids or slurries
Bioventing	Method of treating contaminated soils by drawing oxygen through the soil to stimulate microbial growth and activity
Composting	Aerobic, thermophilic treatment process in which contaminated material is mixed with a bulking agent; can be done using static piles, aerated piles, or continuously fed reactors
Landfarming	Solid-phase treatment system for contaminated soils; may be done in situ or in a constructed soil treatment cell

involve the physical removal of the contaminated material to another area (possibly within the site) for treatment. Bioreactors, landfarming, composting, and some forms of solid-phase treatment are all examples of ex situ treatment techniques. In contrast, *in situ techniques* involve treatment of the contaminated material in place. Bioventing for the treatment of contaminated soils and biostimulation of indigenous aquifer microorganisms are examples of these treatment techniques.

Regardless of the exact nature of the treatment technology, all bioremediation techniques depend on having the right microorganisms in the right place with the right environmental conditions for degradation to occur. The right microorganisms are those bacteria or fungi which have the physiological and metabolic capabilities to degrade the contaminants. In many instances, these organisms will already be present at the site (indigenous microorganisms). In other circumstances, such as bioreactors treating wastes with high concentrations of toxic materials, there may be a need to add exogenous microorganisms to the material. In order for the microorganisms to degrade the contaminants, they must be in close proximity to the contaminants; they must be in the right place. Thus the presence of toluene-degrading microorganisms in the surface soils at a site will be of little use for the remediation of a contaminated aquifer. In many instances, the presence of a contaminant which is biodegradable will have resulted in an enrichment for microorganisms capable of degrading it in the contaminated area. If such populations are not present, then some mechanism must be engineered to bring the microorganisms into contact with the contaminants. This may involve such techniques as flushing the system to transport the contaminants to an above-ground bioreactor, the addition of surfactants to the subsurface to release adsorbed contaminants and render them available to the microorganisms, or the introduction and transport of the microorganisms to the contaminated area. Once the right microorganisms are present in the right place, the environmental conditions must be controlled or altered to optimize the growth and metabolic activity of the microorganisms. Such environmental factors as temperature, inorganic nutrients (primarily nitrogen and phosphorus), electron acceptors (oxygen, nitrate, and sulfate), and pH can be modified to optimize the environment for bioremediation.

Bioremediation offers several advantages over such conventional remediation techniques as landfilling and pump and treat. Table 1-2 summarizes the advantages of bioremediation compared with conventional remediation technologies. Often, bioremediation can be done on site, thereby eliminating transportation costs and liabilities. Because bioremediation can often be applied in an in situ fashion, site disruption can be minimized. In many instances, manufacturing and industrial use of the site can continue while the bioremediation process is implemented. Bioremediation results in the decomposition of the waste materials, often to carbon dioxide and water, thus permanently eliminating the waste and the long-term liability associated with nondestructive treatment methods. Finally, bioremediation can be coupled with other treatment technologies into a treatment train allowing for the treatment of mixed, complex wastes.

Like any treatment technology, bioremediation also has its limitations and disadvantages. Some chemicals, e.g., highly chlorinated

Table 1-2. Advantages of Bioremediation

Can be done on site

Keeps site disruption to a minimum

Eliminates transportation costs and liabilities

Eliminates waste permanently

Eliminates long-term liability

Biological systems, often less expensive, are used

Can be coupled with other treatment techniques into a treatment train

Table 1-3. Chemical Classes and Their Susceptibility to
Bioremediation

Chemical class	Examples	Biodegradability
Aromatic hydrocarbons	Benzene, toluene	Aerobic and anaerobic
Ketones and esters	Acetone, MEK	Aerobic and anaerobic
Petroleum hydrocarbons	Fuel oil	Aerobic
Chlorinated solvents	TCE, PCE	Aerobic (methan-otrophs), anaerobic (reductive dechlorina-tion)
Polyaromatic hydrocarbons	Anthracene, benzo[a]pyrene, creosote	Aerobic
PCBs	Arochlors	Some evidence; not readily degradable
Organic cyanides		Aerobic
Metals	Cadmium	Not degradable; experimental biosorption
Radioactive materials	Uranium, plutonium	Not biodegradable
Corrosives	Inorganic acids, caustics	Not biodegradable
Asbestos		Not biodegradable

compounds and metals, are not readily amenable to biological degradation. Table 1-3 summarizes general categories of contaminants and
their relative susceptibility to biodegradation. In addition, for some
chemicals, microbial degradation may lead to the production of more
toxic or mobile substances than the parent compound. For example,
under anaerobic conditions, trichloroethylene (TCE) undergoes a
series of microbiologically mediated reactions resulting in the sequential removal of chlorine atoms from the molecule. This process is called
reductive dehalogenation. The end product of this series of reactions is
vinyl chloride (VC), a known carcinogen. Thus, if bioremediation is
applied without a thorough understanding of the microbial processes
involved, it could actually lead to a worse situation than already exists
in some cases. Bioremediation is a scientifically intensive procedure
which must be tailored to the site-specific conditions. Therefore, initial
costs for site assessment, characterization, and feasibility evaluation
for bioremediation may be higher than the costs associated with more

Table 1-4. Sources of Information Concerning Bioremediation
Computer Databases

Alternative Treatment Technology Information Center (ATTIC)
EPA database on innovative treatment technologies. Provides information on
treatment technologies, treatability, sources of technical assistance, and a calendar of conferences, etc. relating to hazardous waste treatment. In addition,
ATTIC has a special interest group (SIG) electronic bulletin board dedicated to
bioremediation.
System operator: (301) 670-6294   Online access: (301) 670-3808

Cleanup Information Bulletin Board (CLU-IN)
Formerly known as the Office of Solid Waste and Emergency Response
(OSWER) Bulletin Board. Contains information on innovative technologies for
the remediation of Superfund and RCRA corrective action sites. Numerous
special interest group areas including Ground Water, and Innovative
Technologies.
System operator: (301) 589-8368   Online access: (301) 589-8366

EPA's Computerized On-Line Information Service (COLIS)
Free EPA on-line service for information on site cleanups. Currently consists
of four modules: (1) case history files, (2) library search system, (3) site appli-
cations analysis reports, and (4) RREL treatability database.
System operator: (908) 906-6851   Online access (908) 548-4636   Password
EPA.

Vendor Information System for Innovative Treatment Technologies (VISITT)
Free EPA database ($5\frac{1}{4}$-in or $3\frac{1}{2}$-in diskettes, DOS Version 3.3 or higher).
Contains information on innovative technologies (bioremediation, thermal
desorption, chemical treatment, and soil washing) and companies which supply the technology. Information is based on information supplied by vendors.
Information is not reviewed or certified for accuracy by the EPA. To obtain
copies of the database, call VISITT Hotline: (800) 245-4505 or (703) 883-8448.

---

conventional technologies such as air stripping. As with all remediation technologies, there is the need for extensive monitoring of the site
during implementation of the project. Monitoring requirements may
include some form of microbiological monitoring in addition to the
chemical monitoring associated with physical/chemical remediation
techniques. Finally, there are regulatory constraints which impact on
the implementation of bioremediation. Bioremediation projects must
meet all the requirements of the environmental regulations applicable
to the site and contaminant being treated. Depending on the site, the
Resource Conservation and Recovery Act (RCRA), the Comprehensive
Environmental Response, Compensation and Liability Act (CERCLA),
and the National Pollution Discharge Elimination System (NPDES)

permits as well as applicable state and local permits may be required. In addition, if microorganisms, particularly genetically modified organisms, are to be added to the site, the Toxic Substances Control Act (TSCA) and the Federal Plant Pest Act (FPPA) approval also may be required (Baskt, 1991).

Because bioremediation involves the integration of several scientific and engineering disciplines, information regarding bioremediation is widely scattered throughout the technical literature. Journals and books in the fields of chemical engineering, environmental engineering, and civil engineering, as well as hydrogeology, analytical chemistry, toxicology, and microbiology, are all likely sources of information. In addition to these journals and books, there are a number of scientific meetings (usually with published proceedings) which include information on bioremediation. Finally, several electronic databases are available through the Environmental Protection Agency (EPA) which can supply useful information. Table 1-4 presents information on these sources of information.

References

Baskt, J. S. 1991. Impact of present and future regulations on bioremediation. *J. Indust. Microbiol.* **8:** 13–22.

Eckenfelder, W. W., Jr. 1989. *Industrial Water Pollution Control.* McGraw-Hill Publishing Company, New York.

Kovalick, W. W., Jr. 1991. Removing Impediments to the Use of Bioremediation and Other Innovative Technologies, in *Environmental Biotechnology for Waste Treatment,* G. S. Sayler, R. Fox, and J. W. Blackburn (eds.). Plenum Press, New York, pp. 53–60.

2

Microbiology and Biodegradation

Katherine H. Baker

Environmental Microbiology
Associates, Inc
Harrisburg, Pa.

Diane S. Herson

School of Life and Health Sciences
University of Delaware
Newark, Del.

Bioremediation ultimately depends on the activities of microorganisms. In this chapter we will present a brief overview of microbial physiology, metabolism, genetics, and ecology as they relate to the implementation of bioremediation. This review is by no means inclusive; the reader is referred to standard microbiology texts for further information (5, 8, 30, 116). Because the majority of bioremediation systems currently in use are bacterial, emphasis will be placed on this group of microorganisms.

Degradation of organic materials in natural environments is mediated primarily by two groups of microorganisms: bacteria and fungi. Bacteria represent a widely diverse group of prokaryotic organisms with ubiquitous distribution throughout the biosphere. Bacteria are found in all environments containing living organisms. This does not imply, however, that all strains of bacteria are found in all environments. Bacteria are small (typically between 1 and 10 μm) and morphologically simple, lacking the internal membrane-enclosed organelles typical of eukaryotic organisms such as fungi, protozoa, algae, plants, and animals. Biochemically, however, bacteria show amazing metabolic versatility.

Bacteria have several characteristics which have allowed them to be a successful group of organisms. These characteristics—rapid growth and metabolism, genetic plasticity, and the ability to adjust rapidly to a variety of environments—are also the factors which make microorganisms so useful in bioremediation.

Rapid Growth and Metabolism

Bacteria increase in number using a process termed *binary fission*. In this process, one bacterial cell divides to form two cells. The two daughter cells subsequently divide to form four cells. The amount of time necessary for a bacterial cell to divide, and hence for the population of bacteria to double, is termed the *generation time*. Bacteria have short generation times, typically measured in terms of minutes, hours, days, or weeks. Since the number of bacteria in the population doubles with every generation, bacterial numbers can increase at extraordinarily fast rates. Mathematically, the growth of a bacterial population under nonlimiting conditions can be expressed in terms of an exponential function. If we assume that a bacterial population has a generation time of 30 minutes, starting from a single bacterial cell, numbers could increase to over 10 million cells in 12 hours.

This type of exponential growth of bacterial populations does not continue unchecked for long periods of time. A bacterial population growing in this manner will rapidly alter its environment. It may deplete it of necessary nutrients or produce inhibitory metabolic by-products, ultimately causing population growth to stop. Studies of bacterial growth, determined on the basis of batch culture systems, result in a bacterial growth curve similar to that illustrated in Figure 2-1.

This growth curve can be divided into four phases of growth. Initially, the bacterial population shows a period of little or no increase in bacterial numbers. This phase of growth is called the *lag phase*. During the lag phase, microorganisms are synthesizing the molecules necessary for growth and replication, inducing necessary metabolic enzymes, and adapting to new growth conditions. The length of the lag phase will depend on the prior growth conditions of the bacterial population and on the size of the initial inoculum. The more the microbes

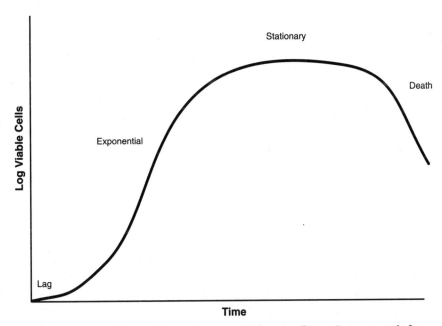

Figure 2-1. Typical bacterial growth curve. This type of growth curve, with four distinct phases, is associated with bacteria growth in batch cultures.

must do to adjust to the new conditions, the longer is the lag. After the lag phase, the population enters the *exponential phase* of growth. Here, the population is increasing at its maximal rate. It is during the exponential phase that the population size will double as a function of the generation time. As microbial numbers increase, exhaustion of the available nutrients or the accumulation of inhibitory by-products slows the growth rate of the population, and it enters the *stationary phase*. During the stationary phase, the growth rate of the population is matched by the death rate of the population, and hence there is no net increase in the number of cells. Finally, as nutrient depletion or the accumulation of toxic metabolic by-products continues, death outstrips growth, and the population numbers and activity decline.

It is impossible to determine the exact growth pattern of bacterial populations in the natural world. In contrast to laboratory-grown batch cultures, microbes in the environment grow under less than ideal conditions. Under such conditions, bacterial population numbers may not increase dramatically over time due to some type of environmental limitation to their growth. Laboratory batch culture studies, however, are important in that they give us an appreciation of the potential for rapid growth.

Microbial growth in the natural environment can be limited by numerous factors. Biotic interactions such as competition between different bacterial strains for the same substrate or predation can function to keep population numbers low (2, 9). For example, Goldstein et al. (51) concluded that protozoan predation on a strain of *Pseudomonas* capable of degrading 2,4-dichlorophenol was responsible for the failure of the introduced organisms to persist in nonsterile sewage.

In addition, microbial population size may be limited by the abiotic environment. Probably the most important abiotic factors limiting bacterial populations from the perspective of bioremediation are (1) water content, (2) temperature, (3) pH, (4) the presence of toxic materials such as metals, (5) the type and amount of organic material (carbon) present, (6) electron acceptors, and (7) inorganic nutrients such as nitrogen and phosphorous.

Like all cells, bacteria are dependent on an adequate supply of water to grow and reproduce. More important than the absolute amount of water present in the environment is the water availability. Water may not be available to microorganisms because it is adsorbed to solid substances and surfaces within the environment (hydrated clays) or because of the presence of high levels of solutes (hypersaline lakes). Therefore, the use of water activity a_w or water potential ψ, which are measures of the physiologically available water, are preferable to measures of total water content in evaluating the environmental effects of water on microorganisms (55, 114). In general, microorganisms need water activities between 0.900 and 1.000 to survive (30). Although there are exceptions, such as the halophile *Halobacter*, which is capable of growing in the Dead Sea ($a_w = 0.700$) (93), water activity below 0.900 typically limits bioremediation.

Biological processes typically increase with increasing temperature up to a maximal temperature above which enzyme denaturation leads to cell inhibition and death. In general, the temperature response of microorganisms shows a decidedly skewed pattern, with maximal activity at a temperature only slightly below the lethal temperature. Bacteria as a group show a wide tolerance range for temperature, with organisms being reported to be able to grow at temperatures from below 0°C to above 100°C (provided water is available) (9). Microorganisms are classified as *psychrophiles* (optimal temperatures between 5 and 15°C), *mesophiles* (optimal temperatures between 25 and 40°C), or *thermophiles* (optimal temperatures between 40 and 60°C). Within the tolerable temperature ranges, microbial activity usually increases by a factor of 2 to 3 for each 10°C increase in temperature up to the optimal temperature of the microorganism.

Although the majority of bioremediation projects have been conducted under mesophilic conditions, the ability to degrade a variety of contaminants has been found in psychrophilic (95, 139), mesophilic (6), and thermophilic (80, 136) microorganisms. For example, a strain of *Corynebacterium* isolated from petroleum-contaminated soils in Antarctica was shown to be capable of active hydrocarbon degradation at 1°C (65). On the other hand, Williams et al. (133) have demonstrated that thermophilic microorganisms can degrade explosives in soil compost systems operating at 55°C. Thus microorganisms capable of degrading at least some types of organic contaminants would be expected to be present in extremely cold environments such as the arctic tundra, in more temperate soils and ground water, and in high-temperature environments such as composting systems.

With the exception of the acidophilic organisms such as *Thiobacillus ferrooxidans*, found in acid mine drainage, microorganisms generally are limited to pH values ranging from 6.0 to 8.0. Dibble and Bartha (38) have reported a soil pH of 7.8 as being optimal for the microbial degradation of petroleum hydrocarbons in soils. Fungi are slightly more tolerant of acidic conditions than are most common soil and aquifer bacteria. Thus, under acidic conditions, soil fungi have an advantage over bacteria. The metabolic pathways for degradation are different in fungi than in bacteria. Fungal decomposition of such contaminants as polyaromatic hydrocarbons may lead to the production of intermediates which are mutagenic (32). Soil liming can be used to increase soil pH, if necessary, favoring bacterial growth.

Perhaps more important than the direct effects of pH on microorganisms, the hydrogen ion concentration of the environment will have profound effects on the soil or ground water chemistry. The availability of macronutrients, especially phosphorous, and the mobility of potentially toxic metals are all highly pH dependent (27). In general, an increase in the pH of the soil environment has been found to decrease the availability of calcium, magnesium, sodium, potassium, ammonia, nitrogen, and phosphorous, while a decrease in soil pH results in decreasing availability of nitrate and chloride (112). Presumably these effects are the result of alterations in the binding of these materials to soil surfaces.

Contaminated sites frequently are impacted by mixtures of waste materials instead of a single compound. In such sites, the presence of high levels of metals in addition to the organic compounds can be inhibitory or toxic to microorganisms. While some metals are necessary as trace elements for microorganisms, elevated levels of metals can disrupt the cell membrane and denature cellular proteins, leading to cell death (20, 43). In addition, microorganisms also can be involved in the transformation and sorption of metals in the environment. Sulfate-reducing bacteria have been implicated in the transformation of inorganic mercury to methylmercury in lake sediments (50). Methylmercury is the form of mercury which accumulates in fish populations and so can be transferred via the food chain to humans, where it is a potent neurotoxin. Lovely and Phillips (78) have shown that *Desulfovibrio desulfuricans* cells can enzymatically reduce uranium in contaminated waste waters, allowing for the recovery of the uranium as a relatively pure precipitate. *Biosorption* involves the sorption of metals to biomass. This is different from enzymatic reduction and precipitation.

In order to grow and reproduce, microorganisms need the proper molecules from which to synthesize new cells. On a total-weight basis, the most common compound in a bacterial cell is water, accounting for between 70 and 85 percent of the cell weight. On a dry-weight basis, approximately 95 percent of the compounds in a bacteria cell are composed of five elements—carbon, oxygen, nitrogen, hydrogen, and phosphorous.

A major element necessary for cell synthesis, in terms of the amount needed, is carbon. In general, microorganisms can either use carbon dioxide as a source of carbon or use organic forms of carbon such as glucose, benzene, or other compounds. Microorganisms which use carbon dioxide as a source of carbon are referred to as *autotrophs*; those which use organic carbon are called *heterotrophs*. In addition to a source of carbon, microorganisms also need a source of energy to power the synthesis of new cellular components.

Organisms which get their energy from photosynthetic reactions with light as the source of energy are called *phototrophs*. Those which get their energy from the oxidation of either inorganic or organic chemicals are called *chemotrophs*. By combining the two requirements (carbon source and energy source), it is possible to classify microorganisms into physiological/nutritional groups. Table 2-1 summarizes this type of classification.

It is the chemoheterotrophic microorganisms which are principally responsible for the degradation of organic contaminants in the environment. Chemoheterotrophs encompass a large group of organisms containing numerous genera and species. Within the chemoheterotrophs, some species of microorganisms are quite limited in the organic compounds which they can metabolize, while others are extremely versatile. For example, some species of microorganisms, such as lactic acid bacteria, are limited to a very small number of sim-

Table 2-1. Nutritional Classification of Microorganisms

Group	Carbon source	Energy source
Photoautotrophs	Carbon dioxide	Light
Photoheterotrophs	Organic carbon	Light
Chemoautotrophs	Carbon dioxide	Inorganic chemical (e.g.,NH_4)
Chemoheterotophs	Organic carbon	Organic carbon

ple organic compounds, while others, such as members of the genus *Pseudomonas,* are capable of metabolizing any of over 90 different organic compounds as a sole source of carbon and energy (116).

Microorganisms break down organic compounds for carbon and energy through a complex series of coupled oxidation-reduction reactions termed *catabolism.* In this process, electrons are removed from and added to intermediates along the pathway. The energy released in these reactions is conserved in the form of the high-energy phosphate bond of ATP for use in fueling biosynthetic reactions. The biochemical processes central to this process (catabolism) can be divided into two major groups: fermentation and respiration. These can be distinguished primarily on the basis of the terminal electron acceptor (the final compound which is reduced) (Table 2-2).

In the case of *fermentation,* the electron donor is an organic compound and the terminal electron acceptor is an organic compound also. The process of fermentation does not lead to a complete oxidation of all the original substrate to carbon dioxide. Fermentative processes are characterized by the production of a mixture of end products, some of which are more oxidized and some of which are more reduced than the initial substrate. While fermentative pathways are extremely

Table 2-2. Major Metabolic Sequences

Type of metabolism	Electron donor	Terminal electron acceptor
Fermentation	Organic compound	Organic compound
Respiration	Organic or inorganic	Inorganic compound
Aerobic respiration		Oxygen (e.g., O_2)
Anaerobic respiration		Nitrate, sulfate (e.g., NO_3, SO_4)

important in industrial microbiology, where they are used for the production of organic acids, fermentative metabolism has not been shown to be significant in the bioremediation of contaminated environments.

Respiratory metabolism is much more important in bioremediation. Respiration involves the oxidation of an organic compound coupled with the terminal reduction of an inorganic compound. When oxygen is used as a terminal electron acceptor, the organisms are said to be carrying out *aerobic respiration*. When other electron acceptors such as nitrate, sulfate, or carbon dioxide are used as terminal electron acceptors, the organisms are said to be carrying out *anaerobic respiration*. Aerobic organisms that can carry out both processes will preferentially use oxygen as a terminal electron acceptor. Upon depletion of the oxygen, the alternative acceptor will be used. The energy obtained by the cell when it is using oxygen as a terminal electron acceptor is greater than that when alternative acceptors are used. This is due to the differences in reduction potentials between the electron donor and the terminal electron acceptor. Oxygen is the most oxidizing, followed by nitrate, sulfate, and carbon dioxide. Because aerobic respiration is generally more efficient than anaerobic respiration, aerobic systems are usually preferred for bioremediation.

Organisms vary in their relationships to oxygen. *Obligate aerobes* require oxygen and use it as a terminal electron acceptor when they carry out respiration. *Obligate anaerobes* cannot grow in the presence of oxygen. Organisms in this group carry out fermentation or anaerobic respiration. *Facultative anaerobes* can grow in the presence or absence of oxygen. Among the facultatives are organisms which carry out aerobic respiration, anaerobic respiration, and fermentation. *Aerotolerant anaerobes,* or *indifferent organisms,* can grow in the presence of oxygen but do not use it as a terminal electron acceptor. *Microaerophilic organisms* require a reduced oxygen partial pressure.

The amount of oxygen present in a soil or ground water system is determined by the difference between those processes which consume oxygen (*sinks*) and those processes which introduce oxygen (*sources*) into the environments. Biological consumption of oxygen is probably the largest sink in these environments, although oxygen also can be removed by abiotic mechanisms such as the formation of iron oxides. Oxygen enters the soil and subsurface environment primarily through the process of diffusion. Typically, diffusion of oxygen into soils is a slow process. At sites where microbial activity is high, biological losses of oxygen may exceed the rate of diffusion, leading to the development of anoxic conditions. In ground water environments, oxygen availability is further limited by the low solubility of oxygen in water.

The energy in the form of ATP obtained from the oxidation-reduction reactions of catabolism is used by the cell for biosynthetic (*anabolic*) reactions. In these reactions, the major types of molecules found in cells—proteins, lipids, carbohydrates, and nucleic acids—are synthesized. In addition to carbon, the elements nitrogen and phosphorous, referred to as *macronutrients*, are necessary in large amounts for the synthesis of some of these molecules.

Nitrogen is necessary for the synthesis of proteins and nucleic acids. Proteins serve both structural and enzymatic (catalytic) functions in the bacterial cell. Nucleic acids (DNA and RNA) function as the genetic information of the cell. In addition, some microorganisms use nitrogen in the form of nitrate as a terminal electron acceptor during the process of anaerobic respiration. Nitrogen is present in the environment in a variety of forms. The majority of nitrogen in the biosphere is in the form of nitrogen gas (N_2). A limited number of bacterial species, called *nitrogen-fixing bacteria*, have the ability to reduce nitrogen gas to ammonia. Nitrogen-fixing bacteria such as *Bradyrhizobium* sp. are found as symbionts in the root nodules of such plants as clover and soybeans.

Microorganisms involved in the degradation of xenobiotics in natural systems are primarily dependent on fixed forms of nitrogen (ammonia, nitrate, nitrite, and organic nitrogen) to meet their nitrogen requirements. These forms of nitrogen are frequently limiting for microbial populations in soils, ground water, and surface waters (7).

Phosphorous is essential to microbial cells for the synthesis of ATP, nucleic acids, and cell membranes. Because of the low water solubility of phosphates, phosphorous is frequently limiting to bacterial growth in both aquatic and soil environments (8).

Using a generalized formula for the composition of a typical bacterial cell, researchers have proposed optimal carbon-to-nitrogen and carbon-to-phosphorous ratios for determining if the concentration of these macronutrients present at a site is sufficient to allow for the complete degradation of the contaminants. Published ratios show a wide variation, with reported values ranging from 200:1 to 10:1 for carbon-to-nitrogen ratios and 1000:1 to 100:1 for carbon-to-phosphorous ratios (38, 103). It is evident from these values that carbon-to-nitrogen and carbon-to-phosphorous ratios should be used with extreme caution in evaluating the inorganic nutrient status of an environment.

Genetic Plasticity

The ability of a microorganism to degrade an organic compound is the result, ultimately, of the genetic makeup of the organism. The chemical reactions involved in metabolism are mediated by enzymes. These molecules are composed of proteins and function as biological catalysts by increasing the rate of chemical reactions in the cell. The range of enzymes which a bacterium has is a reflection of the specific genetic information in that cell.

Genetic information in bacteria, as in all organisms, is stored in the form of DNA. This information is physically present in bacterial cells in two forms—the chromosome and the plasmids. The bacterial chromosome is a single, circular, highly folded double strand of DNA. In addition to chromosomal DNA, a large number of bacteria also have extrachromosomal DNA in the form of plasmids.

Many plasmids contain genes which code for the enzymes necessary for degradative pathways important to bioremediation. Enzymes involved in the degradation of toluene, naphthalene, salicylate, camphor, octane, xylene, 3-chlorobenzoate, 2,4-dichlorophenoxyacetic acid, chloroalkanotes, chlorosalicylate, and p-chlorobiphenyl have been shown to be plasmid encoded (19, 88).

Plasmids are self-replicating. In addition, plasmid-derived DNA can be exchanged between bacteria through the processes of conjugation, transformation, and transduction. Thus DNA from one strain of bacteria can be transferred to another strain, under the appropriate conditions. Such a transfer might confer on the recipient bacterium the genes necessary for degradation of a contaminant. While the rates and extent of such genetic exchanges have not been firmly established in natural systems, genetic exchange has been documented to occur in a variety of natural environments (14, 62, 66, 97, 109).

Plasmids are also important in the development of new microorganisms with enhanced degradative capability. Using molecular biology techniques, it is possible to splice pieces of DNA containing genes for specific degradative pathways into plasmids. These plasmids can then be introduced into a host organism, resulting in a recombinant or genetically engineered microorganism (GEM) with new degradative capabilities (28, 30, 52). These microorganisms may be useful in the bioremediation of contaminated sites either as organisms for bioaugmentation or in bioreactors (41, 79, 100).

Response to the Environment

A bacterial cell contains within its DNA all the information for protein synthesis which it potentially can express. In order for the information encoded in the DNA to be used by the cell, it must first be transcribed into RNA, which is then translated into protein. This transfer of information from DNA to RNA (*transcription*) and from RNA to proteins (*translation*) is called the *central dogma* and was first proposed by Watson and Crick in the 1950s.

Only a portion of the genome will be expressed at any given time. For example, it has been estimated that there cannot be more that 4300 protein-encoding genes in the chromosome (87). When *Escherichia coli* is growing on glucose, approximately 800 enzymes are present in the cell, although the genome, as noted above, has the information for the synthesis of many more (101).

Some proteins, representing what are called *constitutive enzymes*, are constantly synthesized by the cell. The enzymes involved in glucose catabolism fall into this class. Therefore, these enzymes will be synthesized even when glucose is not present in the cell's environment.

This is not the case for many other potential carbon and energy sources whose utilization is ultimately determined by activities at the level of transcription via the action of operons. *Operons* are regions of DNA consisting of genes that code for enzymes and control regions that determine if these genes will be transcribed. In the case of *induction*, the presence of a utilizable substrate in the environment of the cell will result in the synthesis of enzymes, called *inducible enzymes*, which can degrade the substrate. In the case of *repression*, the presence of a specific compound in the environment will serve to stop the transcription of the genes which code for enzymes involved in the biosynthesis of a particular compound such as an amino acid, purine, or pyrimidine. In the case of induction, the genetic information in the cell is normally not expressed. In the case of repression, the genetic information is normally expressed. Environmental conditions may block or allow expression of the genetic information. Thus regulation, while reflecting environmental cues, is a result of the active regulation of gene expression by the cell and allows the cell to respond to its environment.

In both cases it can be seen how these mechanisms ultimately conserve energy for the cell. To continually synthesize enzymes involved in the degradation of a substrate that is usually absent from the cell's environment would be wasteful. It also would be inefficient for the cell to continue to synthesize an amino acid that is provided in the environment.

Enzyme induction is important in the regulation of pathways for the degradation of organic compounds. Enzymes involved in the cleavage of aromatic compounds, octane degradation (alkane hydrolase and alcohol dehydrogenase), naphthalene degradation, and salicylate degradation are regulated by operons (19). Frequently, the initial substrate in the pathway has been found to induce the enzymes necessary for degradation.

Other control mechanisms also exist. In *feedback inhibition,* the activity, as opposed to the synthesis, of enzymes involved in metabolic pathways is inhibited. *Catabolite repression,* like induction and repression, involves transcriptional control. In this control mechanism, the synthesis of a degradative enzyme is inhibited when glucose is present. This can result in diauxic growth, where a cell preferentially uses one growth substrate, such as glucose, over another, such as lactose. Once the glucose has been depleted from the environment, the lactose will be utilized after a short lag period. Cellular activities such as these may come into play in field situations. For example, Swindoll et al. (119) have reported that diauxic growth may be important in regulating the microbial degradation of xenobiotic compounds. They found that the addition of glucose or amino acids to aquifer samples contaminated with toluene, ethylene dibromide, phenol, and *p*-nitrophenol inhibited mineralization of the contaminant, presumably as a result of preferential use of the easily degradable substrate.

The rate of degradation of organic compounds by microbial populations and communities is increased when the microorganisms have been preexposed to the chemical. This phenomenon, called *adaptation,* has been shown to occur in a variety of natural environments, including freshwater, soils, and ground water systems (1, 15, 99, 115, 134). The exact mechanism responsible for adaptation is not known. Such factors as enzyme induction, genetic alteration of the population, or physiological acclimation to stress conditions have all been proposed as important in the phenomenon of adaptation (110, 131,137). Since industrial sites often have a history of chronic exposure to low-levels of contaminants, it is likely that adaptation of the microorganisms at the site will have already occurred prior to initiation of a bioremediation project. Therefore, the indigenous microorganisms should be capable of a more rapid degradation of contaminants than microorganisms from a pristine site.

Metabolic Pathways for the Degradation of Xenobiotics

As mentioned previously, bacteria break down complex organic compounds through a series of coupled chemical reactions termed *catabolism*. For respiratory bacteria, the central pathways involved in metabolism are glycolysis (Fig. 2-2), the Krebs cycle (Fig. 2-3), and the electron transport system (Fig. 2-4). Figure 2-5 summarizes the relationship between these three pathways. The overall generalized reaction is

$$CH_2O \text{ (complex carbon)} + O_2 \rightarrow CO_2 + H_2O + \text{energy}$$

This equation represents the complete conversion of the complex carbon compound, also called the *substrate,* to carbon dioxide and water, a process called *mineralization.* Energy attained in the form of ATP is used for a variety of cellular activities, including synthesis of new cell components and motility.

Microbial degradation of complex organic compounds does not always result in mineralization. Incomplete degradation, also called *transformation,* of the compound may occur as a result of microbial activity. In some instances, microbial activities can transform the parent compound into one that is a greater environmental concern than the original compound. As a example of this type of metabolism, the anaerobic transformations of trichloroethylene (TCE) can lead to the accumulation of vinyl chloride, a carcinogen, in the environment. Microbial transformation of chlorinated organic compounds such as lindane has recently been proposed as a contributor to elevated levels of halogenated anisoles in the atmosphere (127). In addition, microorganisms may produce surface-active substances, called *bioemulsifiers* or *surfactants,* which do not degrade the contaminants but which may increase their mobility and bioavailability. This is of particular concern when dealing with such compounds as polyaromatic hydrocarbons (e.g., benzo[*a*]pyrene) which are typically tightly bound to soil humic and other organic fractions. Because of these partial transformations and influences on compound mobility and solubility, studies on biodegradation which measure metabolism only through parent compound disappearance may lead to misleading results.

In order for a compound to be mineralized, it must be able to be converted into one of the compounds involved in the central metabolic pathways. The diversity of mechanisms by which bacteria can achieve this is astonishing. In many instances, bacterial mineralization of a compound is known to occur, but the exact reactions in the pathway have not been fully elucidated. In addition, an appreciation has

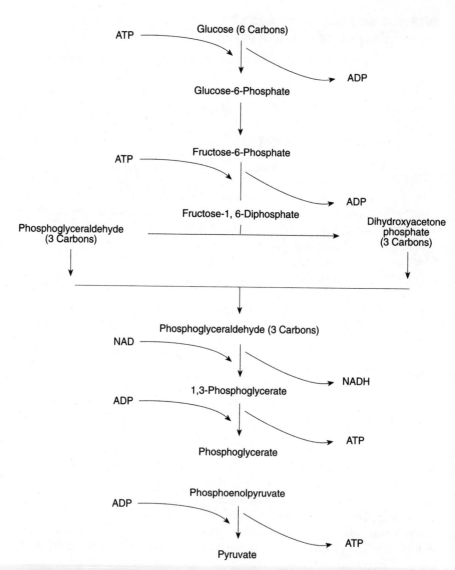

Figure 2-2. Glycolysis. Glycolysis, or the Embden-Meyerhof pathway as it is also termed, results in the conversion of one molecule of glucose into two molecules of pyruvate. In addition, there is a gain to the cell of two ATP and two NADH. ATP is generated by the conversion of ADP to ATP. During this process, energy is stored in the form of a high-energy phosphate bond. This energy subsequently can be used by the cell for biosynthetic reactions. NADH, more correctly written as NADH + H⁺, functions to carry hydrogens and electrons. These can be used for biosynthetic reactions or carried to the electron transport system, where additional ATP is generated.

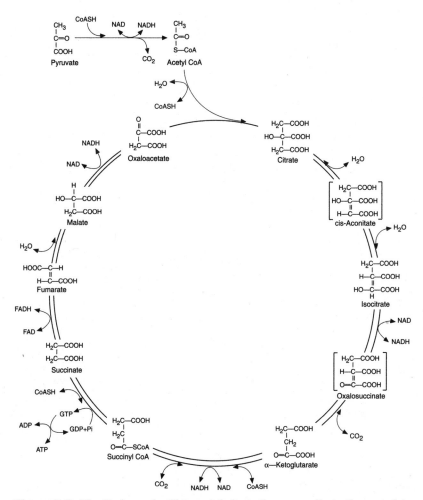

Figure 2-3. The Krebs cycle. This metabolic pathway results in the complete oxidation of one molecule of pyruvate to three molecules of carbon dioxide. Energy, in the form of high-energy phosphate bonds, is stored during the conversion of ADP to ATP. In addition, reduced coenzymes NADH and FADH are generated. Conversion of a xenobiotic compound into any of the intermediates (e.g., succinate, citrate, etc.) involved in the Krebs cycle allows for the complete oxidation of the compound.

recently developed for the role of mixed microbial populations (consortia) in degradation in the natural environment. Thus, while a single strain of microorganism may not be capable of carrying out the mineralization of a compound, it may be able to partially transform the

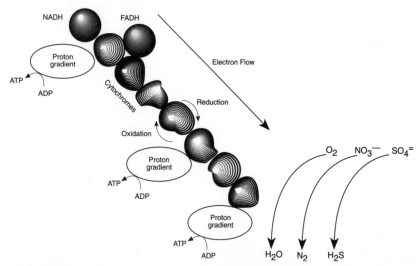

Figure 2-4. Electron transport system. During electron transport, hydrogens (one proton and one electron) from reduced coenzymes are donated to a series of carrier molecules (e.g., cytochromes) which are sequentially reduced and oxidized. During this process, protons are translocated to the outside of the cell membrane. Reentry of the protons into the cell is coupled with the conversion of ADP to ATP. The electrons originally donated by the reduced coenzymes are ultimately transferred to a terminal electron acceptor which is reduced. Under aerobic conditions, oxygen functions as the terminal electron acceptor and is reduced to water. Under anoxic or anaerobic conditions, other inorganic compounds such as nitrate or sulfate can function as terminal electron acceptors. The specific carrier molecules in the electron transport system and the terminal electron acceptor depend on the specific bacterium.

compound into a product that can serve as a metabolic substrate for a second, third, or fourth group of microorganisms.

Recent evidence has demonstrated that for many compounds of environmental concern, cometabolic reactions are important in degradation. *Cometabolism,* or *co-oxidation* as it is sometimes termed, is the transformation of a nongrowth substrate by a microorganism. For example, microorganisms which are not capable of growing on chlorinated aromatic compounds may be able to carry out limited transformations of these compounds in the presence of the nonchlorinated analogues. This phenomenon is most easily explained as a case of fortuitous degradation resulting from a low specificity of degradative enzymes. Thus the structural similarity between the xenobiotic compound (which is cometabolized) and the metabolic substrate may be sufficient for enzymes with low specificity to use

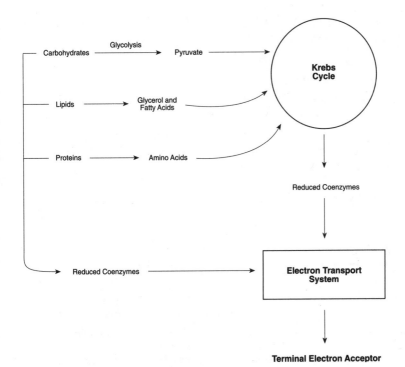

Figure 2-5. Central metabolic pathways. The interrelationships between the central metabolic pathways—glycolysis, Krebs cycle, and electron transport system—are illustrated here. The combined functioning of these pathways results in the complete oxidation of organic compounds to carbon dioxide, water, and inorganic compounds (such as chloride in the case of chlorinated solvents). In addition, energy in the form of ATP is stored by the cell for the production of additional biomass.

both as a substrate, resulting in the transformation and partial degradation of the xenobiotic (37). Cometabolic reactions are particularly important in the transformation and degradation of halogenated organic compounds.

In this section, the major pathways involved in the degradation of several of the classes of chemicals which are frequently of environmental concern will be reviewed. The reader interested in more detail concerning the biochemistry and microbial ecology of degradation is referred to several recent books (4, 33, 46, 71, 74, 105) and review articles (3, 6, 11, 34, 35, 53, 72, 73, 83, 98, 117), as well as the current microbiological literature.

Petroleum Hydrocarbons

Petroleum and petroleum products such as gasoline, fuel oils, and diesel fuels are complex mixtures of organic compounds. Gasoline, for example, contains over 100 different substances. The majority of the compounds found in petroleum products are hydrocarbons. Table 2-3 summarizes the major components of different petroleum products.

Aliphatic hydrocarbons are straight or branched chains of carbon

Table 2-3. Composition of Petroleum Products

Product	Major components
Gas	Normal and branched-chain alkanes. One to five carbons in length. Examples: ethane, propane.
Gasoline	Normal and branched hydrocarbons between 6 and 10 carbons in length. Cycloalkanes and alkylbenzenes are also present.
Kerosene/diesel fuel no.1	Primarily 11 and 12 carbon hydrocarbons. Both normal and branched hydrocarbons are present. Normal alkanes usually predominate. Cycloalkanes and aromatic and mixed aromatic cycloalkanes also present. Generally contain low to nondetectable levels of benzene and polyaromatic hydrocarbons. Jet fuel oils have a similar composition.
Light gas oils	Twelve to 18 carbon hydrocarbons. Lower percentage of normal alkanes than kerosene. Cycloalkanes. Aromatic and mixed aromatic cycloalkanes. Olefins and mixed aromatic olefins such as styrenes. These products include diesel and furnace fuel oils (no. 2 fuel oil).
Heavy gas oils and light lubricating oils	Hydrocarbons between 18 and 25 carbons long.
Lubricants	Hydrocarbons between 26 and 38 carbons in length.
Asphalts	Heavy polycyclic compounds.

SOURCE: Nyer and Skladany, 1989.

atoms with sufficient hydrogen to satisfy the valency requirements of the carbons. They have the empirical formula C_nH_m. Depending on the degree of saturation (number of carbon-carbon bonds), aliphatic hydrocarbons can be classified as alkanes, alkenes, and alkynes (102). Figures 2-6a, 2-6b, and 2-6c illustrate typical hydrocarbons. Aromatic compounds represent a separate class of hydrocarbons. These compounds are based on the familiar benzene ring structure C_6H_6. Before

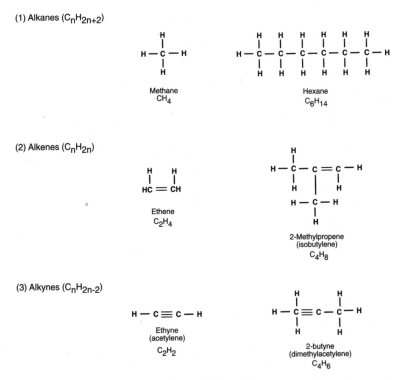

Figure 2-6a. Structures of hydrocarbons. Hydrocarbons are a class of organic compounds composed of carbon and hydrogen. There are three major categories of hydrocarbons: aliphatic hydrocarbons, alicyclic hydrocarbons, and aromatic hydrocarbons. Aliphatic hydrocarbons (a) are straight or branched chains of carbon atoms with sufficient hydrocarbons to satisfy the valency of the carbon atoms. Aliphatic hydrocarbons are further subclassified into alkanes (a-1), alkenes (a-2), and alkynes (a-3) depending on the presence of double and triple bonds within the carbon chain.

(b) Alicyclic Hydrocarbons

(1) Cycloalkanes (C_nH_{2n})

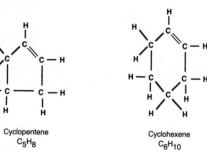

Cyclopropane
C_3H_6

Cyclohexane
C_6H_{12}

(2) Cycloalkenes (C_nH_{2n-2})

Cyclopentene
C_5H_8

Cyclohexene
C_6H_{10}

(3) Cycloalkynes (C_nH_{2n-4})

Cyclopropyne
C_3H_2

Figure 2-6b. (*Continued*) Structures of hydrocarbons. Hydrocarbons are a class of organic compounds composed of carbon and hydrogen. There are three major categories of hydrocarbons: aliphatic hydrocarbons, alicyclic hydrocarbons, and aromatic hydrocarbons. Alicyclic hydrocarbons (*b*) are characterized by the presence of a carbon ring structure. As with aliphatic hydrocarbons, they are further subdivided into cycloalkanes (*b*-1), cycloalkenes (*b*-2), and cycloalkynes (*b*-3).

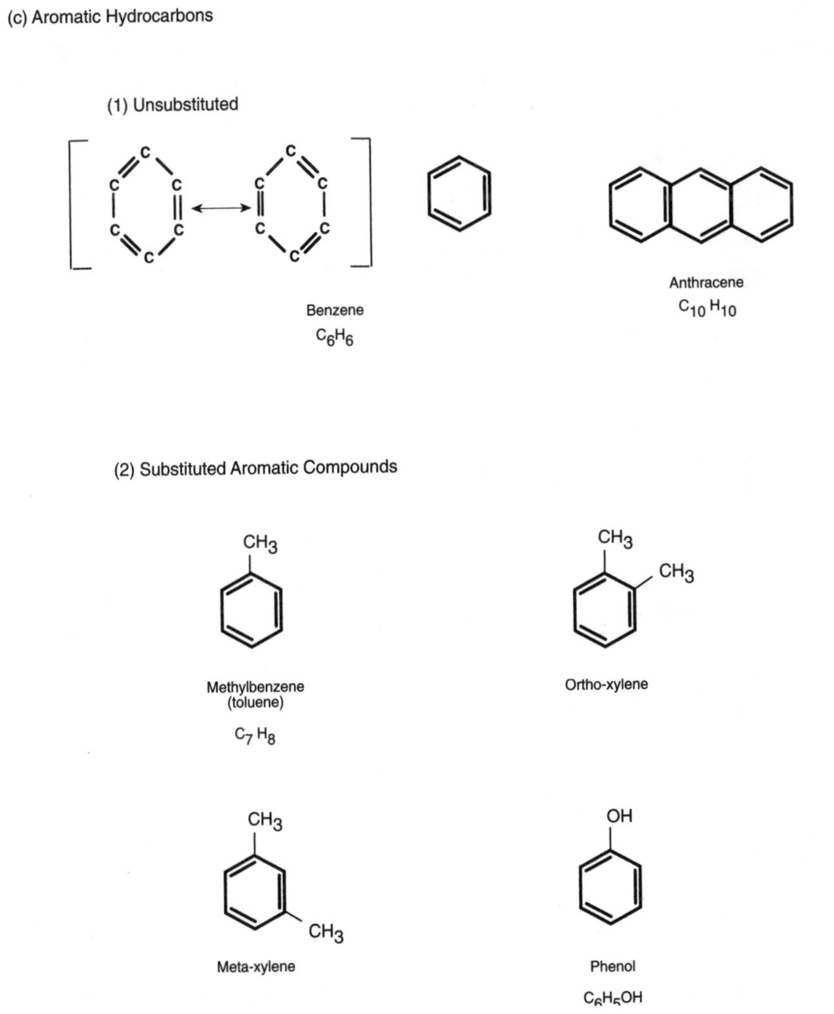

(c) Aromatic Hydrocarbons

(1) Unsubstituted

Benzene
C_6H_6

Anthracene
$C_{10}H_{10}$

(2) Substituted Aromatic Compounds

Methylbenzene
(toluene)
C_7H_8

Ortho-xylene

Meta-xylene

Phenol
C_6H_5OH

Figure 2-6c. (*Continued*) Structures of hydrocarbons. Hydrocarbons are a class of organic compounds composed of carbon and hydrogen. There are three major categories of hydrocarbons: aliphatic hydrocarbons, alicyclic hydrocarbons, and aromatic hydrocarbons. Aromatic compounds (*c*) are characterized by the presence of one or more resonance-stabilized six-carbon rings. Unsubstituted aromatic compounds include benzene and polyaromatic hydrocarbons such as anthracene (*c*-1). In addition, a number of substituted aromatic compounds (*c*-2) are important pollutants.

reviewing the pathways involved in the degradation of these compounds, there are several generalizations which should be kept in mind. These can serve as useful "rules of thumb" in a first-order evaluation of the biodegradability of a petroleum hydrocarbon.

1. Aliphatic hydrocarbons are generally easier to degrade than aromatic compounds.

2. Straight-chain aliphatic hydrocarbons are easier to degrade than branched-chain hydrocarbons. The introduction of branching into the hydrocarbon molecule hinders biodegradation.

3. Saturated hydrocarbons are more easily degraded than unsaturated hydrocarbons. The presence of carbon-carbon double or triple bonds hinders degradation.

4. Long-chain aliphatic hydrocarbons are more easily degraded than short-chain hydrocarbons. Hydrocarbons with chain lengths of less than 9 carbons are difficult to degrade because of their toxicity to microorganisms. Some specialized microorganisms (methanotrophs) can degrade these short-chain hydrocarbons. The optimal chain length for biodegradation appears to be between 10 and 20 carbons (29).

Straight-chain alkanes are degraded primarily through oxidation of the terminal methyl group, followed by cleavage of the molecule between the second and third carbon in the chain (β cleavage). Other pathways, such as the subterminal oxidation by methane monooxygenase found in *Ps. methanica,* also have been documented (29). Figures 2-7a and 2-7b illustrate hydrocarbon degradation by terminal methyl group oxidation. The initial reaction in the degradation involves the direct addition of oxygen to the terminal carbon of the hydrocarbon. Hence there is a requirement for the presence of molecular oxygen in order for the reaction to occur. This reaction is mediated by a class of enzymes called *oxygenases.* The addition of oxygen to the terminal carbon results in the formation of a primary alcohol which is subsequently oxidized to the corresponding aldehyde and finally to the acid, forming a fatty acid. A two-carbon-long fragment is ultimately cleaved from the fatty acid, providing the intermediate (acetyl-CoA) which can enter the central metabolic pathways and a hydrocarbon of chain length C_{n-2}. Sequential repetition of this series of reactions ultimately results in complete oxidation of the hydrocarbon molecule.

The presence of branching within the alkane molecule will prevent

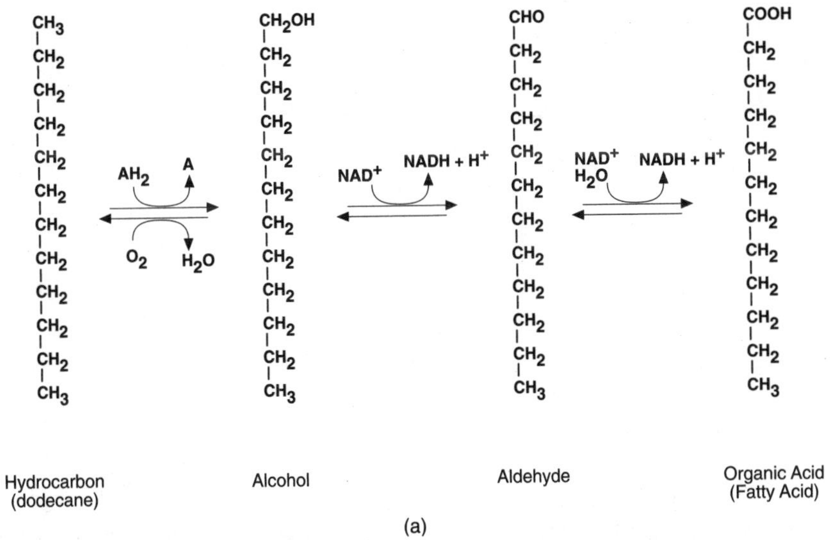

Hydrocarbon (dodecane)			Alcohol			Aldehyde			Organic Acid (Fatty Acid)
CH_3			CH_2OH			CHO			COOH
CH_2			CH_2			CH_2			CH_2
CH_2			CH_2			CH_2			CH_2
CH_2			CH_2			CH_2			CH_2
CH_2	AH_2	A	CH_2	NAD^+	$NADH + H^+$	CH_2	NAD^+	$NADH + H^+$	CH_2
CH_2			CH_2			CH_2	H_2O		CH_2
CH_2			CH_2			CH_2			CH_2
CH_2	O_2	H_2O	CH_2			CH_2			CH_2
CH_2			CH_2			CH_2			CH_2
CH_2			CH_2			CH_2			CH_2
CH_2			CH_2			CH_2			CH_2
CH_3			CH_3			CH_3			CH_3

(a)

Figure 2-7a. Degradation of *n*-alkanes. (*a*) Straight-chain (*n* or normal) alkanes are most commonly degraded by initial oxidation of the terminal methyl (CH_3) group to an organic (fatty) acid. This initial oxidation is the result of a three-step reaction sequence in which the hydrocarbon is first oxidized to form an alcohol, the alcohol is oxidized to an aldehyde, and finally, the aldehyde is converted into an organic (fatty) acid. Molecular oxygen (O_2) is required for the first reaction in this sequence.

the β cleavage reactions and render the molecule refractory to biodegradation. Some highly branched alkanes such as pristane (2,6,10,14-tetramethylpentadecane) are extremely resistant to microbial degradation and are frequently used as markers in biodegradation studies, although recent evidence has indicated that there are pathways by which the oxidation of branched alkanes can occur (29). For example, Rontani and Giusti (106) demonstrated direct oxidation of the third carbon atom leading to the degradation of branched alkanes.

Oxidation of unsaturated aliphatic hydrocarbons is not as straightforward as degradation of the alkanes. Most studies on alkene degradation have focused on molecules which contain a double bond in the C-1 position. Little is known about the pathways for degradation of molecules with internal double bonds. The presence of a double bond at the C-1 position on the molecule allows for several mechanisms of

(b)

Figure 2-7b. (*Continued*) Degradation of *n*-alkanes. (*b*) The fatty acid formed is then degraded via a metabolic pathway known as ß-oxidation, in which two carbon units are cleaved sequentially from the molecule, resulting in the formation of acetyl-CoA and an fatty acyl-CoA molecule which is two carbon atoms shorter than the parent molecule. The acetyl-CoA formed can be further metabolized via the Krebs cycle. The fatty acyl-CoA molecule serves as a substrate for further ß-oxidation.

attack, and multiple modes of alkane degradation within the same organism have been demonstrated (29).

Evidence for the anaerobic degradation of aliphatic hydrocarbons is equivocal. Although anaerobic degradation of aliphatic hydrocarbons has been reported, it is generally considered to be of limited significance in the degradation of petroleum hydrocarbons in the natural environment (6, 7).

Aromatic hydrocarbons such as benzene, toluene, ethylbenzene, and xylenes are found predominantly in light petroleum fractions such as gasoline, although they may be present in trace amounts in any type of petroleum product. In addition, these compounds are widely used as industrial solvents and intermediates in chemical production. These compounds are relatively soluble in water (1780 ppm at 20°C for benzene, for example) and hence can be transported away from spill sites, resulting in extensive soil and ground water contamination. In addition, because of their small molecular size and low boiling point, these compounds can volatilize at environmental temperatures and as such may be a source of air as well as soil and ground water contamination. The aromatic hydrocarbons are highly toxic, and in addition, they have mutagenic and hence potentially carcinogenic properties.

Aerobic degradation of aromatic hydrocarbons by bacteria was first demonstrated in the early 1900s (48). Since these pioneering studies, numerous investigators have examined the biochemistry and genetics of aromatic hydrocarbon degradation. These studies have demonstrated the ability to aerobically degrade aromatic compounds to be widespread among bacterial (and fungal) genera. There are a large number of different pathways which can be used by bacteria in the degradation of aromatic compounds.

Benzene is degraded by conversion first into either catechol or protocatechuate (Figs. 2-8a and 2-8b). The aromatic nucleus in these com-

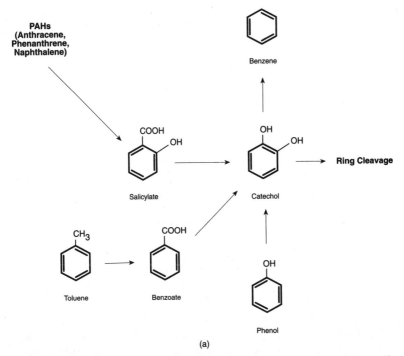

(a)

Figure 2-8a. Initial degradation of benzene. Under aerobic conditions, the majority of aromatic compounds (benzenes and substituted benzenes) are converted into catechol (a) or protocatechuate (b) before oxidative ring cleavage. Other pathways such as the homogentisate and gentisate pathways (not illustrated) may occur; however, these are not as common as the initial transformations to catechol and protocatechuate.

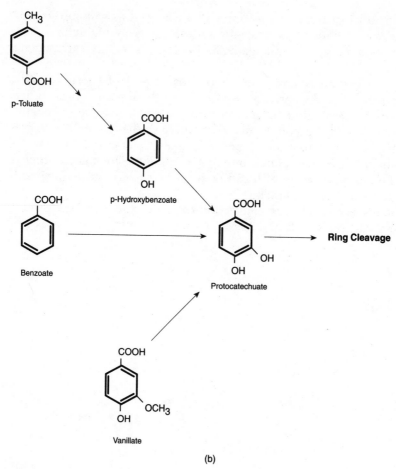

Figure 2-8b. (*Continued*) Initial degradation of benzene.

pounds is subsequently opened by one of two pathways—the *ortho-cleavage* (3-oxoadipate pathway) or the *meta-cleavage pathway*. Figures 2-9 and 2-10 summarize these pathways.

The ortho pathway involves cleavage of the aromatic nucleus of catechol or protocatechuate between the two hydroxyl groups. This leads to the formation of the respective muconates and muconolactones, which are further metabolized to 4-oxoadipate enol-lactone and then to 3-oxoadipate (β-ketoadipate). Metabolism finally proceeds to the Krebs cycle intermediates acetyl-CoA and succinate. In meta cleavage, on the other hand, initial ring cleavage occurs adjacent to the two

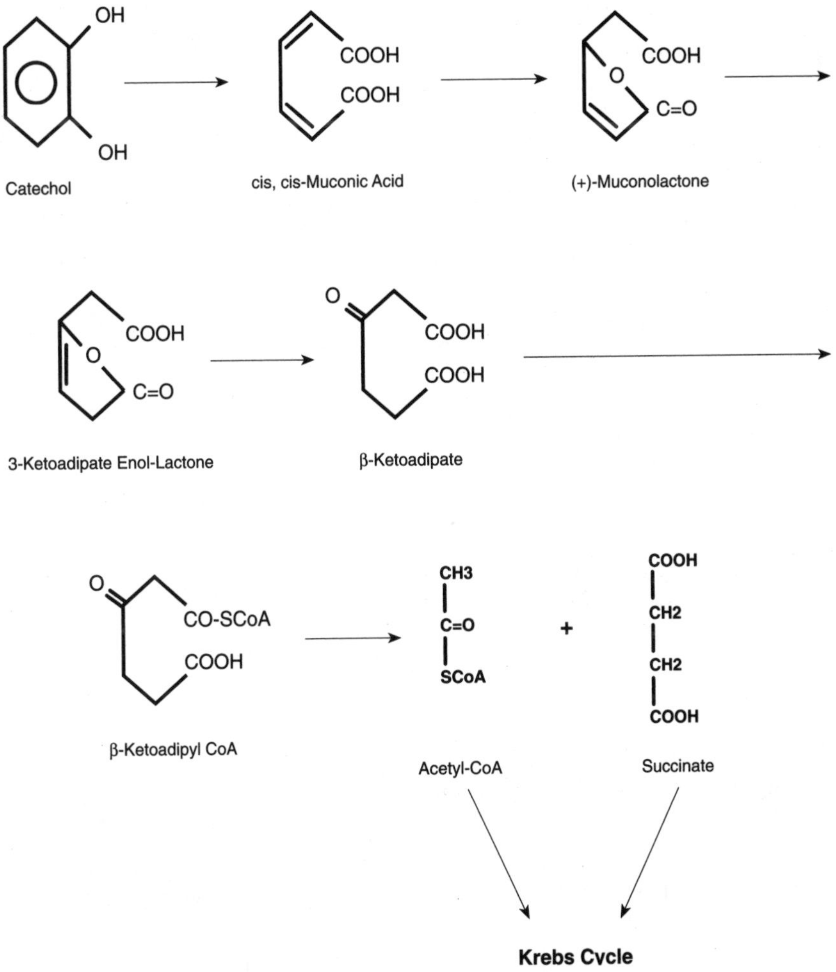

Figure 2-9. Aromatic ring cleavage—ortho cleavage. The aromatic ring of catechol is opened as a result of the introduction of molecular oxygen into the ring structure. Cleavage of the aromatic nucleus occurs between the two carbons bearing hydroxyl groups. The acetyl-CoA and succinate produced as a result of ring cleavage can be further oxidized via the Krebs cycle and the electron transport system. A similar pathway for ortho cleavage also exists for protocatechuate (not illustrated).

hydroxyl groups, forming a 2-hydroxy-muconic semialdehyde as the initial ring cleavage product. Subsequent metabolism results in the formation of pyruvate, formate, and acetaldehyde, which feed into the Krebs cycle.

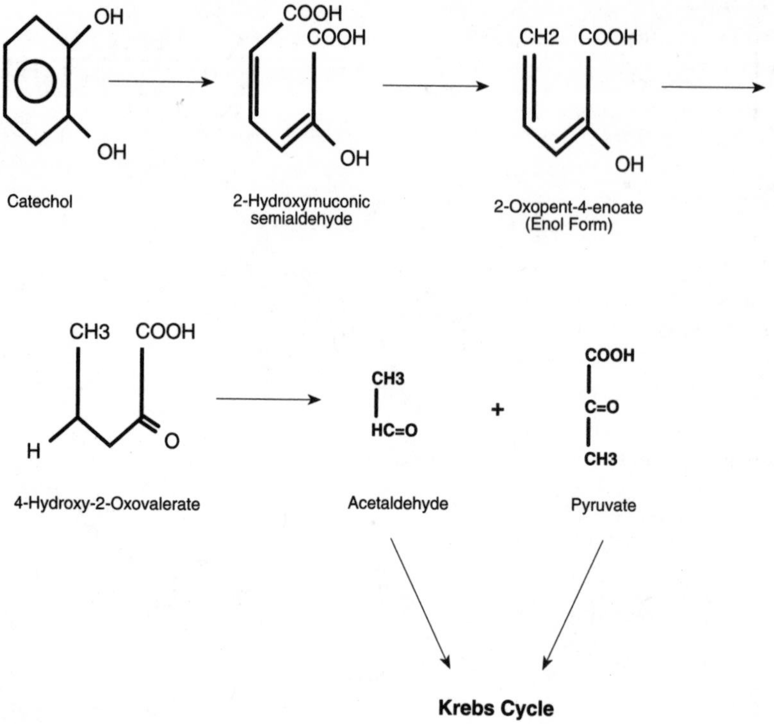

Figure 2-10. Aromatic ring cleavage—meta cleavage. As is the case with the ortho-cleavage pathway, the aromatic ring is opened as a result of the introduction of molecular oxygen into the ring structure. Cleavage of the aromatic ring occurs between an hydroxylated carbon and the adjacent unsubstituted carbon. The final products of this pathway, acetaldehyde and pyruvate, can be further oxidized via the Krebs cycle and the electron transport system. A similar pathway for meta cleavage also exists for protocatechuate (not illustrated).

In both the ortho- and meta-cleavage pathways, the initial reaction of the pathway involves the direct addition of oxygen to the aromatic nucleus to cleave catechol or protocatechuate. These reactions are catalyzed by a group of enzymes called *oxygenases*. Molecular oxygen serves as a source of oxygen for these reactions; hence degradation of aromatic compounds by either of these pathways is obligately aerobic, with oxygen serving as both a terminal electron acceptor and a reactant in the metabolism of aromatic compounds.

Alkyl-substituted benzenes (toluene and ethylbenzene) are initially oxidized using one of several different pathways. If the initial attack is on the aromatic ring, alkyl-catechols are formed which can be cleaved using either the ortho- or meta-cleavage pathways discussed previously. If the alkyl group is oxidized initially, however, aromatic carboxylic acids are formed. For example, the initial oxidation of the alkyl substitution of p-xylene leads to the formation of toluic acid. These aromatic carboxylic acids can be converted into homogentisate or gentisate. Cleavage of the aromatic ring subsequently occurs with degradation ultimately to intermediates such as fumarate and acetoacetate which can feed into the Krebs cycle. Gibson (47) has recently reviewed the major pathways from microbial degradation of aromatic hydrocarbons such as toluene.

Until recently, microbial metabolism of aromatic compounds was considered to be limited to aerobic environments. Within the last 10 years, however, numerous studies have demonstrated the ability of anaerobic microorganisms to degrade this class of compounds. Table 2-4 summarizes some of the aromatic compounds for which anaerobic degradation has been demonstrated. Many of these studies have used enrichment cultures of mixed microbial populations. Thus, while anaerobic degradation can be demonstrated to occur with these cultures, the exact metabolic pathways responsible are not as well established as with pure culture studies of aerobic degraders. This does not negate the validity of the anaerobic studies; rather, the results indicate that in many cases anaerobic decomposition may be the result of the activities of a microbial consortium.

Anaerobic degradation of aromatic compounds has been shown to occur under denitrifying, sulfate-reducing, and methanogenic conditions (53, 117). In addition, ferric iron and manganese oxide have been shown to serve as alternate electron acceptors for the anaerobic degradation of aromatic compounds. The terminal electron acceptor used for the degradative reactions will depend to a large extent on the redox conditions of the environment. Table 2-5 summarizes the optimal redox conditions for various microbial processes.

Under anaerobic conditions, the initial steps of aromatic degradation are significantly different from those found in the aerobic degradation pathways. The initial step in the aromatic degradative sequences involves the hydrogenation of the benzene ring. This hydrogenation serves to destabilize the aromatic ring structure. The ring is then cleaved through a hydration reaction during which oxygen is attached to the ring structure. Ring cleavage yields aliphatic hydrocarbons which are then further metabolized to Krebs cycle intermediates via β-oxidation. Figure 2-11 illustrates the reaction sequence

Table 2-4. Anaerobic Degradation of Aromatic Compounds

Compound	Organism	Culture conditions	Reference
Benzoate	*Pseudomonas stutzeri*	Denitrifying	132
	Desulfonema magnum	Sulfate reducing	130
	Not identified	Methanogenic	60
Catechol	Not identified	Methanogenic	13
	Desulfobacterium catecholicum	Sulfate reducing	121
p-Cresol	Mixed culture	Denitrifying	22
	Not identified	Sulfate reducing	67
	Not identified	Methanogenic	58
Phenol	Not identified	Denitrifying	12
	Not identified	Sulfate reducing	49
	Desulfobacterium phenolicum		10
	Not identified	Methanogenic	60
Toluene	Mixed culture	Denitrifying	40
	Not identified		70
	Mixed culture	Methanogenic	54
o-Xylene	Mixed culture	Denitrifying	40

Table 2-5. Optimal Redox Potential for Microbial Metabolism

Type of metabolism	Oxygen requirements of microbes involved	Redox potential (mV)
Aerobic respiration	Aerobes	700–500
Denitrification	Facultative	300
Sulfate reduction	Obligate anaerobes	− 200
Methanogenesis	Obligate anaerobes	− 200

for the cleavage of benzoate (21, 53). It is essential to note that in anaerobic degradation, water serves as an oxygen source for the metabolic reactions, in contrast to aerobic degradation, where molecular oxygen is an essential reactant.

In addition to the reductive mechanism outlined above, recent research has demonstrated that in some cases, particularly for molecules containing electron-withdrawing substitutes on the aromatic ring (e.g., *p*-cresol), an initial oxidative reaction occurs. In this case, the substituted aromatic compound is oxidized to the corresponding alcohol, which is

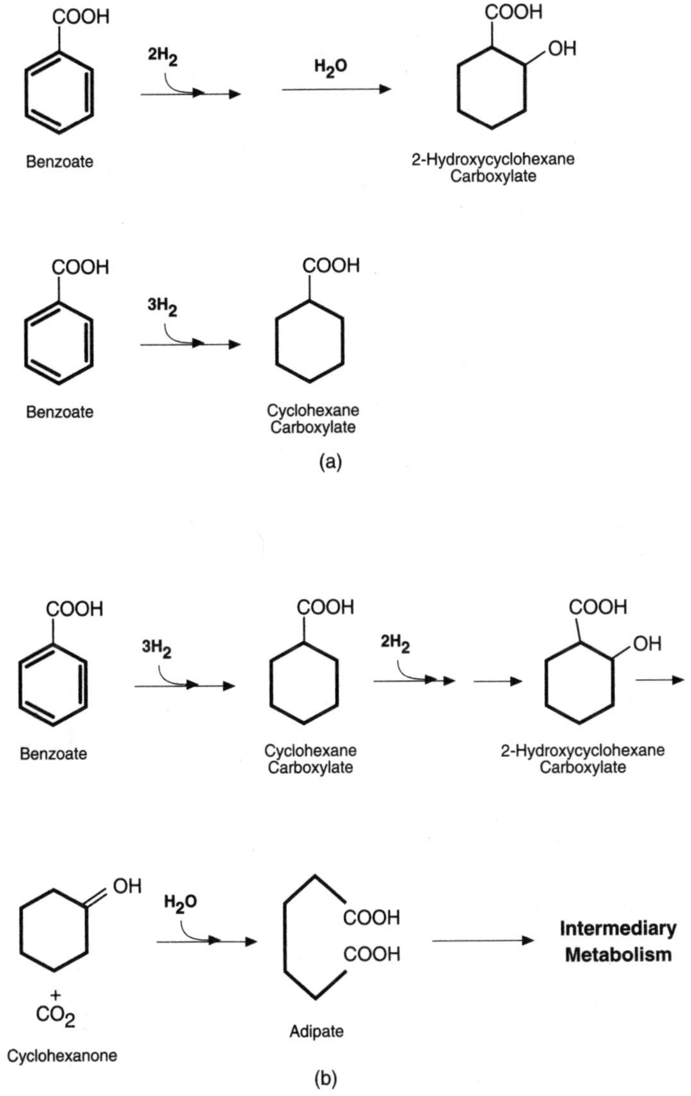

Figure 2-11. Anaerobic degradation of aromatic compounds: Initial reductive transformations. (a) Benzoate and related compounds can be reduced to the corresponding alicyclic compounds under anoxic and anaerobic conditions. Note that in contrast to the initial transformations of aromatic compounds under aerobic conditions, molecular oxygen is not involved. (b) Reductive pathway for the degradation of benzoate by *Moraxella* sp. Not all intermediates are shown. Note that the oxygen in the pathway is derived from water, not molecular oxygen. (*Adapted from Refs. 21, 53, and 137.*)

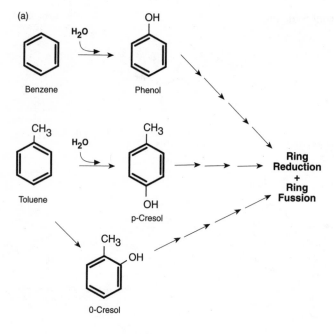

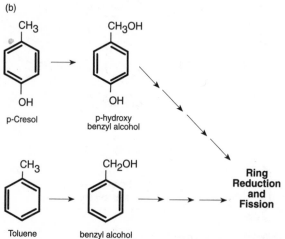

Figure 2-12. Anaerobic degradation of aromatic compounds: Initial oxidative transformations. Oxidation is accomplished initially through the addition of a hydroxyl group derived from a molecule of water. Oxidation can occur either through (*a*) the addition of the hydroxyl group to the aromatic nucleus or by (*b*) oxidation of a methyl substitution on the aromatic structure. In both cases, an aromatic alcohol is produced. This is subsequently converted to the aromatic aldehyde and finally to the aromatic acid before ring reduction and fission to form aliphatic products. Note that in all cases the oxygen added to the molecule is derived from water, not molecular oxygen. (*Adapted from Refs. 53, 54, and 58.*)

subsequently dehydrogenated to the aldehyde and acid (Fig. 2-12). The oxygen for this initial reaction is derived from water (58).

Polycyclic Aromatics
(Polyaromatic Hydrocarbons)

Polycyclic aromatic compounds, also called *polyaromatic hydrocarbons* (PAHs) and *polynuclear aromatics* (PNAs), contain two or more fused aromatic rings. Figure 2-13 illustrates some PAHs commonly found as environmental contaminants. Present in small amounts in heavy petroleum products such as fuel oils and crude oils, polyaromatic hydrocarbons are also a major component in creosote. Extensive environmental contamination with these compounds has resulted from the use of creosote in wood preservation. In a study of ground water contamination at wood treatment sites, Rosenfeld and Plumb (107) reported that such PAHs as acenaphthene, naphthalene, fluorene, phenanthrene, fluoranthene, pyrene, and anthracene were present in over 20

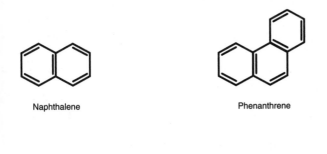

Naphthalene   Phenanthrene

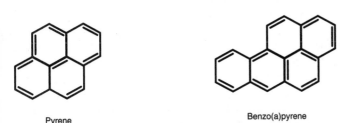

Pyrene   Benzo(a)pyrene

Figure 2-13. Structure of common polyaromatic hydrocarbons. Polyaromatic hydrocarbons are compounds composed of two or more fused aromatic nuclei. Compounds with four or five aromatic rings (pyrene, benzo[*a*]pyrene, etc.) are typically resistant to biodegradation, although there have been reports of microbial cultures capable of transforming these compounds.

percent of the sites tested. Additional PAHs (benzo[*a*]anthracene, chrysene, acenaphthylene, benzo[*b*]fluoranthene, and benzo[*a*] pyrene) also were detected at wood treating sites but with less frequency. PAH contamination of soils and ground water is of particular concern because many of these compounds are known or suspected carcinogens.

Degradation of two- and three-ring aromatic compounds (e.g., naphthalene, anthracene, and phenanthrene) has been shown to be widespread among aerobic bacteria. As is the case with the aerobic degradation of monoaromatic compounds discussed in the preceding section, the initial reactions of PAH degradation involve the introduction of molecular oxygen into the ring structure (48) with the ultimate production of intermediates that feed into the Krebs cycle.

Susceptibility of PAHs to microbial degradation is inversely correlated with the number of rings in the PAH. Thus naphthalene is more easily degraded than anthracene or phenanthrene. In part, the decreased degradation of three-, four-, and five-ring PAHs is a result of the low water solubility and hence low bioavailability of these compounds.

Traditional techniques for isolating and culturing microorganisms depend on the presence of soluble carbon sources. Despite these difficulties, recent studies have been able to demonstrate the existence of aerobic heterotrophic bacteria, both pure cultures and consortia, capable of degrading the majority of these high-molecular-weight PAHs. For example, Mueller et al. (84) reported on a mixed microbial community containing seven morphologically distinct bacterial strains which is capable of degrading fluorene, anthracene, anthraquinone, fluoranthene, and pyrene. No doubt microbial consortia are important in the degradation of other PAHs and complex mixtures of PAHs (e.g., creosote) at contaminated sites. Table 2-6 summarizes bacterial degradation of selected PAHs.

Biodegradation of polycyclic aromatic hydrocarbons under strictly anaerobic conditions has not been reported. Recently, however, Mihelicic and Luthy (81) demonstrated that degradation of naphthalene could occur under denitrifying conditions, although no degradation of naphthalene or anthracene was found under strictly anaerobic (sulfate-reducing or methanogenic) conditions. In contrast, naphthol, which contains a hydroxyl group substitution, was degraded under both denitrifying and strictly anaerobic conditions. Thus the presence of hydroxyl substitutions may render polycyclic aromatic materials susceptible to anaerobic degradation.

Table 2-6. Aerobic Degradation of Selected Polycyclic Aromatic Hydrocarbons

Compound	No. of rings	Microorganism(s)	Reference
Naphthalene	2	*Rhodococcus* sp.	56
		Soil microorganisms	85, 86, 129
		Marine *Pseudomonas*	123
Anthracene	3	Soil microorganisms	75
		Marine microorganisms	18
Acenaphthene	3	Soil microorganisms	129
Fluorene	3	Soil microorganisms	138
Phenanthrene	3	Estuarine enrichments	57
		Soil microorganisms	138, 94
Fluoranthene	4	Soil microorganisms	129
Pyrene	4	Soil microorganisms	128, 129
		Mycobacterium sp.	61
		Rhodococcus sp. UWI	127
Benzo[*a*]anthracene	4	Soil microorganisms	76, 85, 86
Benzo[*a*]pyrene	5	Soil microorganisms	76, 82, 129
		Estuarine microorganisms	113

In addition to bacterial degradation, eukaryotic organisms such as fungi and algae are also known to transform and degrade PAHs under aerobic conditions. The mechanism of oxidation of PAHs in fungi and other eukaryotes is fundamentally different from that in bacterial systems, typically involving a reaction sequence called the *NIH shift* in the initial stages of transformation (48). Fungal degradation of PAHs ranging from two to five rings (naphthalenebenzo[*a*]pyrene), as well as benzene and substituted benzenes (e.g., phenol) has been demonstrated to occur, although the role of fungi in the degradation of PAHs in the natural environment is unclear.

One group of fungi, commonly called the *white-rot fungi,* are known to have extensive biodegradative capabilities associated with the production of enzymes involved in lignin biodegradation. Such organisms as *Phanerochaete chrysosporium* produce a family of enzymes called *lignin peroxidases* or *ligninases.* Production of ligninase enzymes by the white-rot fungi is dependent on the nutritional status of the organism and optimal under conditions of inorganic nutrient limitations (N starvation). Lignin degradation represents a type of secondary metabolism for the fungi; therefore, it is dependent on the presence of a primary growth substrate such as glucose or cellulose. Ligninases catalyze the initial degradation of the aromatic structures in lignin.

Lignin is a highly complex and variable molecule; thus it is understandable that the enzymes involved in lignin degradation show low substrate specificities. As a result of this low specificity, ligninases are also capable of catalyzing the initial oxidation of the aromatic nucleus of PAHs. Bumpus (31) has demonstrated that *Phanerochaete chrysosporium* is capable of degrading many of the polycyclic aromatic hydrocarbons found in anthracene oil, a coal tar derivative. Based on disappearance of the PAHs from the initial culture media, he concluded that degradation of 22 different PAHs could be attributed to the metabolic activities of the fungus. Sanglard et al. (108) have demonstrated that *Phanerochaete chrysosporium* is capable of mineralizing benzo[a]pyrene, although the intermediates in this pathway have not been determined. Sutherland et al. (118) have presented evidence indicating that the initial oxidation of phenanthrene (and presumably other PAHs) is not dependent solely on the presence of ligninases. Rather, the activities of fungal monooxygenases and epoxide hydrolases are proposed to catalyze the initial formation of unstable PAH arine oxides which are subsequently isomerized to phenols or hydrated to form *trans*-dihydrodiols.

Halogenated Aliphatic Hydrocarbons

Halogenated hydrocarbons are widely used as industrial solvents and degreasers. These compounds are substituted hydrocarbons in which one or more of the hydrogens has been replaced with a halogen, most commonly chlorine. Figure 2-14 illustrates some of the halogenated hydrocarbons of environmental concern. Contamination of the environment with these compounds is now widespread. Recent studies have indicated that trichloroethylene (TCE), for example, is the most common contaminant in ground water.

Halogenated hydrocarbons can be degraded under both anaerobic and aerobic conditions. Anaerobically, degradation proceeds through a process called *reductive dechlorination*. In this process, halogen atoms are sequentially removed from the molecule and replaced with hydrogen. In this type of reaction, the halogenated hydrocarbon is used as

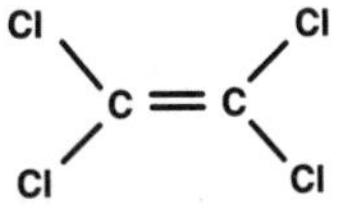

Tetrachloroethene
(PCE)

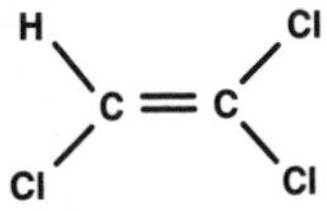

Trichloroethene
(TCE)

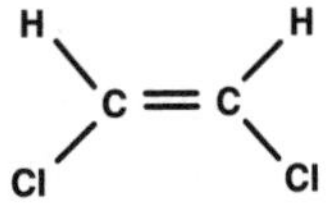

Cis-1, 2 dichloroethene
(cis-DCE)

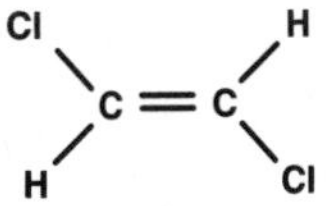

trans-1, 2 dichloroethene
(trans-DCE)

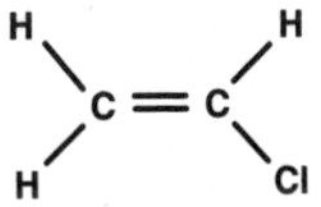

Vinyl Chloride
(VC)

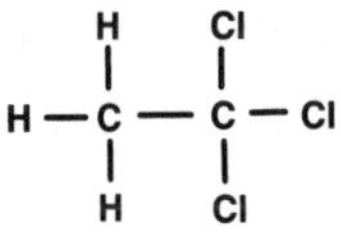

1, 1, 1 Trichloroethane
(TCA)

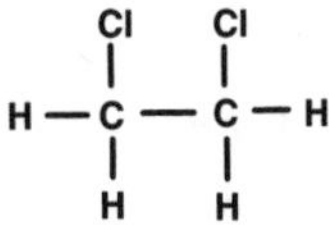

1,2-Dichloroethane
(1,2-DCA)

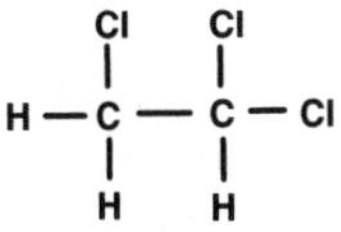

1, 1, 2 Trichloroethane
(1,1,2 TCA)

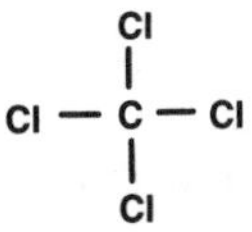

Carbon Tetrachloride
(CT)

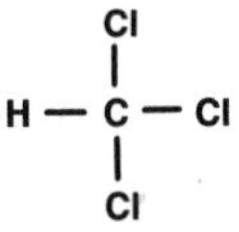

Chloroform
(CF)

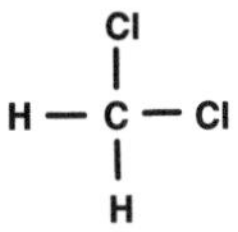

Methylene Chloride

Figure 2-14. Chlorinated hydrocarbons. Chlorinated hydrocarbons are widely used as industrial solvents.

an electron acceptor, not as a source of carbon. Therefore, reductive dechlorination can only proceed in the presence of an appropriate source of carbon for microbial growth.

Dehalogenation depends on the redox potential of the molecule, which is determined primarily by the strength of the halogen-carbon bond. The higher the bond strength, the less likely it is that the halogen will be removed. Bond strength is dependent on both the type and number of halogen atoms present and the degree of saturation of the halogenated molecule. In general, bromine and iodine substitutions, which have a lower bond strength than chlorine, are easier to remove. Fluorine-carbon bonds have a high bond strength and are generally harder to remove than chlorine (124). As the degree of saturation of the molecule decreases, bond strength increases. Thus saturated compounds (alkanes) are generally more susceptible to reductive dehalogenation than unsaturated compounds (alkenes and alkynes).

Until the early 1980s, biological degradation of chlorinated ethenes, such as perchloroethylene (PCE) and trichloroethylene (TCE), was not thought to occur. Bouwer et al. (25) and Bouwer and McCarty (23, 24) were the first to demonstrate conclusively that biological transformations of these compounds not only could occur but occurred at a rate faster than abiotic transformations. They demonstrated that microbial consortia, enriched under methanogenic conditions with acetate as a source of carbon, could transform C_1 and C_2 halocarbons (TCE, carbon tetrachloride, 1,1,1 TCA) into carbon dioxide and methane. Only carbon tetrachloride was transformed using enrichments made under denitrifying conditions (23, 24). Biologically mediated reductive dechlorination was proposed as one of the mechanisms responsible for these transformations, although the entire pathway for transformation remained unclear. Since these initial observations, numerous studies have confirmed the ability of anaerobic consortia to transform C_1 and C_2 halocarbons. Table 2-7 summarizes some of these findings.

Initially, it was thought that transformation and degradation of halogenated aliphatic compounds was limited to anaerobic organisms. Recently, however, several groups of microorganisms have been shown to aerobically degrade chlorinated aliphatic compounds, particularly TCE. Wilson and Wilson (135), Fogel et al. (42), and Henson et al. (63) have demonstrated degradation of chlorinated aliphatics in soils exposed to methane or natural gas. Because the prior exposure of the samples to methane stimulated degradation, they concluded that a specialized group of aerobic microorganisms, the *methanotrophs*, was responsible. Little et al. (77) have recently isolated a methanotrophic (methane-oxidizing) bacterium which

Table 2-7. Anaerobic Degradation of Selected Aliphatic Halocarbons

Compound	Microorganisms	Reference
Chloroform	Methanogenic enrichment	124
Carbon tetrachloride	Methanogenic enrichment	23, 26
	Methanobacterium sp.	39
	Desulfobacterium sp.	39
	Clostridium sp.	45
	Denitrifying enrichment	104
	Pseudomonas (denitrifying)	36
1, 2 Dichloroethane (DCA)	Anaerobic consortia	23, 24
Dichloromethane (DCM)	Anaerobic consortia	64
	Acetogenic consortia	44
Freons	Denitrifying enrichment	111
1, 1, 1-Trichloroethane (TCA)	Denitrifying enrichment	111
	Sulfate-reducing enrichment	68
	Methanogenic enrichment	68
	Clostridium sp.	45
Trichloroethylene (TCE)	Anaerobic consortia	69
		17
Tetrachloroethylene (PCE)	Methanogenic enrichment	23
		96
		126
Vinyl chloride (VC)	Anaerobic consortia	16

degrades TCE in pure culture. In addition to the methanotrophs, Nelson et al. (89, 90, 91) have isolated a strain of *Pseudomonas* (G-4) which can degrade TCE in the presence of aromatic compounds such as phenol. Finally, Vannelli et al. (125) have isolated an ammonia-oxidizing bacterium, *Nitrosomonas europaea*, which can degrade chlorinated aliphatic compounds. In general, susceptibility to degradation was inversely correlated with the number of chlorine atoms on the molecule. Thus, in contrast to anaerobic systems, vinyl chloride is most susceptible to aerobic degradation, while tetrachloroethylene (PCE) is most resistant.

The understanding of the metabolic pathways involved in the aerobic degradation of chlorinated aliphatics is incomplete. Aerobic degradation of chlorinated aliphatic compounds involves an initial oxidation of the molecule. For example, under aerobic conditions, TCE is oxidized to an epoxide prior to dechlorination (34). This oxidation is presumed to be mediated by mono- and dioxygenase enzymes. These are the same oxygenase enzymes which are involved in the degradation of aliphatic and aromatic hydrocarbons by the microorganisms. The presence of the metabolizable substrate (methane, toluene/phenol, or ammonia) is necessary for degradation to occur. Thus aerobic degradation of chlorinated alkanes seems to involve cometabolism of the chlorinated alkanes and is the result of the low specificity of the mono- and dioxygenase enzymes.

Halogenated Aromatic Compounds

Halogenated aromatic compounds represent an immense group of organic chemicals structurally related to benzene and substituted benzene molecules such as toluene and phenol. Table 2-8 summarizes the most environmentally important groups of halogenated aromatic compounds. Because of the diversity of this group of compounds, a thorough review of the microorganisms and metabolic pathways involved in degradation is not possible.

Degradation of halogenated aromatic compounds proceeds through many of the pathways described for degradation of nonhalogenated aromatic compounds. Chlorobenzenes, for example, are first converted to the analogous chlorocatechol. The aromatic nucleus is then broken, followed by dechlorination of the ring cleavage products (83). In some cases, such as the chlorobenzoic acids, dechlorination may occur either before or after ring cleavage. Halogenated polyaromatic compounds such as chlorinated biphenyls are usually degraded by cleavage of the nonsubstituted ring followed by degradation of the resultant chlorobenzoate.

Although pure cultures of bacteria have been isolated with the ability to degrade many of the halogenated aromatic hydrocarbons, degradation is frequently the result of a consortium of microorganisms. For example, Sylvestre et al. (120) found mineralization of 4-chlorobiphenyl to be dependent on the interaction of two bacterial populations. The first of these populations transformed 4-chlorobiphenyl into 4-chlorobenzoate, while the second mineralized the 4-chlorobenzoate to carbon dioxide.

Susceptibility to degradation for all the halogenated aromatic hydrocarbons depends on the nature of the halogen substitution, the number of substitutes, and the position of the substitutes, (stereospecificity). As with the halogenated aliphatic hydrocarbons, com-

Table 2-8. Halogenated Aromatic Compounds

Class	Example	Sources
Chlorobenzoic acids	2, 3, 6-Trichlorobenzoic acid	Degradative products of polychlorinated biphenyls; herbicides, plant growth regulators
Chlorobenzenes	Chlorobenzene Pentachloronitrobenzene	Industrial solvents, diluents for PCBs, paint solvents, by-products of textile dyeing, fungicides
Chlorophenols	2, 3, 4, 6-Tetrachlorophenol Pentachlorophenol (PCP)	Antifungal agents, wood perservatives, degradation products of chlorophenoxy herbicides
Chlorophenoxy and chlorophenyl herbicides	2, 4-D 2, 4, 5-T	Herbicides
Phenylamide herbicides	Phenyl ureas	Herbicides
Chlorinated biphenyls	PCBs; Aroclors	Dielectric fluids in capacitors and transformers, gas turbines, and hydraulic systems; fire retardants; plasticizers
Chlorinated dioxins and furans	2, 3, 7, 8-Tetrachlorodibenzo-p-dioxin (TCDD)	By-products during the manufacturing of chemicals; pyrolysis of chlorophenol salts, heat exchange fluids, and hydraulic fluids; pyrolysis of PCBs and polychlorinated diphenyl ethers

pounds containing bromine or iodine substitutions are more susceptible to degradation than those containing chlorine substitutions, which in turn are more susceptible than molecules with fluorine substitutions. Susceptibility to degradation generally decreases as the number of substitutions increases; however, even highly chlorinated compounds such as pentachlorophenol have been shown to be susceptible to aerobic degradation. Finally, since the dioxygenase enzymes involved in ring cleavage require the presence of two adjacent hydroxyl group–carrying carbon atoms (for ortho cleavage, the predominant mechanism), the stereochemistry of the compound will play a significant role in determining its susceptibility to degradation. Any substitution which blocks potential sites for ring cleavage will decrease the compound's susceptibility to cleavage.

Anaerobic degradation of chlorinated aromatic compounds such as chlorobenzenes, chlorophenols, and PCBs has been found to occur in a wide variety of environments. As is the case with anaerobic degradation of halogenated aliphatic hydrocarbons, degradation generally proceeds through reductive dechlorination reactions, and highly substituted compounds are more easily dehalogenated than monohalogenated compounds. PCBs, which are mixtures of congeners rather than pure compounds, show evidence of reductive dechlorination, particularly of the more highly substituted congeners, resulting in an enrichment for the congeners containing a low number of chlorines. The placement of the halogen atoms on the molecule has profound effects of the degradability of the compound. Thus reductive dechlorination was found to occur preferentially at the ortho and para positions in PCP, while dechlorination of the meta position was less frequent. In contrast, only meta-chlorines were removed from chlorobenzoates (122). Typically, degradation is the result of the activities of a consortium of microorganisms rather than an individual strain. For example, Teidje et al. (122) discuss a nine-membered anaerobic consortium, enriched from sewage sludge, which degrades 3-chlorobenzoate to carbon dioxide and methane.

The majority of the studies on anaerobic degradation of halogenated aromatic compounds have used methanogenic enrichment cultures. At the present time, information concerning degradation of these compounds under sulfate-reducing conditions is limited.

Häggblom and Young (59) have demonstrated that five chlorophenols (2-chlorophenol, 3-chlorophenol, 4-chlorophenol, 2,4-chlorophenol, and 2, 6-chlorophenol) can be degraded under sulfate-reducing conditions. Colberg (35) has reviewed recent research on degradation of chlorinated aromatic hydrocarbons by sulfate-reducing microorganisms. There are no thermodynamic barriers to the degradation of chlorinated aromatic hydrocarbons by sulfate-reducing organisms.Therefore, the lack of documented cases of degradation may reflect the amount of research which has been conducted in this area rather than a limitation of the microorganisms.

References

1. Aelion, C. M., C. M. Swindoll, and F. K. Pfaender. 1987. Adaptation to and biodegradation of xenobiotic compounds by microbial communities from a pristine aquifer. *Appl. Environ. Microbiol.* **53**: 2212–2217.

2. Alexander, M. 1971. *Microbial Ecology.* John Wiley and Sons, New York.

3. Atlas, R. M. 1981. Microbial degradation of petroleum hydrocarbons: An environmental perspective. *Microbiol. Rev.* **45**: 180–209.

4. Atlas, R. M. 1984. *Petroleum Microbiology.* Macmillan Publishing Company, Inc., New York.

5. Atlas, R. M. 1984. *Microbiology: Fundamentals and Applications.* Macmillan Publishing Company, Inc., New York.

6. Atlas, R. M. 1988. Biodegradation of Hydrocarbons in the Environment, in *Environmental Biotechnology Reducing Risks from Environmental Chemicals through Biotechnology,* G. S. Omen (ed.). Plenum Press, New York, pp. 211–222.

7. Atlas, R. M. 1991. Bioremediation of Fossil Fuel Contaminated Soils, in *In Situ Bioreclamation,* R. E. Hinchee and R. F. Olfenbuttel (eds.) Butterworth-Heinemann, Boston. pp. 14–32.

8. Atlas, R. M., and R. Bartha. 1972. Degradation and mineralization of petroleum in seawater: Limitation by nitrogen and phosphorous. *Biotechnol. Bioeng.* **14**: 308–318.

9. Atlas, R. M., and R. Bartha. 1987. *Microbial Ecology: Fundamentals and Applications.* Benjamin/Cummings Publishing Company, Menlo Park, Calif.

10. Bak, F., and F. Widdel. 1986. Anaerobic degradation of phenol and phenol derivatives by *Desulfobacterium phenolicum* sp. nov. *Arch. Microbiol.* **146**: 177–180.

11. Baker, K. H., and D. S. Herson. 1990. In situ bioremediation of contaminated aquifers and subsurface soils. *Geomicrobiol. J.* **8**: 133–146.

12. Bakker, G. 1977. Anaerobic degradation of aromatic compounds in the presence of nitrate. *Microbiol. Lett.* **1**: 103–108.

13. Balba, M. T., and W. C. Evans. 1980. The methanogenic biodegradation of catechol by a microbial consortium: Evidence for the production of phenol through *cis*-benzenediol. *Biochem. Soc. Trans.* **8**: 452–453.

14. Bale, M. J., J. C. Frey, and M. J. Dale. 1988. Transfer and occurrence of large mercury resistance plasmids in river epilithon. *Appl. Environ. Microbiol.* **54**: 972–978.

15. Barkay, T., and P. H. Pritchard. 1988. Adaptation of aquatic microbial communities to pollutants. *Microbiol. Sci.* **5**: 165–169.

16. Barrio-Lage, G. A., F. Z. Parsons, R. M. Narbaitz, and P. A. Lorenzo. 1990. Enhanced anaerobic biodegradation of vinyl chloride in ground water. *Environ. Toxicol. Chem.* **9**: 403–415.

17. Barrio-Lage, G. A., F. Z. Parsons, R. S. Nassar, and P. A. Lorenzo. 1987. Biotransformation of trichloroethane in a variety of subsurface materials. *Environ. Toxicol. Chem.* **6**: 571–578.

18. Bauer, J. E., and D. G. Capone. 1988. Effects of co-occurring aromatic hydrocarbons on degradation of individual polycyclic aromatic hydrocarbons in marine sediment slurries. *Appl. Environ. Microbiol.* **54**: 1649–1655.

19. Bayly, R. C., and M. G. Barbour. 1984. The Degradation of Aromatic Compounds by the Meta and Gentisate Pathways, in *Microbial Degradation of Organic Compounds*, D. T. Gibson (ed.). Marcel Dekker, Inc., New York, pp. 253– 294.

20. Beppu, I., and K. Arima. 1969. Induction by mercuric ion of extensive degradation of cellular ribonucleic acid in *E. coli. J. Bacteriol.* **98**: 888–897.

21. Berry, D. F., A. J. Francis, and J.-M. Bollag. 1987. Microbial metabolism of homocyclic and heterocyclic aromatic compounds under anaerobic conditions. *Microbiol. Rev.* **51**: 43–59.

22. Bossert, I. D., and L. Y. Young. 1986. Anaerobic oxidation of *p*-cresol by a denitrifying consortium. *Appl. Environ. Microbiol.* **52**: 1117–1122.

23. Bouwer, E. J., and P. L. McCarty. 1983. Transformations of 1- and 2-carbon halogenated aliphatic organic compounds under methanogenic conditions. *Appl. Environ. Microbiol.* **45**: 1286–1294.

24. Bouwer, E. J., and P. L. McCarty. 1983. Transformations of halogenated organic compounds under denitrifying conditions. *Appl. Environ. Microbiol.* **45**: 1295–1299.

25. Bouwer, E. J., B. E. Rittman, and P. L. McCarty. 1981. Anaerobic degradation of halogenated 1- and 2-carbon organic compounds. *Environ. Sci. Technol.* **15**: 596–599.

26. Bouwer, E. J., and J. P. Wright. 1988. Transformation of trace halogenated aliphatics in anoxic biofilm columns. *J. Contaminant Hydrol.* **2**: 155–169.

27. Brady, N. C. 1974. *The Nature and Properties of Soils*. Macmillan Publishing Co., Inc., New York.

28. Brand, J. M., D. L. Cruden, G. J. Zylstra, and D. T. Gibson. 1992. Stereospecific hydroxylation of indole by *Escherichia coli* containing the cloned toluene dioxygenase genes from *Pseudomonas putida* F1. *Appl. Environ. Microbiol.* **58**: 3407–3409.

29. Britton, L. N. 1984. Microbial Degradation of Aliphatic Hydrocarbons, in *Microbial Degradation of Organic Compounds*, D. T. Gibson (ed.). Marcel Dekker, Inc., New York pp. 89–129.

30. Brock, T. D., and M. T. Madigan. 1991. *Biology of Microorganisms*, 6th ed. Prentice-Hall, Englewood Cliffs, N.J.

31. Bumpus, J. S. 1989. Biodegradation of polycyclic aromatic hydrocarbons by *Phanerochaete chrysosporium*. *Appl. Environ. Microbiol.* **55**: 154–158.

32. Cerniglia, C. E., R. L. Herbert, R. H. Dodge, P. J. Szaniszlo, and D. T. Gibson. 1979. Some Approaches to Studies on the Degradation of Aromatic Hydrocarbons by Fungi, in *Microbial Degradation of Pollutants in Marine Environments*, A. L. Bourquin and H. Pritchard (eds.) EPA Report No. EPA-600/9-79-012. EPA, Washington, D.C.

33. Chakrabarty, A. M. 1982. *Biodegradation and Detoxification of Environmental Pollutants*. CRC Press, Inc., Boca Raton, Fla.

34. Chaudhry, G. R., and S. Chapalamadugu. 1991. Biodegradation of halogenated organic compounds. *Microbiol. Rev.* **55**: 59–79.

35. Colberg, P. J. S. 1990. Role of sulfate in microbial transformation of environmental contaminants: Chlorinated aromatic compounds. *Geomicrobiol. J.* 8:147-165.

36. Criddle, C. S., J. T. Dewitt, D. Grbíc-Galic, and P. L. McCarty. 1990. Transformation of carbon tetrachloride by *Pseudomonas* sp. strain KC under denitrifying conditions. *Appl. Environ. Microbiol.* **56**: 3240–3246.

37. Dagley, S. 1984. Introduction, in *Microbial Degradation of Organic Compounds*, D. J. Gibson (ed.)., Marcel Dekker, Inc., New York, pp. 1–11.

38. Dibble, J. T., and R. Bartha. 1979. Effect of environmental parameters on the biodegradation of oil sludge. *Appl. Environ. Microbiol.* **37**: 729–739.

39. Elgi, C., R. Scholtz, A. M. Cook, and T. Lieslinger. 1987. Anaerobic dechlorination of tetrachloromethane and 1,2-dichloroethane to degradable products by pure cultures of *Desulfobacterium* sp. and *Methanobacterium* sp. *FEMS Microbiol. Lett.* **43**: 257–261.

40. Evans, P. J., D. T. Mang, and L. Y. Young. 1991. Degradation of toluene and *m*-xylene and transformation of *o*-xylene by denitrifying enrichment cultures. *Appl. Environ. Microbiol.* **57**: 450–454.

41. Focht, D. D. 1988. Performance of Biodegradative Microorganisms in Soil: Xenobiotic Chemicals as Unexploited Metabolic Niches, in *Environmental Biotechnology: Reducing Risks for Environmental Chemicals through Biotechnology*, G. S. Omenn (ed.). Plenum Press, New York, pp. 15–30.

42. Fogel, M. M, A. R. Taddeo, and S. Fogel. 1986. Biodegradation of chlorinated ethanes by a methane-utilizing mixed culture. *Appl. Environ. Microbiol.* **51**: 720–724.

43. Foster, T. J. 1983. Plasmid determined resistance to antimicrobial drugs and toxic ions in bacteria. *Microbiol. Rev.* **47**: 361–409.

44. Freedman, D. L., and J. M. Gossett. 1991. Biodegradation of Dichloromethane in a Fixed-Film Reactor under Methanogenic Conditions, in *On Site Bioreclamation*, R. E. Hinchee and R. F. Olfenbuttel (eds.). Butterworth-Heinemann, Boston, pp. 113–134.

45. Galli, R., and P. L. McCarty. 1989. Biotransformation of 1,1,1-

trichloroethane, trichloromethane, and tetrachloromethane by *Clostridium* sp. *Appl. Environ. Microbiol.* **55**: 837–844.

46. Gibson, D. T., (ed.). 1984. *Microbial Degradation of Organic Compounds.* Marcel Dekker, Inc., New York.

47. Gibson, D. T.. 1988. Microbial Metabolism of Aromatic Hydrocarbons and the Carbon Cycle, in *Microbial Metabolism and the Carbon Cycle*, S. R. Hagedron, R. S. Hanson, and D. A. Kunz (eds.). Harwood Academic Publisher, Chur, Switzerland, pp. 33–58.

48. Gibson, D. T., and V.. Subramanian. 1984. Microbial Degradation of Aromatic Hydrocarbons, in *Microbial Degradation of Organic Compounds*, D. T. Gibson, (ed.). Marcel Dekker, Inc., New York, pp. 182–252.

49. Gibson, S. A., and J. M. Suflita. 1986. Extrapolation of biodegradation results to groundwater aquifers: Reductive dehalogenation of anaerobic compounds. *Appl. Environ. Microbiol.* **52**: 681–688.

50. Gilmour, C. C., E. A. Henry, and R. Mitchell. 1992. Sulfate stimulation of mercury methylation in freshwater sediments. *Environ. Sci. Technol.* **26**: 2281–2287.

51. Goldstein, R. M., L. M. Mallory, and M. Alexander. 1985. Reasons for possible failure of inoculation to enhance biodegradation. *Appl. Environ. Microbiol.* **50**: 977–983.

52. Golovleva, L. A., R. N. Pertsova, A. M. Boronin, V. G. Grishchenkov, B. P. Baskunov, and A. V. Polyakova. 1988. Degradation of polychloroaromatic insecticides by *Pseudomonas aeruginosa* containing biodegradative plasmids. *Microbiology* **51**: 772–777.

53. Grbíc-Galíc, D. 1990. Methanogenic transformation of aromatic hydrocarbons and phenols in groundwater aquifers. *Geomicrobiol. J.* **8**: 167–200.

54. Grbíc-Galíc, D., and T. M. Vogel. 1987. Transformation of toluene and benzene by mixed methanogenic cultures. *Appl. Environ. Microbiol.* **53**: 254–260.

55. Griffin, D. M. 1981. Water and Microbial Stress, in *Advances in Microbial Ecology*, vol. 5, M. Alexander (ed.). Plenum Press, New York, pp. 91–136.

56. Grund, E., B. Denecke, and R. Eichenlaub. 1992. Naphthalene degradation via salicylate and gentisate by *Rhodococcus* sp. strain B4. *Appl. Environ. Microbiol.* **58**: 1874–1877.

57. Guerin, W. F. 1989. Phenanthrene degradation by estuarine surface microlayer and bulk water microbial populations. *Microbial Ecol.* **17**: 89–104.

58. Häggblom, M. M., M. D. Rivera, I. D. Bossert, J. E. Rogers, and L. Y. Young. 1990. Anaerobic biodegradation of *p*-cresol under three reducing conditions. *Microb. Ecol.* **20**: 141–150.

59. Häggblom, M. M., and L. Y. Young. 1990. Chlorophenol degradation coupled to sulfate reduction. *Appl. Environ. Microbiol.* **56**: 3255–3260.

60. Healy, J. B., Jr., and L. Y. Young. 1978. Catechol and phenol degradation by a methanogenic population of bacteria. *Appl. Environ. Microbiol.* **35**: 216–218.

61. Heitkamp, M. A., J. P. Freeman, D. W. Miller, and C. E. Cerniglia. 1988. Pyrene degradation by a *Mycobacterium* sp.: Identification of ring oxidation and ring fission products. *Appl. Environ. Microbiol.* **54**: 2556–2565.

62. Henschke, R. B., and F. R. Schmidt. 1990. Plasmid mobilization from genetically engineered bacteria to members of the indigenous soil microflora in situ. *Curr. Microbiol.* **20**: 105–110.

63. Henson, J. M., M. V. Yates, J. W. Cochran, and D. L. Shackleford. 1988. Microbial removal of halogenated methanes, ethanes, and ethylenes in an aerobic soil exposed to methane. *FEMS Microbial. Ecol.* **53**: 193–201.

64. Hughes, J. B., and G. F. Parkin. 1991. The effect of electron donor concentration on the biotransformation of chlorinated aliphatics, in *On Site Bioreclamation*, R. E. Hinchee and R. F. Olfenbuttel (eds.). Butterworth-Heinemann. Boston, pp. 59–76.

65. Kerry, E. 1990. Microorganisms colonizing plants and soil subjected to different degrees of human activity, including petroleum contamination, in the Vestfold Hills and MacRobertson Land, Antarctica. *Polar Biol.* **10**: 423–430.

66. Khanna, M., and G. Stotzky. 1992. Transformation of *Bacillus subtilis* by DNA bound on montmorillonite and effect of DNase on the transforming ability of bound DNA. *Appl. Environ. Microbiol.* **58**: 1930–1939.

67. King, G. M. 1988. Dehalogenation in marine sediments containing natural sources of halophenols. *Appl. Environ. Microbiol.* **54**: 3079–3085.

68. Klecka, G. M., S. J. Gonsior, and D. A. Markham. 1990. Biological transformations of 1,1,1-trichloroethane in subsurface soils and ground water. *Environ. Toxicol. Chem.* **9**: 1437–1451.

69. Kleopfer, R. D., D. M. Easley, B. B. Haas, Jr., T. G. Deihl, D. E. Jackson, and C. J. Wurrey. 1985. Anaerobic degradation of trichloroethylene in soil. *Environ. Sci. Technol.* **19**: 277–280.

70. Kuhn, E. P., J. M. Suflita, M. D. Rivera, and L. Y. Young. 1989. Influence of alternate electron acceptors on the metabolic fate of hydroxybenzoate isomers in anoxic aquifer slurries. *Appl. Environ. Microbiol.* **55**: 590–598.

71. Lal, R. 1984. *Insecticide Microbiology*. Springer-Verlag, Berlin.

72. Leahy, J. G., and R. R. Colwell. 1990. Microbial degradation of hydrocarbons in the environment. *Microbiol. Rev.* **54**: 305–315.

73. Lee, M. D., J. M. Thomas, R. C. Borden, P. B. Bedient, C. H. Ward, and J. T. Wilson. 1988. Biorestoration of aquifers contaminated with organic compounds. *CRC Crit. Rev. Environ. Control* **18**: 29–89.

74. Leisinger, Th., A. M. Cook, R. Hütter, and J. Nüesch. 1981. *Microbial Degradation of Xenobiotics and Recalcitrant Compounds*. Academic Press, London.

75. Lieberman, M. T., and J. A. Caplan. 1988. Biological Treatment of Petroleum Hydrocarbons and Polynuclear Aromatic Hydrocarbons in Waste Lagoons, in *Environmental Biotechnology: Reducing Risks from Environmental Chemicals through Biotechnology*, G. S. Omenn (ed.). Plenum Press, Inc., New York, p. 453.

76. Litchfield, C. D., D. S. Herson, K. H. Baker, R. Krishnamoorthy, and W. J. Zeli. (1992). Biotreatability Study and Field Pilot Design for a Wood Treating Site. Presented at IGT Symposium on Gas, Oil and Environmental Biotechnology, September 21–23, 1992, Chicago, Ill.

77. Little, C. D., A. V. Palumbo, S. E. Herbes, M. E. Lidstrom, R. L. Tyndall, and P. J. Gilmer. 1988. Trichloroethylene biodegradation by a methane-oxidizing bacterium. *Appl. Environ. Microbiol.* **54**: 951–956.

78. Lovely, D. R., and E. J. P. Phillips. 1992. Bioremediation of uranium contamination with enzymatic uranium reduction. *Environ. Sci. Technol.* **26**: 2228–2234.

79. McClure, N. C., A. J. Weightman, and J. C. Fry. 1989. Survival of *Pseudomonas putida* UWC1 containing cloned catabolic genes in a model activated-sludge unit. *Appl. Environ. Microbiol.* **55**: 2627–2634.

80. Merkel, G. J., S. S. Stapleton, and J. J. Perry. 1978. Isolation and peptido-glycan of gram-negative hydrocarbon-utilizing thermophilic bacteria. *J. Gen. Microbiol.* **109**: 141–148.

81. Mihelcic, J. R., and R. C. Luthy. 1988. Degradation of polycyclic aromatic hydrocarbon compounds under various redox conditions in soil-water systems. *Appl. Environ. Microbiol.* **54**: 1182–1187.

82. Miller, R. M., G. M. Singer, J. D. Rosen, and R. Bartha. 1988. Photolysis primes biodegradation of benzo[a]pyrene. *Appl. Environ. Microbiol.* **54**: 1724–1730.

83. Morgan, P., and R. J. Watkinson. 1989. Microbiological methods for the cleanup of soil and ground water contaminated with halogenated organic compounds. *FEMS Microbiol. Rev.* **63**: 277–300.

84. Mueller, J. G., P. J. Chapman, and P. H. Pritchard. 1989. Action of a fluoranthene-utilizing bacterial community on polycyclic aromatic hydrocarbon components of creosote. *Appl. Environ. Microbiol.* **55**: 3085–3090.

85. Mueller, J. G., S. E. Lantz, B. O. Blattmann, and P. J. Chapman. 1991. Bench-scale evaluation of alternative biological treatment processes for the remediation of pentachlorophenol- and creosote-contaminated materials: Solid-phase bioremediation. *Environ. Sci. Technol.* **25**: 1045–1055.

86. Mueller, J. G., S. E. Lantz, B. O. Blattmann, and P. J. Chapman. 1991. Bench-scale evaluation of alternative biological treatment processes for the remediation of pentachlorophenol- and creosote-contaminated materials: Slurry-phase bioremediation. *Environ. Sci. Technol.* **25**: 1055–1061.

87. Neidhardt, F. C., J. L. Ingraham, and M. Schaechter. 1990. *Physiology of the Bacterial Cell.* Sinauer Associates, Sunderland, Mass.

88. Neilson, A. H. 1990. The biodegradation of halogenated organic compounds. *J. Appl. Bacteriol.* **69**: 445–470.

89. Nelson, M. J. K., S. O. Montgomery, E. J. O'Neill, and P. H. Pritchard. 1986. Aerobic metabolism of trichloroethylene by a bacterial isolate. *Appl. Environ. Microbiol.* **52**: 383–384.

90. Nelson, M. J. K., S. O. Montgomery, W. R. Mahaffey, and P. H. Pritchard. 1987. Biodegradation of trichloroethylene and involvement of an aromatic biodegradation pathway. *Appl. Environ. Microbiol.* **53**: 949–954.

91. Nelson, M. J. K., S. O. Montgomery, and P. H. Pritchard. 1988. Trichloroethylene metabolism by microorganisms that degrade aromatic compounds. *Appl. Environ. Microbiol.* **54**: 604–606.

92. Nyer, E. K., and G. J. Skladany. 1989. Relating the physical and chemical properties of petroleum hydrocarbons to soil and aquifer remediation. *Ground Water Monit. Rev.* **Winter**: 54–60.

93. Oren, A. 1988. The Microbial Ecology of the Dead Sea, in *Advances in Microbial Ecology,* vol. 10, K. C. Marshall (ed.). Plenum Press. New York, pp. 193–229.

94. Park, K. S., R. C. Sims, R. R. Dupont, W. J. Doucette, and J. E. Matthews. 1990. Fate of PAH compounds in two soil types: Influence of volatilization, abiotic loss and biological activity. *Environ. Toxicol. Chem.* **9**: 187–195.

95. Parr, J. F., L. J. Sikora, and W. D. Burge. 1983. Factors affecting the degradation and inactivation of waste constituents in soil, in *Land Treatment of Hazardous Wastes,* J.E. Parr, P. B. Marsh and J. M. Kla (eds.). Noyes Publishing, Park Ridge, N.J., pp. 20–49.

96. Parsons, F. Z., P. R. Wood, and J. Demarco. 1984. Transformation of tetrachloroethene and trichloroethane in microcosms and groundwater. *J. Am. Water Works Assoc.* **76**: 56–59.

97. Paul, J. H., M. E. Frischer, and M. J. Thurmond. 1991. Gene transfer in marine water column and sediment microcosms by natural plasmid transformation. *Appl. Environ. Microbiol.* **57**: 1509–1515.

98. Perry, J. J. 1977. Microbial metabolism of cyclic hydrocarbons and related compounds. *Crit. Rev. Microbiol.* **5**: 387–412.

99. Pfaender, F. K., R. J. Shimp, and R. J. Larson. 1985. Adaptation of estuarine ecosystems to the biodegradation of nitrilotriacetic acid: Effects of preexposure. *Environ. Toxicol. Chem.* **4**: 587–593.

100. Phillips, S. J., D. S. Dalgarn, and S. K. Young. 1989. Recombinant DNA in wastewater: pBR322 degradation kinetics. *J. Water Pollut. Control Fed.* **61**: 1588–1595.

101. Prescott, L. M., J. P. Harley, and D. A. Klein. 1990. *Microbiology.* Wm. C. Brown and Company, Inc., Dubuque, Iowa.

102. Reusch, W. H. 1977. *An Introduction to Organic Chemistry.* Holden-Day, Inc., San Francisco.

103. Riser-Roberts, E. 1992. *Bioremediation of Petroleum Contaminated Sites.* CRC Press, Inc., Boca Raton, Fla.

104. Rittmann, B. E., A. J. Valocchi, J. E. Odencrantz, and W. Bae. 1988. In-Situ Bioreclamation of Contaminated Groundwater. Illinois Hazardous Waste Research and Information Center, HWRIC RR 031.121, Chicago.

105. Rochkind-Dubinsky, M. L., G. S. Sayler, and J. W. Blackburn. 1987. *Microbiological Decomposition of Chlorinated Aromatic Compounds.* Marcel Dekker, Inc., New York.

106. Rontani, J. F., and G. Giusti. 1986. Study of the biodegradation of poly-branched alkanes by a marine bacterial community. *Mar. Chem.* **20:** 197–205.

107. Rosenfeld, J. K., and R. H. Plumb, Jr. 1991. Groundwater contamination at wood treatment facilities. *Ground Water Monit. Rev.* **Winter:** 133–140.

108. Sanglard, D., M. S. A. Leisola, and A. Fiechter. 1986. Role of extracellular ligninases in biodegradation of benzo[a]pyrene by *Phanerochaete chrysosporium. Enzyme Microb. Technol.* **8:** 209–212.

109. Saye, D. J., O. Ogunseitan, G. S. Sayler, and R. V. Miller. 1987. Potential for transduction of plasmids in a natural freshwater environment: Effect of plasmid donor concentration and a natural microbial community on trans-duction in *Pseudomonas aeruginosa. Appl. Environ. Microbiol.* **53:** 987–995.

110. Schmidt, S. K., K. M. Scow, and M. Alexander. 1987. Kinetics of *p*-nitro-phenol mineralization by a *Pseudomonas* sp.: Effects of second substrates. *Appl. Environ. Microbiol.* **53:** 2617–2623.

111. Semprini, L., G. D. Hopkins, P. V. Roberts, and P. L. McCarty. 1991. In Situ Biotransformation of Carbon-Tetrachloride, Freon-113, Freon-11, and 1,1,1-TCA under Anoxic Conditions, in *On Site Bioreclamation,* R. E. Hinchee and R. F. Olfenbuttel (eds.). Butterworth-Heinemann. Boston, pp. 41–58.

112. Sharpley, A. N. 1991. Effect of soil pH on cation and anion solubility. *Commun. Soil Sci. Plant Anal.* **22:** 827–841.

113. Shiaris, M. P. 1989. Seasonal biotransformation of naphthalene, phenan-threne and benzo[a]pyrene in surficial estuarine sediments. *Appl. Environ. Microbiol.* **55:** 1391–1399.

114. Skujins, J. 1984. Microbial Ecology of Desert Soils, in *Advances in Microbial Ecology,* vol. 7, K. C. Marshall (ed.). Plenum Press, New York, pp. 49–91.

115. Spain, J. C., P. H. Pritchard, and A. W. Bourquin. 1980. Effects of adapta-tion on biodegradation rates in sediment/water cores from estuarine and freshwater environments. *Appl. Environ. Microbiol.* **40:** 726–734.

116. Stanier, R. Y., J. L. Ingraham, M. L. Wheelis, and P. R. Painter. 1986. *The Microbial World,* 5th ed. Prentice-Hall, Englewood Cliffs, N.J.

117. Suflita, J. M., and G. W. Sewell. 1991. Anaerobic Biotransformation of Contaminants in the Subsurface. U.S. EPA Environmental Research Brief. EPA/600/M-90/024. EPA, Washington, D.C.

118. Sutherland, J. B., A. L. Selby, J. P. Freeman, F. E. Evans, and C. E. Cerniglia. 1991. Metabolism of phenanthrene by *Phanerochaete chrysosporium. Appl. Environ. Microbiol.* **57**: 3310–3316.

119. Swindoll, C. M., C. M. Aelion, and F. K. Pfaender. 1988. Influence of inorganic and organic nutrients on aerobic biodegradation and on the adaptation response of subsurface microbial communities. *Appl. Environ. Microbiol.* **54**: 212–217.

120. Sylvestre, M., R. Masse, C. Ayotte, F. Messier, and J. Fauteaux. 1985. Total biodegradation of 4-chlorobiphenyl (4CB) by a two-membered bacterial culture. *Appl. Microbiol. Biotechnol.* **21**: 192–195.

121. Szewzyk, R., and N. Pfennig. 1987. Complete oxidation of catechol by the strictly anaerobic sulfate-reducing *Desulfobacterium catecholicum* sp. nov. *Arch. Microbiol.* **147**: 163–168.

122. Teidje, J. M., S. A. Boyd, and B. Z. Fathepure. 1987. Anaerobic degradation of chlorinated aromatic hydrocarbons. *Dev. Ind. Microbiol.* **27**: 117–127.

123. Valdes-Garcia, E., E. Cozar, R. Rotger, J. Lalucat, and J. Ursing. 1988. New naphthalene-degrading marine *Pseudomonas* strains. *Appl. Environ. Microbiol.* **54**: 2478–2485.

124. van Beelen, P., and F. van Keulen. 1990. The kinetics of the degradation of chloroform and benzene in anaerobic sediment from the river Rhine. *Hydrobiol. Bull.* **24**: 13–21.

125. Vannelli, T., M. Logan, D. M. Arciero, and A. B. Hooper. 1990. Degradation of halogenated aliphatic compounds by the ammonia-oxidizing bacterium *Nitrosomonas europaea. Appl. Environ. Microbiol.* **56**: 1169–1171.

126. Vogel, T. M., and P. L. McCarty. 1985. Biotransformation of tetrachloroethylene to trichloroethylene, dichloroethylene, vinyl chloride, and carbon dioxide under methanogenic conditions. *Appl. Environ. Microbiol.* **49**: 1080–1083.

127. Walter, B., and K. Ballschmiter. 1991. Biohalogenation as a source of halogenated anisoles in air. *Chemosphere* **22**: 557–567.

128. Wang, X., X. Yu, and R. Bartha. 1990. Effect of bioremediation on polycyclic aromatic hydrocarbon residues in soil. *Environ. Sci. Technol.* **24**: 1086–1089.

129. Werner, P. 1991. German Experiences in the Biodegradation of Creosote and Gaswork-Specific Substances, in *In Situ Bioreclamation: Applications and Investigations for Hydrocarbon and Contaminated Site Remediation*, R. E. Hinchee and R. F. Olfenbuttel (eds.). Butterworth-Heinemann, Boston, pp. 496–516.

130. Widdel, F. 1988. Microbiology and Ecology of Sulfate and Sulfur-

Reducing Bacteria, in *Biology of Anaerobic Microorganisms*, A. J. B. Zehnder (ed.). John Wiley and Sons, Inc., New York, pp. 469–585.

131. Wiggins, B. A., S. H. Jones, and M. Alexander. 1987. Explanations for the acclimation period preceding the mineralization of organic chemicals in aquatic environments. *Appl. Environ. Microbiol.* **53**: 791–796.

132. Williams, R. J., and W. C. Evans. 1973. Anaerobic metabolism of aromatic substrates by certain microorganisms. *Biochem. Soc. Trans.* **1**: 186–187.

133. Williams, R. T., P. S. Ziegenfuss, and P. J. Marks. 1988. *Field Demonstration: Composting of Explosives-Contaminated Sediments at Louisiana Army Ammunition Plant (LAAP)*. Report No. AMXTH-Ir-TE-88242. Final Report. U.S. Army Toxic and Hazardous Materials Agency, Aberdeen Proving Ground, Md.

134. Wilson, J. T., J. F. McNabb, J. W. Cochran, T. H. Wang, M. B. Tomson, and P. B. Bedient. 1985. Influence of microbial adaptation on the fate of organic pollutants in ground water. *Environ. Toxicol. Chem.* **4**: 721–726.

135. Wilson, J. T., and B. H. Wilson. 1985. Biotransformation of trichloroethylene in soil. *Appl. Environ. Microbiol.* **49**: 242–243.

136. Woodward, R. 1990. *Evaluation of Composting Implementation: A Literature Review*. Report No. AMXTH-IR-TE-88242. Final Report. U.S. Army Toxic and Hazardous Materials Agency. Aberdeen Proving Ground, Md.

137. Wyndham, R. C. 1986. Evolved aniline catabolism in *Acinetobacter calcoaceticus* during continuous culture in river water. *Appl. Environ. Microbiol.* **51**: 781–789.

138. Yong, R. N., L. P. Tousignant, R. Leduc, and E. C. S. Chan. 1991. Disappearance of PAHs in a Contaminated Soil from Mascouche, Québec, in *In Situ Bioreclamation: Applications and Investigations for Hydrocarbon and Contaminated Site Remediation*, R. E. Hinchee and R. F. Olfenbuttel (eds.). Butterworth-Heinemann, Boston, pp. 377–395.

139. ZoBell, C. E., and J. Agosti. 1972. Bacterial oxidation of mineral oil at sub-zero Celsius, in *Abstracts 72nd Annual Meeting of the American Society for Microbiology*, Abstract E11.

3

Hydrogeologic Information and Ground Water Modeling

Craig Rosenbaum-Wilkinson

Global Basins Research Network
LaMont-Doherty Earth Observatory
Columbia University

The hydrogeology of a site is a first-order factor determining both the potential for a contamination problem and the difficulty encountered in remediation. This knowledge is used for testing alternative schemes for bioremediation and for evaluating the results of long-term monitoring before, during, and after bioremediation. The limited scope of this chapter precludes a detailed introduction to the subject; however, several excellent references exist (DeMarsily, 1986; Freeze and Cherry, 1979). This chapter is not for hydrogeologists but for those who may be working closely with one during a bioremediation project.

The basic goal of a hydrogeologic investigation is to understand the movements of fluids (water and organics), dissolved materials, and possibly microorganisms in the porous subsurface. This review concerns the planning, execution, and evaluation of ground water studies and modeling of fluid flow and solute and microbial transport.

The term *model* is used commonly to refer to several distinct things: (1) a theoretical treatment of some physical process, (2) a mathematical rendering of this theory, (3) a computer code which will yield approximate solutions to the mathematical equations, or (4) the results of the implementation of the computer code using some site-specific data input. An appropriate model uses a formulation and assumptions which are valid for the site in question. The more sophisticated the model, the more extensive (and expensive) are the data requirements. The fewer poorly defined parameter values used, the more defensible the modeling conclusions will be. In addition, while all models assume some sort of homogeneity in properties on some scale, nature is stubbornly heterogeneous. A skilled hydrogeologist and modeler are required to know when certain assumptions are valid.

Site-specific data are necessary for meaningful modeling but are rarely available before the study begins. A sampling program is often the only way to adequately define values for many of the physical parameters required by the models.

A site reconnaissance stage is critical both to identify possible processes that are site-specific and to evaluate any problems that will be associated with the execution of the sampling program. Despite the costs, planners are encouraged to make the sampling plan as comprehensive as possible. When reconnaissance is completed and possible site processes are identified, a detailed sampling plan is designed and carried out. This plan must aim to prove the presence or absence of the various proposed processes. It is also likely that the plan will need to be revised in progress as the initial results are examined. Field measurements are invaluable in revealing trends because they yield immediate results, although they often lack the accuracy of laboratory determinations. Field data should be compiled and reviewed from day to day to ensure that the sampling program yields as much useful data as possible.

When the laboratory analysis of samples is complete, the data set is assembled, and some standard manipulations carried out (Mazor, 1991) to characterize the general nature of the site. The values of parameters are estimated, and the required input data are used to execute an appropriate computer model. The resulting model predictions are compared with the field data. Contradictions may be resolved with further sampling. It is likely that the data set will indicate interesting trends that need to be investigated further.

How close to reality is the model? This should be determined before the model results are relied on, and *rely* must be used in a very restricted sense. Few models can claim to be more than predictive approximations, albeit useful ones for many purposes.

Validation of a model is only relative; by showing accuracy and error estimates, the validity of results can be evaluated by all parties involved, and other interpretations can be considered. In some cases, as in the study by Mercer et al. (1983), a range of possible velocities and the associated errors were estimated for transport at a site. This approach avoids some of the subjective aspects of a "single-number" result. Mangold and Tsang (1991) discuss several validation issues. The model formulation and the solution must be mathematically correct, and the model should be demonstrated to make good predictions of actual simple processes. This step is usually accomplished to some degree by the creator of the model code. It is the user's responsibility to determine whether the model in use is appropriate for the job at hand and that the site conditions lie within the range of conditions for which the model has already been proven to yield accurate results. A model must be validated with respect to a specific process before it may be used with confidence. A model that has been validated with respect to a given site may not yield accurate results at another site.

Basic Physical Concepts and Terms of Subsurface Flow

The *unsaturated zone* refers to the soil and rock between the surface and the water table. The unsaturated zone may be a few meters to hundreds of meters thick. The top 2 m is commonly marked by high clay content, abundant natural organic matter, and extensive microbial activity. The *water table* is the surface below which virtually all the pore space is full of water ("saturated"). All the rock or soil below the water table (in the saturated zone) is commonly assumed to be saturated with ground water. Water at the surface moves through the *pores* (spaces between the solid rock particles) in the soil or rock (*matrix*) and flows through the unsaturated zone toward the water table. Once the water reaches the saturated zone, it can, in general, move much more freely and rapidly.

The most important parameters that need to be defined in a subsurface bioremediation project are those relating to the subsurface flow of water. Many of these parameters vary in three-dimensional space, which is commonly divided into x, y, and z directions, with z representing depth. For example, hydraulic conductivity commonly exhibits two different values at one point corresponding to the conductivity in the x and z directions.

The fundamental equation used to describe the flow of water through a porous media is *Darcy's law*. This equation, first stated by the French engineer Henry Darcy, is an empirical law which was developed from experiments using tubes of a known cross section A filled with various soils. Basically, Darcy's law states that the amount of a fluid (water or any other liquid) which can be forced through a given area of soil Q is related to the driving force behind it (dh/dx, the hydraulic pressure gradient, or just hydraulic gradient) and the nature of the material through which it flows (the hydraulic conductivity of the medium). Mathematically, Darcy's law (for flow in the x direction) is expressed as

$$Q = - KA(dh/dx) \qquad (3\text{-}1)$$

where Q = flow (m³/s)
   K = hydraulic conductivity (m/s)
   dh/dx = hydraulic gradient (m/m)
   A = cross-sectional area (m²)

Specific discharge q, (also called *Darcy velocity* or *Darcy flux*) is equal to Q/A. Thus Darcy's law (for the x direction) becomes

$$q_x = - K_x(dh/dx) \qquad (3\text{-}2)$$

The hydraulic gradient term represents the change in hydraulic head dh over a small distance dx. Hydraulic head, usually written as h is classically defined as the measurable physical quantity that results in flow occurring from higher values of head to lower values (Freeze and Cherry, 1979). There are two components to the hydraulic head— the pressure head (usually written as ψ) and the elevation head (written as z). The relationship between hydraulic head, elevation head, and pressure head can be described by the simple equation

$$h = z + \psi \qquad (3\text{-}3)$$

Elevation head z is the difference in elevation between the drop of water under consideration and some reference datum, usually sea level or the ground surface. Pressure head ψ is due to the forces exerting pressure on the fluid. While not an important component of the hydraulic head in surface water, the pressure head does become important in deep artesian aquifers, where water is under large pressures. Pressure head is also important at the top of the saturated zone, where negative pressures result from capillary action. Hydraulic gradient is a measure of the difference in the hydraulic head of a fluid at two different points within an aquifer or, in its simplest form, the difference in hydraulic head at two points divided by the distance between the points. Where large vertical gradients exist, pressure head dominates the hydraulic head, resulting in upward or downward flow of ground water.

Hydraulic head is measured under field conditions using a piezometer. A piezometer is simply a very small diameter well which is open to the atmosphere at the top and terminates within the aquifer to be measured at depth. The height of the standing water in the piezometer relative to a standard reference point is the total hydraulic head at that site. It is important to note that the hydraulic head will vary in three dimensions in the ground. Thus a single piezometer reading will tell you only the total hydraulic head for a single specific point in three-dimensional space. When numerous piezometric measurements of hydraulic head are plotted and contoured (connecting points with equal head), a *water table map* can be developed. This is the usual result for measurements in *unconfined aquifers*, which are shallow aquifers whose top is the ground surface (or more correctly, the water table or the top of the saturated zone). In a confined aquifer, where the top boundary is some semipermeable material, pressures can build up which are not normally possible near the ground surface. Piezometers which terminate within a confined aquifer will define a hydraulic head surface which may be largely unrelated to the water table surface. This surface is often referred to as a *potentiometric surface,* although there are problems with this usage (Freeze and Cherry, 1979).

The hydraulic conductivity K of a system is a measure of the ease with which water (or any other fluid) passes through a given medium and is a function of both the medium and the fluid. The proportionality constant K in Darcy's law is defined as $k\rho g/\mu_f$. Density ρ_f and viscosity μ_f are the most important properties of the fluid affecting K. The term g is the gravitational acceleration. Permeability k, a characteristic of the medium alone, is the most important property of the medium influencing K. It reflects both the amount of pore space within the formation and the degree of interconnection within the pore structure. Exact permeability values are extremely difficult to determine. Laboratory measurements such as constant or falling head permeameter tests are limited by the size of the sample which can be evaluated. Field tests such as pump tests are limited by incomplete knowledge of the subsurface environment. As a result of these limitations, "order of

Table 3-1. Hydraulic Conductivities

Medium	K (m/d)
Clay soils (surface)	10^{-2}–0.2
Deep clay beds	10^{-8}–10^{-12}
Loam soils (surface)	0.1–1.0
Fine sand	1.0–5.0
Medium sand	5.0–20.0
Coarse sand	10^2–10^3
Sand and gravel mixes	5.0–10^2
Glacial till	10^{-3}–10^{-1}
Sandstone	10^{-3}–1.0
Shale	10^{-7}
Dense solid rock	10^{-5}
Fractured or weathered rock	10^{-3}–10^1

SOURCE: Bouwer, 1984.

magnitude" estimates of permeability or hydraulic conductivity frequently must be relied on for hydrogeologic planning and modeling. Field and laboratory studies have shown that such estimates can vary by more than 13 orders of magnitude; within a single rock type (e.g., quartz sandstones), conductivity commonly spans 2 to 3 orders of magnitude. Table 3.1 summarizes typical hydraulic conductivities for geologic media.

Darcy's law alone is insufficient for all but the simplest flow calculations. Fluid flow modeling requires the conservation of fluid mass:

$$- [\delta(\rho q_x)/\delta x + \delta(\rho q_y)/\delta y + \delta (\rho q_z)/\delta z] + \rho W = \delta(\rho \phi)/\delta t \qquad (3\text{-}4)$$

where ρ = fluid density
  q = specific discharge in the $x, y,$ or z direction
  W = source/sink rate term
  ϕ = porosity of the medium
  t = time

Simply stated, the conservation of fluid mass requires that the sum of (1) fluid mass entering or leaving the control volume in the $x, y,$ or z

directions (term in brackets) and (2) the creation or destruction of free fluid mass within the volume must equal (3) the change in fluid mass within the volume during the time step. When Eq. (3-4) is combined with Eq. (3-2) (Darcy's law), changes in hydraulic head with respect to distance are related to changes in fluid flux with respect to time. This is the basis of fluid flow modeling.

The conservation of energy is required when temperature calculations are coupled with fluid flow models:

$$\rho_f c_f [q_x \delta T / \delta x + q_y \delta T / \delta y + q_z \delta T / \delta z] - \lambda_m [\delta^2 T / \delta^2 x +$$
$$\delta^2 T / \delta^2 y + \delta^2 T / \delta^2 z] = \rho_m c_m (\delta T / \delta t) \qquad (3\text{-}5)$$

where ρ = density of the fluid (f) or matrix (m)
  c = specific heat capacity of fluid (f) or matrix (m)
  q = specific discharge in the x, y, or z direction
  T = temperature of fluid and matrix
  λ_m = thermal conductivity of the matrix
  t = time

Simply stated, the conservation of energy requires that the sum of (1) heat energy entering the control volume carried by moving fluid (first term) and (2) the heat leaving the control volume by heat conduction through the solid grains (second term) must equal (3) the change in heat content within the control volume during the time step. Through Eq. (3-5), which contains specific discharge q, fluid flux and changes in the temperature field are related.

Data Collection and Synthesis

The hydrogeologist attempts to provide a description of the subsurface system and its properties as they affect the bioremediation project, including the geologic, hydrologic, and contaminant transport properties of the rock and soil, as well as the chemical composition of the ground water.

Common goals of a field investigation include

1. A geologic interpretation, including structure (faults, strata, deformation, lithologies), soil maps, an interpretation of the depth to bedrock ("basement"), paleogeology (buried river channels, allu-

vium, glacial deposits, karst), and synthesis of regional and local data (geologic, structural, borehole)

2. A hydrologic interpretation, including the seasonal behavior of the water table, local water courses, rainfall, infiltration, and flow in wells

3. A hydrogeologic interpretation, including subsurface hydraulic conductivity structure from laboratory measurements, borehole logging, hydraulic packer experiments (with or without tracers), surface pumping tests, and fracture mapping

4. A hydrochemical interpretation, including characterization of the subsurface chemical conditions, which should be consistent with the hydrogeologic interpretation

5. A historical perspective on the site and its problems

6. Establishment of multiple working hypotheses

7. Resolution of conflicting evidence, with retesting, if necessary

8. Preliminary interpretation of the "real" hydrogeologic situation

9. Application of one or more validated mathematical models to this system to test interpretations

10. Predictions and recommendations based on tested models

To accomplish these goals, a number of projects may be implemented:

Historical Data Search. Data may exist in the form of historical records (state and local agencies, industry, etc.) and chemical analyses if you are lucky and know where to look. Drillers' and well operators' records are often quite useful in constructing a historical picture of pumping and disposal practices as a function of time. Old measurements can be useful for seeing long-term changes in chemistry or water table location. Regional or local historical development data can be used to understand (1) the local history of human land use, (2) the number of wells and pumping rates as a function of time in the area, and (3) the history of the development or introduction of chemical compounds into the region.

Mapping. Mapping of a site may include creating or using geologic base maps; air photographs; field surveys of soil, rocks, and vegetation; land use and development maps; surface water maps (extent of watershed, river stage, recharge/discharge zones); or soil moisture maps to characterize the site.

Seismic Refraction Profiles.   Shallow seismic methods may be used for mapping depth to basement and delineating buried river channels, alluvial fans, and glacial deposits.

Hydrologic Monitoring.   Long-term hydrologic monitoring of rainfall, infiltration, evaporation, and depth to water table in wells and flow rates in water courses is useful in understanding the hydrologic budget and its seasonal changes.

Sampling.   Samples of soil, water, gases, and plant material are collected and analyzed to characterize some of the properties of the solid/liquid system. Soil and rock chemistry are needed for making sense of chemical and isotopic data for ground water mapping. Samples are obtained from water wells, surface water, tanks, and industrial equipment. Historical samples may be available.

Borehole Geophysical Well Logging.   These methods can provide insight into subsurface geology, structure, physical properties, and some hydrogeologic properties (hydraulic conductivity, fluid velocity from flowmeters) and include conventional pumping tests and downhole fluid sampling. Wells may be drilled or may exist already, but they are invaluable in that they give access to the subsurface. Water wells, disposal wells, exploration wells, special-purpose (scientific) wells, mines, and quarries can be useful.

The basic logging tools measure gamma ray (natural potassium, uranium, and thorium) activity, formation electrical resistivity, borehole caliper, and sonic velocity, which are interpreted for rock type, stratigraphic boundaries, formation fluid properties, and clay (shale) content. More specialized tools such as the borehole televiewer and the multichannel sonic logging tool are used for fracture mapping. Many references exist for these tools and their uses (Ellis, 1987; Verdier, 1986), and logging services may be contracted for easily in many areas. The larger companies sell only logging services (rather than the tools themselves) as a rule. Basic logging tools may be purchased from one of the smaller manufacturers, who often advertise at professional conferences.

Other useful tools for hydrogeologic investigations are the hydraulic packer, the temperature tool, and the downhole flowmeter.

The hydraulic packer is designed for pumping/injection tests on isolated intervals. It can be modified for fluid sampling from specific formation intervals. The temperature tool is a self-contained microprocessor-based temperature profiling instrument which can detect very small differences in ground water temperature that are often related to fluid flow, while the downhole flowmeter can be positioned at any desired depth to record fluid flow.

Physical Properties Measurements.   Measurements of temperature, pH, and major ion values using field test kits can all be made quickly and repeatedly. Hydraulic conductivity determinations and various borehole tool calibrations are made in the laboratory using cores (if available) or cuttings.

Chemical Methods.   These include chemical analyses of soil, rock, and water. Although the exact analyses depend on the site-specific conditions, general measurements include salinity, pH, dissolved solids, organics, inorganic nutrients and electron acceptors, and some trace elements (iron). Radioisotopes (e.g., oxygen 18, carbon 14, and tritium) are used to determine the age of the ground water.

Pumping Tests.   These are used to determine hydrologic properties such as formation hydraulic conductivity and cross-hole connectivity. Well tests involve the addition or removal of water from one or more wells and the measurement of the water table response over time. Either single or multiple well tests are used. Multiple well tests are usually more expensive because of the need to drill additional monitoring wells, the additional data interpretation required, and the larger quantities of contaminated water which must be disposed of. The data, however, are usually of better quality than available from single well tests. Information on the design and analysis of pump tests is available from the National Water Well Association (NWWA), the United States Geological Survey (USGS; *National Handbook of Recommended Methods for Water-Data Acquisition*), the American Institution of Professional Geologists (AIPG), or any of a number of standard reference books (Freeze and Cherry, 1979; McWhorter and Sunada, 1977).

Ground-Penetrating Radar.   This relatively new technique is used for the detection of the water table, organics floating on the water table, and buried tanks and pipes.

Processes That Are Modeled

Ground water modeling commonly follows one of three approaches: (1) single-phase (water-only) saturated or unsaturated flow models, (2) dissolved substance transport models, or (3) multiphase (water and organics) flow models. These are discussed very briefly below. For greater detail, see Bear and Verruijt (1987), National Research Council (1990), DeMarsily (1986), Huyakorn and Pinder (1983), Javandel et al. (1984), and Wang and Anderson (1982).

Saturated/Unsaturated Ground Water Flow

Ground water flows in response to a hydraulic potential gradient due either to potential differences on the boundaries of the flow cells, heterogeneity in the hydraulic conductivity structure, or the presence of sources or sinks (withdrawing or injecting wells) within the boundaries of the flow system. Steady-state models assume that hydraulic potential is constant with time, while transient models assume that potential may change with time.

Dissolved Substance Transport and Geochemical Reactions

Advection, the process wherein a dissolved substance moves with the moving water, is often the dominant transport process. *Diffusion,* the migration of a substance in response to a concentration gradient, results in mass mixing. *Dispersion* commonly operates at a fine scale and results in a mixing fringe between a contaminant plume and fresh ground water. Dispersion has both a transverse and a longitudinal component. The effects of dispersion are commonly much greater than those of diffusion. Other transport models include processes such as *radioactive decay, sorption* (wherein substances attach to surfaces of minerals, organic particulate matter, clay minerals, metal oxides, or noncharged organic molecules), *precipitation/dissolution, complexation, hydrolysis/substitution, oxidation/reduction,* or *biological transformation reactions.*

Multiphase Flow

Multiphase flow refers to the combined flow of water and other immiscible fluids such as organic liquids. As in ground water flow, the individual potential fields for these liquids separately determine the flow direction and rate for each. Since the hydraulic potentials of organics and water may be different, flow directions may be different. The relative hydraulic conductivity of the medium (matrix) for each of these liquids increases as the percentage saturation increases; in other words, fluids like to flow with themselves. The style of migration of an organic liquid depends on several factors, including its density relative to water. Organics can flow within the *capillary fringe*, which is a transition zone between the unsaturated and saturated zones just above the water table. If the organic liquid is lighter than water, it can float on the water table. Such a compound is often referred to as a *light non-aqueous-phase liquid* (LNAPL). If it is heavier (*dense non-aqueous-phase liquid, DNAPL*), it will sink through the water table to the next boundary. The migration of organic liquid through a porous medium often leaves behind an immobile residue.

Models

Ground water models are based on mathematical simplifications of physical processes and make use of what is known about the geologic structure, the physical properties of the porous media, and the distribution of potential for all the phases involved. A prospective user of a given model should keep in mind the following questions:

1. What specific hydrologic, chemical, or physical processes have been incorporated into the formulation?
2. What are the simplifying assumptions of the model?
3. What input information is required, and how well constrained are these data?
4. What does the computer code actually do?
5. What is the output information?

Computer-Based Numerical Solutions of Mathematical Models

A recent review of hydrologic and hydrochemical models by Mangold and Tsang (1991) provides an excellent summary of some of the properties of various models currently available. Those considering the selection of a model for their particular needs are referred to this work.

Models of ground water flow are based on numerical approximations to the exact solution of Laplace's equation for fluid flow in a porous medium. A model commonly includes one or more differential equations. Solutions to these equations are found at numerous points on a grid which represents (in a schematic way) the cross-sectional physical and/or chemical properties of the site under consideration. In general, other values required by the model include various parameter values estimated from field data, the boundary conditions and initial conditions for the model (values of temperature, pressure, fluid, and/or heat flux at the boundaries, initial values throughout for temperature, pressure, and all physicochemical parameters, variables, constants, and functions), and some parameters that concern the operation of the code itself (time steps, grid size). An example of the transition from the geologic "reality" (based on field work) to the simplified "model" is shown in Figures 3-1 and 3-2. The locations of the boundaries of different strata are approximated by sets of some type of discrete element, in this figure rectangles (often triangles or squares).

The governing equations of interest may include (1) the conservation of fluid mass (continuity equation), (2) the conservation of energy, (3) the conservation of dissolved mass, and possibly, (4) various relationships concerning specific chemical, physical, or biological processes such as microbial reactions. In the absence of sources/sinks for fluid or mass, the sum of the fluid volume and dissolved mass over the entire model domain must remain constant (or very close to it) at all stages of the calculations. The boundary conditions apply to the edges of the grid. The three basic types of boundary conditions which may be used are (1) the *Dirichlet type,* which means that a specified value of some parameter (e.g., temperature, pressure, or chemical concentration) exists at a particular boundary, (2) the *Neumann type,* which means a specific gradient or flux exists, or (3) a *variable flux* of value as a function of some other parameter. Fluid sources or sinks, such as pumping, injection, mineral dehydration reactions, and infiltration from rainfall, are examples of boundary conditions. Chemical and biological reactions, contaminant sources, and dissolved substances in atmospheric recharge water are mass sources or sinks.

For multiphase flow, one continuity equation is required for each phase. For dissolved components, the required data may include advective flux, dispersion coefficient, rate constants for radioactive decay, and coefficients (retardation factors) for sorption processes. For a single dissolved ion, the multicomponent velocity field, transverse

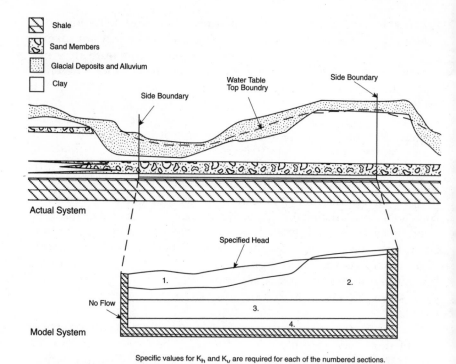

Figure 3-1. A simplified geologic representation (*a*) serves as the basis for a model cross section (*b*) with the approximate shape, hydrogeologic properties, and boundary conditions of the actual system. (*Adapted from National Research Council Committee on Groundwater Modeling, 1990.*)

and longitudinal dispersivities, and decay rates for any reactions are required. Data requirements increase as the complexity of the system of equations increases. In addition, some complex reactions are not well understood.

Solving the Equations

It is rare for a set of useful differential equations to have an exact solution. Exact solutions generally exist only for simple cases. In general, the governing differential equations are expressed as approximations in algebraic form (sums over discrete intervals instead of integrals over continuous segments). The result is a set of related similar equations for each volume element (each section on the grid), which can be

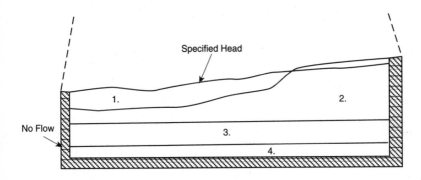

Specific values for K_h and K_v are required for each of the numbered sections.

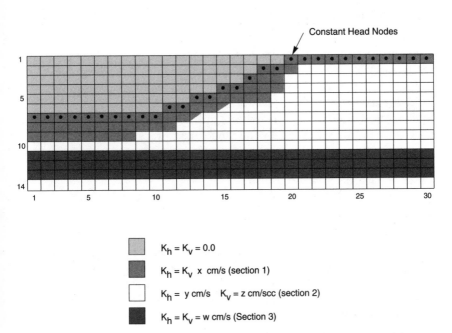

$K_h = K_v = 0.0$

$K_h = K_v \ x$ cm/s (section 1)

$K_h = y$ cm/s $\quad K_v = z$ cm/scc (section 2)

$K_h = K_v = w$ cm/s (Section 3)

Figure 3-2. A grid of rectangular elements is superimposed on the model cross section from Figure 3-1 *b* to obtain a set of volume elements. The governing equations are solved with reference to this set of elements. (*Adapted from National Research Council Committee on Groundwater Modeling, 1990.*)

assembled conveniently into a numerical matrix and solved by standard numerical methods (Huyakorn and Pinder, 1983). Once the governing equations have been formulated and combined, the remaining issues largely concern which numerical techniques will be used to manipulate this system of equations to produce an approximated solution for each point on the grid. This solution must satisfy the energy and continuity equations. The solutions method commonly used is either simultaneous solution by direct means (e.g., matrix inversion) or iterative methods which rely on successive approximations and refinements based on the use-specified allowable error level. The numerical accuracy of the solution to these equations (be they right or wrong conceptually) at this stage depends on grid fineness, the size of time steps, and the numerical method utilized. These numerical methods may be finite-difference or finite-element methods or some other equally well established numerical approximation technique.

Current Modeling Capabilities

Some processes have been addressed for many years and are very accurately modeled. Others are still in the research stage and are not commonly available (or reliable enough) for commercial modeling efforts. Models exist for saturated, unsaturated, and fracture flow, saturated and unsaturated multiphase transport, and geochemical and biological processes, but they display a wide range of accuracy and reliability.

Flow in the Saturated Zone

This type of modeling is well established for fully saturated porous media. The equations are almost linear, and simple mathematical models give good results generally. One- or two-dimensional models are often sufficient, although three-dimensional models are becoming more common. Assumptions of steady-state flow, spatial homogeneity, and the applicability of Darcy's law (invalid for high-velocity flow near well screens) are common. Estimates of flow velocities based on spatial variations in hydraulic potential are obtained. A good bit of input data (soil, rock, and fluid physical parameters) are required even for these simple models. Parameter estimation is a critical step but still not a routine matter. It is important to decide whether a one-, two-, or three-dimensional model is required. The physical boundary conditions must be known or estimated. The user must be certain that the steady-state flow approximation is suitable for this site. Training in the use of widely accepted programs such as the USGS's MOD-FLOW is available through professional groups such as the NWWA.

Flow in the Unsaturated Zone

Modeling of flow in the unsaturated zone has become relatively common. Water flow in the unsaturated zone is transient because of the transient nature of recharge and fluid transport from the surface to the water table. Infiltration at the surface is governed by soil properties and recent soil moisture history. Water flows through the unsaturated zone seeking a (quasi)equilibrium distribution of potential. When the transition to equilibrium is complete, the zone is hydraulically static, and the only sources and sinks are evapotranspiration and chemical or biological reactions. The pores in the unsaturated zone are filled with water, gases, and possibly non-aqueous-phase liquids (NAPLs). The degree of saturation depends on soil properties and surface sources/sinks. Hydraulic conductivity is a nonlinear variable in unsaturated soils; conductivity is proportional to the degree of saturation for the phase in question, and flow of a particular phase is proportional to the potential gradient of that phase. Models such as SUTRA (USGS/Voss, 1984) or TRACR3D (Los Alamos National Lab/Travis, 1984) are well-verified and flexible.

Flow in Fractures

Modeling of flow in fractured rocks is still in the research stage. Fractures (joints, faults, shear zones) result from a wide range of geologic processes. They are quite common and vary a great deal in size, shape, distribution, and hydraulic conductivity. The porosity created by fracturing is superimposed on the original matrix porosity and may exceed it by several orders of magnitude.

Flow in (open) fractures may approximate conduit flow, and here the applicability of Darcy's law may come into question. The fracture is often modeled as a simple channel between parallel plates. In fact, fracture surfaces are found to be rough rather than smooth, with non-parallel faces (Scholz, 1990). The fracture plane is a rough, sinuous surface in reality. In addition, fractures may be either open or closed as a function of the chemical or physical transport processes and of pressure or depth.

The current efforts at fracture flow modeling include (1) single parallel-plate fracture models, which predict flow through a network of discrete fractures (of specific location, orientation, and property), (2) dual-porosity models, which calculate flow in fractures and flow in the porous matrix separately and apply a transfer function to the surface area between the two "phases," and (3) models which approximate the fractured rock using an equivalent porous medium with a modified (but spatially homogeneous) hydraulic conductivity.

Several problems exist for the fracture flow modeler. On some scale, the medium must be treated as homogeneous. The value of this characteristic scale length must be determined. The *representative volume element* (RVE) is that volume which is neither so small nor so large that geologic or hydraulic heterogeneities are exaggerated or obscured. In practice, this results in a subjective judgment. The discrete network approach requires lots of input data and lots of computer effort. In addition, it is unlikely that the exact geometry of the real fracture network is known.

Transport Models

Modeling of transport in the saturated and unsaturated zones is relatively common, while transport modeling in fractured rocks is still in the research stage. Naymick (1987) presents an excellent review of transport modeling. Transport models seek to predict the concentration of substances as a function of time and space, and a good fluid flow model is essential to a good transport model. In fact, advection often dominates the transport of contaminants to such a degree that the influence of other processes becomes negligible. The equations used are simplified approximations to real reactions. Linear or equilibrium relationships are straightforward to model. Nonlinear reactions are more difficult. It is also difficult to get the numerous and varied field data needed to calibrate these models. The compositional variability of the medium is as much an issue here as hydraulic heterogeneity is for ground water flow modeling. Several more specific parameters are required for each immiscible liquid phase. The longitudinal and transverse dispersion coefficient, for example, must be determined or estimated. Unfortunately, this information is rarely known in any detail, and some of the determinations are subject to considerable error. Detailed chemical field studies require time, money, and experience.

No one model is appropriate over a wide range of conditions. A false assumption of steady-state flow will result in error, since dispersion increases with variations in velocity (as a function of time). Dispersivity is an approximate linear function of transport distance, not a constant, and so its relative importance depends on the scale of modeling.

The case of nonconservative solutes (those which undergo reactions along the flow path) is generally considered either as an equilibrium concentration problem or as a kinetic formulation depending on the rate of flow. Redox, acid-base, and radioactive decay processes are relatively well characterized. Complexation, substitution, hydrolysis, and dissolution reactions are characterized by parameters whose values are less accurately known.

Transport in the unsaturated zone is marked by advection with accompanying retardation and attenuation, as well as vapor-phase transport. Since flow within this zone is transient (a function of infiltration rate at the surface), solute transport is also transient. The physical properties of the soil and water at the surface control the flow events and the reactions in the unsaturated zone. The idea that this zone acts as a buffer for the saturated zone is only partly correct. Under certain circumstances, contaminants may remain immobile within the unsaturated zone or at least move very slowly. In terms of remediation, this represents a potential danger in that the unsaturated zone can act as a temporary sponge or repository. This could delay the prompt transport of contaminants to the water table (and likewise to the remediation wells) and result in the release of contaminants to the ground water over a long time period.

Saturated solute transport models include those with two-dimensional capabilities, such as PTC (Lai et al., 1986), KONBRED (Konikow and Bredehoeft, 1978), or the Grove/Galerkin model (Grove, 1977), and those with three-dimensional capabilities, such as CFEST (Gupta et al., 1982), HST3D (Kipp, 1987), or PORFLOW (Runchal, 1985). Unsaturated solute transport models include the two-dimensional FLOWS (Noorishad and Mehran, 1982) and SUTRA (Voss, 1984) and the three-dimensional CHAMP (Narasimhan and Alavi, 1986), GS3 (Davis and Segol, 1985), and SWANFLOW (Faust, 1985).

A common problem with all these models is the large computer effort required. Three-dimensional models for transport in a homogeneous medium represent the most advanced models which are still practical to solve currently.

Multiphase Transport

Organic fluid transport in the unsaturated zone depends on the fluid potential, fluid retardation, and relative hydraulic conductivity of the air, water, organics, and solids. Truly multiphase systems require the ability to model the transient flow in the unsaturated zone. The equations of flow for multiphase organic transport modeling are reviewed in Pinder and Abriola (1986). The migration of miscible contaminants in water can be modeled using the standard solute transport models.

Equations are required for fluid conservation from each of the phases, fluid retention, and relative hydraulic conductivity relationships for each phase. If organic partitioning into water and gas phases is modeled, partition coefficients are needed. This dramatically increases the model complexity.

Geochemical Models

Common equilibrium-type models include MINTEQ (Felmy et al., 1983), PHREEQE (Parkhurst et al., 1980), SOILCHEM (Sposito and Coves, 1988), and WATEQ4F (Ball et al., 1987). Kinetic models, such as EQ6 (Delany et al., 1986), are far less common.

Models that couple substance transport and chemical or biological reactions (e.g., DYNAMIX, Narasimhan et al., 1986; CHEMTRN, Miller and Benson, 1983; and TRANQL, Cedarberg et al., 1985) are complicated and require a large computer effort. Potentially large source/sink terms are present in phase transfer processes, yet for these to be modeled successfully, a rate term must be supplied in addition to the subsurface composition.

Simpler (and faster) models such as FASTCHEM (Hostetler and Erikson, 1989) solve for flow, transport, and equilibrium chemistry for major species only.

Field testing and calibration are critical for all transport and geochemical models. The combination of field and laboratory tests with model verification is invaluable, though not common.

Microbial Transport Models

Microorganisms differ from particles mainly in their ability to grow in the subsurface. Microbes are commonly observed attached to the surface of the matrix grains, where they use nutrients dissolved in the ground water. The characterization of bacterial growth and substance utilization in a porous medium is complex.

Successful microbial transport modeling requires a wide range of chemical, geologic, hydraulic, and microbial measurements to adequately characterize the flow/transport system. Many field studies of microbial transport in the subsurface are based on incomplete hydrogeologic data sets; future modeling projects should include collection of soil and aquifer hydrogeologic data to increase their accuracy.

Processes affecting microbes in the subsurface include

1. Advection, the transport of microbes with moving fluid

2. Dispersion, composed of mechanical mixing and diffusion

3. Adsorption, a function of the organism and soil properties

4. Filtration, including straining and sedimentation, both functions of the organism, fluid velocity, soil type, size of the population, and water quality

5. Decay/growth, a function of the organism and ambient conditions

Microbial transport models (MTMs) are commonly developed from solute transport models (STMs), with several specific assumptions. Two basic approaches to mass and energy balances in modeling are (1) the *eulerian approach,* commonly used for fluid flow modeling, where the calculated balances are relative to volume elements which have constant sizes and spatial locations, and (2) the *lagrangian approach,* commonly used for particle tracking, where a fixed unit of mass is tracked as it moves through space and time. The transport of a solute particle through a porous medium involves many factors (pore size and geometry, path lengths, degree of saturation, turbulence, etc.). Since the lagrangian approach produces a very complex modeling problem, the eulerian method is a commonly used approximation for deriving STMs.

In the case of MTMs, the advective and dispersive transport of microorganisms is calculated from the average flow velocity of the ground water that carries them. The hydrodynamic advection-dispersion equation (ADE) is the basis for the advective-dispersive microbial transport models (ADMTMs) based on STMs:

$$\theta(\delta C / \delta t) = div\,(\theta D_h \cdot gradC) - q(gradC) - \theta\mu C \qquad (3\text{-}6)$$

where θ = moisture content
C = mass concentration of solute
t = time
D_h = coefficient of dispersion
q = specific discharge (Darcy velocity or Darcy flux)
μ = first-order decay constant

The ADE states that the time rate of change in concentration of a solute in a volume element is equal to movement through the volume, solute sources and sinks (solution/deposition) within the volume, and losses due to decay. This equation is valid for both saturated and unsaturated conditions.

For saturated flow (where θ is constant), a simpler version holds:

$$\delta C/\delta t = div(D_h \cdot gradC) - v_{gw} (gradC) - \mu C \qquad (3\text{-}7)$$

where v_{gw} is the ground water flow velocity. The ADE and the mass/energy conservation equations (Eqs. 3-2, 3-4, and 3-5) have this variable in common. They are coupled equations and must be solved together.

Various transport models try to make use of equations which attempt to describe mathematically the processes affecting microbes in porous media. Each of these equations can be thought of as a small model to predict one particular process. Limited understanding of the exact nature of these processes severely limits their usefulness. The parameters in these equations must be measurable in some way in order to be used in a transport model. In practice, parameter determination is often very difficult.

Microbial advection equations generally assume that microbes travel with the fluid as it moves through pores. Under favorable nutrient conditions, however, some microbes are observed to penetrate porous media in the absence of fluid flow simply by growth and propagation into pores. The self-propulsion exhibited by some microbes (rates approaching 1 m/d) can be important when fluid velocities are low.

Microbial dispersion equations attempt to account for the mechanical mixing and dissemination of microbes as they travel away from a source point through a porous medium with some measured or inferred hydraulic properties. In reality, the porous matrix often includes secondary pore structures (macropores and fractures) which can radically alter the hydraulic character of the medium. Dual-porosity models (Mills et al., in press) are one result of attempts to understand the hydraulic influence of fractures on transport. This approach treats the fractured aquifer as two separate compartments with different properties, and flow between the two is governed by an exchange coefficient. Permeability is an important factor controlling microbial penetration into rock. In media where grain size is small relative to cell dimensions, straining of cells by the medium can result in permeability reductions. Colloid transport theory predicts that increases in fluid velocity tend to concentrate colloids into faster flow paths. When permeability is large, the pore distribution tends to large pore sizes, and penetration of microbes into the medium is large. When permeability is low and pores are small, penetration is small.

Microbial source/sink equations try to account for the gain or loss of microbial biomass from the subsurface. Adsorption, described by colloidal filtration theory, is a function of the balance of attractive and repulsive forces affecting the microbe and the matrix surface. Deadsorption can occur with changes in ionic strength, composition, or pH of the fluid. Growth and decay functions are formulated in several ways as a function of the organism(s) and water chemistry.

In order to derive equations for growth and substrate utilization, some assumptions are commonly made about the location and distribution of the bacterial biomass. While many models of the adsorption of microbial biomass assume that a biofilm forms on grain surfaces, this may not be the best approach. Baveye and Valocchi (1989) discuss three competing conceptual approaches used in developing mathematical models of bacterial growth and transport of biologically reacting solute in a saturated porous medium. In the first approach (e.g., Borden and Bedient, 1986), pore-scale processes are ignored. No assumption is made about the microscopic configuration or spatial distribution of bacteria in the medium. The second approach (Molz et al., 1986) considers the bacterial mass to be distributed in discrete microcolonies fixed on the particle surfaces. These may be considered to grow as a result of substrate consumption or may be treated as having uniform and constant size. The third approach (e.g., Bouwer and Cobb, 1987) assumes that the surface of the matrix particles is completely covered by a continuous sheet of bacteria, called a *biofilm*. The transport of nutrients into the biofilm is kinetically controlled, while movement of compounds within the film is diffusion controlled. Baveye and Valocchi (1989) demonstrate that these three formulations differ only in the way the equations for transport parameters are expressed, and thus prefer the first approach because it makes no unwarranted assumptions about the spatial distribution of biomass.

Equations for growth/decay, filtering by the medium, adsorption, and many other mechanisms of potential significance to MTMs have been derived by Corapcioglu and Haridas (1984, 1985). Detailed descriptions of the mechanisms are presented, but their use in MTMs requires the estimation of many parameters which are site-specific and difficult to measure.

Many difficulties remain with ADMTMs. In reality, microbes are transported as colloids rather than solutes, but colloid transport theory is complex and incomplete. Some of the current models tend to underpredict transport because they assume homogeneity of the medium. The undetected field-scale heterogeneities can be responsible for significant transport. Many transport parameters either are not directly measurable or else vary in space. Hydraulic and transport properties of porous media (conductivity, hydraulic gradient, porosity, soil characteristic curve, etc.) are not always measured or reported. Local changes in conductivity or porosity due to clogging with biomass are neglected.

Field testing suggests that the standard ADM is inadequate for field-scale prediction due not only to undetected heterogeneities but also to the fact that dispersivity is a function of travel distance. Pickens and Grisak (1981) incorporated a distance-dependent dispersivity function in their models. The main result of distance-dependent dispersivity is that solutes (microbes) are seen to travel farther but at lower concentrations than predicted by standard dispersivity calculations.

Two independent data sets are required to establish the validity of a new model code (one to calibrate and one to verify). Because microbial transport data are currently scarce, two appropriate data sets are not always available for establishing new models.

Some working models such as SUMATRA-1 (Van Genuchten, 1978) and WORM (Van Genuchten, 1986) model unsaturated one-dimensional microbial transport using colloid transport theory. These models work well conceptually, but the parameter values are not well known. The BIOPLUME model (Borden and Bedient, 1986) is applied to a bacterial transport problem in the following section.

Examples of Hydrogeologic Modeling

An excellent and detailed review of current modeling capabilities and issues has been compiled by the National Research Council's Committee on Ground Water Modeling Assessment (1990). Flow and transport processes, development of models, regulatory issues, and research needs are all discussed in much greater scope than is possible here. In particular, Chapter 5, entitled "Experience with Contaminant Flow Models in the Regulatory System," gives examples of model/data interaction and its limitations. Particularly appropriate to our discussion are the cases of (1) the Snake River plain aquifer in Idaho (dissolved transport from a point source) and (2) the S-Area Landfill at Niagara Falls, N.Y. (multiphase organics transport). A final example of biological reaction modeling connected with bioremediation concludes the section.

Example 1. The Snake River basin aquifer in Idaho was the subject of field investigations and modeling of the transport of dissolved species (chloride, tritium, strontium) in ground water (Robertson, 1974). This case is notable in that follow-up field testing was performed several years later (Lewis and Goldstein, 1982) so that the true field concentrations could be compared with the predictions.

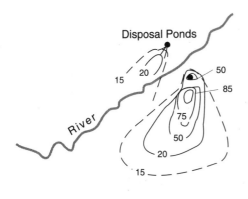

Disposal Ponds

0   1   2 Miles

0   1   2 Kilometers

●   Disposal Well

— 20 —   Line of equal chloride concentration in milligrams per liter: interval varies

1972 Actual

Figure 3-3. Measured distribution (1972) of chloride in ground water in the Snake River plain aquifer. (*Adapted from Robertson, 1974.*)

The area of interest is the site of discharge of diluted chemical waste in liquid form into the subsurface via pond seepage and well disposal. The geologic basin is filled with volcanics, alluvium, and lake deposits. The highly permeable basalts permit high ground water flow velocities. The hydraulic conductivity structure is very complex.

Robertson's solute transport model used a calibration based on 20 years of disposal data from 45 local wells. Chloride and tritium concentrations were examined, and a large lateral dispersion tread was evident (Fig. 3-3). The assumptions made in modeling included steady disposal rates (fixed at modern levels) and recharge from the river in alternate years (an approximation based on hydrologic data). The results of this modeling (Fig. 3-4) were compared with actual concentration values measured in the field several years later by Lewis and Goldstein (1982). They found a general agreement in the direction, speed, and degree of lateral dispersion (Fig 3-5).

Unfortunately, predictions of the location of the contaminant front are 1 to 2 miles in error in general. The shape of the contoured plume

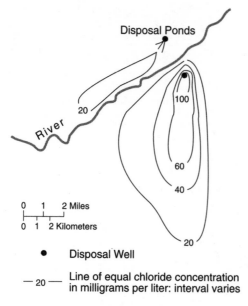

0   1   2 Miles

0   1   2 Kilometers

●   Disposal Well

— 20 —   Line of equal chloride concentration
in milligrams per liter: interval varies

1980 Predicted

Figure 3-4. Model-predicted distribution
(1980) of chloride in ground water in the
Snake river plain aquifer. (*Adapted from
Robertson, 1974.*)

is skewed in the opposite direction to that predicted by the model. Spreading is greater than predicted on the east, less than predicted on the west, and the plume is broader in an east-west sense than predicted. The smaller plume on the north side of the river was not detected in the later sampling.

Why did the model predictions fail? It was noted that actual rainfall and recharge in the years 1977–1980 were lower than expected and chloride input increased over the time period, thus violating two of the model assumptions. In addition, the grid may be insufficiently fine, and parameter values for hydrologic and transport processes may be poorly determined. The areal flow model itself calculated only horizontal flow components, ignoring the effects of vertical flow within the formation. To improve the accuracy of the prediction, the original model could be coupled with a two-dimensional (cross-sectional) model for further study. The lack of field data may cause prediction errors; few wells (measurement points)

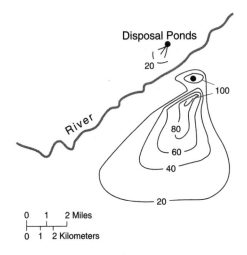

Disposal Ponds

20

100

River

80

60

40

20

```
0   1   2 Miles
├───┼───┤
0   1   2 Kilometers
```

●   Disposal Well

— 20 —   Line of equal chloride concentration
          in milligrams per liter: interval varies

1980 Actual

Figure 3-5. Measured distribution (1980) of chloride in ground water in the Snake River plain aquifer. (*Adapted from Lewis and Goldstein, 1982.*)

were available and may have been poorly positioned or improperly screened to ensure representative water samples for testing this model's hypothesis.

The relative contributions of each of these possible sources of error could be evaluated by rerunning the model with some modifications. It is almost certain that no unique solution exists; however, it is possible to identify dominant sources of variation. This interaction between field testing and model revision increases the usefulness of the model and defines locations in the field site which require greater sampling density. The opportunity to compare predictions of two different models gives an invaluable insight into which trends are shared by both models.

Example 2.   The S-Area Landfill in Niagara Falls, New York, has been the topic of extensive remediation modeling (Mercer et al., 1983; Guswa, 1985; Faust et al., 1989). A huge volume of chemical waste is present on the site, and a large amount of organic NAPLs, including chloroben-

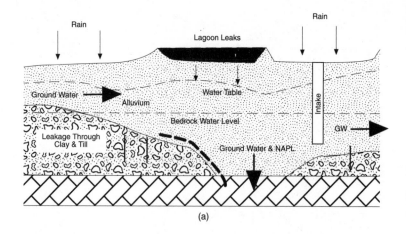

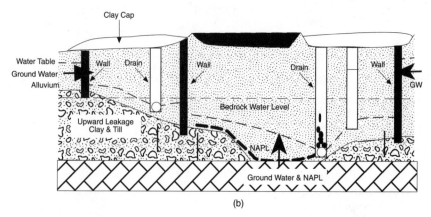

Figure 3-6. Conditions at S-Area Landfill (*a*) before and (*b*) after remediation. By pumping to lower the water table, the downward flow of organics can be arrested. (*Adapted from Cohen et al., 1987.*)

zenes and trichloroethylene, has been transported from the landfill to bedrock via gaps in the clay liner (Fig. 3-6).

Models of ground water flow and contaminant transport are presented by Mercer et al. (1985). The ground water flow and transport of NAPLs are of primary concern at the site. The potential for downward flow of NAPLs is of interest especially. The effectiveness of clay caps and barriers in controlling the downward flow of NAPLs is investigated. The modeled geologic column consists of 20 ft of fine sand ($K = 1E - 5$ cm/s), 1 ft of clay ($K = 1E - 7$ cm/s), and 2 ft of dolomite ($K = 1E - 3$ cm/s).

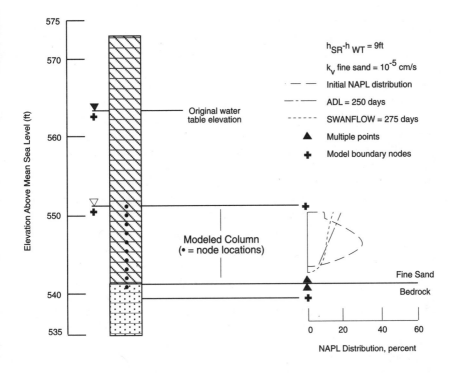

NAPL saturation profiles at one time for the two-layer simulation.
Source: National Research Council, 1990.

Figure 3-7. Predicted NAPL saturation profiles in a modeled column (one-dimensional of sand, clay, and dolomite) at the S-Area Landfill. The results indicate that the water table must be lowered by 9 ft before NAPL migration stops. (*Adapted from National Research Council Committee on Ground Water Modeling, 1990.*)

The investigation at this site is remarkable for the fact that both one-dimensional and two-dimensional two-phase flow models are undertaken and compared. While the model formulations differ in some respects and conventions, the same basic equations are being solved. The one-dimensional two-phase flow model incorporates hydraulic conductivity and hydraulic gradient variability and the influence of lithology on capillary pressure functions (Guswa, 1985). A finite-difference model solves flow equations for water and one immiscible NAPL phase.

The two-dimensional two-phase flow model is used to check the one-dimensional model predictions. The code SWANFLOW (GeoTrans, 1985) is used and the predictions compared (Faust et al., 1989). Good agreement is obtained between the two models (Fig. 3-7).

Some small differences exist between the model predictions partly because different gridding methods and different time steps were used for the calculations. Both models predict that a head difference of 9 ft (between the water table outside the landfill barriers and the water table inside) may be sufficient to keep NAPLs out of bedrock and the ground water system.

Example 3. Many organic contaminants in the subsurface can be degraded by microbial processes if sufficient oxygen and/or other inorganic nutrients are available. Bitton and Gerba (1984) show that active microbial populations, consisting primarily of bacteria, exist at depths much greater than previously thought. Subsurface bioremediation requires some models to predict bacterial effects on organic concentrations.

Several good examples of microbial transport modeling concern an abandoned wood creosoting site in Conroe, Texas. Wastes were disposed of in unlined pits, and as a result, the shallow aquifer beneath the site is contaminated. The low fluid potential gradient in the area produces ground water flow that carries a chloride plume southward at 17 ft/yr. Polynuclear aromatics (PNAs) are the major contaminant present and are found exclusively within the plume. The site is well characterized both hydrogeologically and microbiologically as a result of numerous studies (e.g., Bedient et al., 1984). The microbe population at the site is dominated by bacteria, which appear to be well adapted to the waste material. The geology of the site is very heterogeneous, with many rapid changes between relatively clean, well-sorted sand and sandy clay. Two distinct water-bearing zones are identified: (1) a low-hydraulic-conductivity unconfined zone and (2) a semiconfined zone with slightly higher hydraulic conductivity. These are separated by a thin semiconfining layer of medium-fine clayey sand. The organic carbon content of the soil is very low.

A three-well tracer test (Bordon and Bedient, 1987) suggested that in situ biodegradation rather than adsorption onto the formation accounts for the creosote (PNA) plume being smaller than the chloride plume, although they originate from the same source pit.

Borden et al. (1989) modeled the oxygen-limited biodegradation at this site. A USGS-sponsored finite-difference two-dimensional saturated solute transport model, KONBRED (Konikow and Bredehoeft, 1978), is used. This model has been verified and applied to several field problems, as noted by Mangold and Tsang (1991). Using guidelines presented by Tracy (1982), this program was modified by Borden and Bedient (1986), who developed equations for simulating the growth, decay, and transport of microbes and the transport and removal of hydrocarbons and oxygen. The growth of microorganisms and removal of organics and oxygen are simulated with a modified Monod function to model oxygen limitation. This is combined with an advection-dispersion equation (Bear, 1979) for a solute undergoing linear instantaneous adsorption. The movement of microorganisms is modeled with a simple retardation factor. Parameter values are obtained from field and laboratory tests or from literature values. The differential equations are solved with a standard stiff matrix solver package, yielding concentration values for hydrocarbon, oxygen, non-reactive tracer, and microbial biomass. This modified model, referred to as BIOPLUME in Borden et al. (1989), was tested on a field application with good results. This model assumes that an adapted microbial population is present, oxygen is required for the degradation, and the reaction between hydrocarbons and oxygen is instantaneous. If the contaminant is a slowly degradable compound, this last assumption may be inappropriate. The BIOPLUME model is used to aid in the interpretation of field push-pull well tests on the semiconfined zone which aim to determine the effect of oxygen addition on the in situ biotransformation of PNAs. The model version used in this study simulates one-dimensional radial flow away from the injection well, accounts for the apparent minimum degradable hydrocarbon concentration, and incorporates oxygen demand during both cell growth and decay. Transport of microorganisms within the aquifer was assumed to be negligible.

A model such as this requires many educated guesses about calibrations and various parameter values. The main transport parameters are estimated by calibrating the solute transport model to the chloride plume at the site. Oxygen is assumed to be the only electron receptor. Four main PNAs are assumed to comprise 70 percent of all hydrocarbons present. The ratio of oxygen to hydrogen consumed in cell synthesis is estimated from a reaction equation. Oxygen demand due to cell decay was calculated assuming complete mineralization to carbon dioxide and water. The initial microbial population, the maximum hydrocarbon utilization rate, the hydrocarbon half-saturation constant, the oxygen half-saturation constant, and the minimum degradable hydrocarbon concentration were estimated. Since no data are available from the site on the microbial cell yield or decay coefficient, estimates of these values are obtained from literature sources.

The simulated total PNA concentrations closely matched the observed values for both aerobic and anoxic push-pull tests. Results clearly show that oxygen is the primary factor which limits the biodegradation of hydrocarbons at this site. When oxygen was added to the injection water, degradation was rapid. Without oxygen, no significant degradation occurred.

References

Ball, J. W., D. K. Nordstrom, and D. W. Zachmann. 1987. WATQ4F—A Personal Computer FORTRAN Translation of the Geochemical Model WATEQ2 with Revised Database. U.S. Geological Survey Open File Report 87-50.

Baveye, P., and A. Valocchi. 1989. An evaluation of mathematical models of the transport of biologically reacting solutes in saturated soils and aquifers. *Water Resources Res.* **25**(6): 1413–1421.

Bear, J. 1979. *Hydraulics of Ground Water.* McGraw-Hill Publishing Company, New York.

Bear, J., and A. Verruijt. 1987. *Modeling Ground Water Flow and Pollution.* Reidel, Dordrecht.

Bedient, P. B., A. C. Rodgers, T. C. Bouvette, M. B. Tomson, and T. H. Wang. 1984. Ground-water quality at a creosote waste site. *Ground Water* **22**:318–329.

Bitton, G., and C. P. Gerba (eds.). 1984. *Groundwater Pollution Microbiology.* John Wiley and Sons, Inc., New York.

Borden, R. C., and P. B. Bedient. 1986. Transport of dissolved hydrocarbons influenced by oxygen-limited biodegradation, 1. Theoretical development. *Water Resources Res.,* **22**(13):1973–1982.

Borden, R. C., and P. B. Bedient. 1987. In situ measurement of adsorption and biotransformation at a hazardous waste site. *Waste Resources Bull.,* **23**(4):629–636.

Borden, R. C., M. D. Lee, J. M. Thomas, P. B. Bedient, and C. H. Ward. 1989. In situ measurement and numerical simulation of oxygen limited biotransformation. *Ground Water Monit. Rev.* **Winter**: 83–91.

Bouwer, E. J. 1984. Elements of Soil Science and Groundwater Hydrology, in *Groundwater Pollution Microbiology,* G. Bitton and C. P. Gerba (eds.). John Wiley and Sons, Inc., New York, pp. 9–38.

Bouwer, E. J., and G. D. Cobb. 1987. Modeling of biological processes in the subsurface. *Water Sci. Technol.* **19**:769–779.

Cedarberg, G. A., R. L. Street, and J. O. Leckie. 1985. A groundwater mass transport and equilibrium chemistry model for multicomponent systems. *Water Resources Res.* **21**: 1095–1104.

Corapcioglu, M. Y., and A. Haridas. 1984. Transport and fate of microorganisms in porous media: A theoretical investigation. *J. Hydrol.* **72**: 149–169.

Corapcioglu, M. Y., and A. Haridas. 1985. Microbial transport in soils and ground water: A numerical model. *Adv. Water. Res.* **8**: 188–200.

Davis, L. A., and G. Segol. 1985. *Documentation and User's Guide: GS2 and GS3—Variably Saturated Flow and Mass Transport Models*. Report NUREG/CR-3901 and WWL/TM-1791-2. U.S. Nuclear Regulatory Commission, Washington, D.C.

Delany, J. M., I. Puigdomenech, and T. J. Wolry. 1986. *Precipitation Kinetics Option of the EQ6 Geochemical Reaction Path Code*. Lawrence Livermore National Laboratory Report UCRL-53642, Livermore, Calif.

DeMarsily, G. 1986. *Quantitative Hydrogeology*. Academic Press, Orlando, Fla.

Ellis, D. V. 1987. *Well Logging for Earth Scientists*. Elsevier, New York.

Faust, C. R. 1985. Transport of immiscible fluids within and below the unsaturated zone: a numerical model. *Water Resources Res.* 21: 587–596.

Faust, C. R., J. H. Guswa, and J. W. Mercer. 1989. Simulation of three-dimensional flow of immiscible fluids within and below the unsaturated zone. *Water Resources Res.* **25**: 2449–2464.

Felmy, A. R., D. Girvin, and E. A. Jenne. 1983. *MINTEQ: A Computer Program for Calculating Aqueous Geochemical Equilibria*. Report USEPA, Washington, D.C.

Freeze, R. A., and J. A. Cherry. 1979. *Ground Water*. Prentice-Hall. Englewood Cliffs, N.J.

GeoTrans, Inc. 1985. *SWANFLOW: Simultaneous Water, Air and Non-Aqueous-Phase Flow*, Version 1.0. Documentation prepared for the U.S. Environmental Protection Agency.

Grove, D. B. 1977. *The Use of Galerkin Finite Element Methods to Solve Mass Transport Equations*. U.S. Geological Survey Water Resources Investigation 77–49.

Gupta, S. K., C. T. Kincaid, P. R. Meyer, C. A. Newbill, and C. R. Cole. 1982. *A Multidimensional Finite Element Code for the Analysis of Coupled Fluid, Energy, and Solute Transport (CFEST)*. Report PNL-4260, Pacific Northwest Laboratories, Richland, Wash.

Guswa, J. H. 1985. Application of Multi-Phase Flow Theory at a Chemical Waste Landfill, Niagara Falls, New York, in *Proceedings of the Second International Conference on Groundwater Quality Research*. National Center for Groundwater Research, Stillwater, Okla., pp. 108–111.

Hostetler, C. J., and R. L. Erikson. 1989. *FASTCHEM Package*, vol. 5: *User's Guide to the EICM Coupled Geohydrochemical Transport Code*. Report EA-5870-CCM, Electric Power Research Institute, Palo Alto, Calif.

Huyakorn, P. S., and G. F. Pinder. 1983. *Computational Methods in Subsurface Flow*. Academic Press, New York.

Javandel, L., C. Doughty, and C. F. Tsang. 1984. *Ground Water Transport: Handbook of Mathematical Models*. Water Resources Monograph 10, American Geophysical Union, Washington, D.C.

Kipp, K. L., Jr. 1987. *HST3D: A Computer Code for Simulation of Heat and Solute Transport in Three-Dimensional Groundwater Systems*. U.S. Geological Survey Water Resources Investigation 86-4095.

Konikow, L. F., and J. D. Bredehoeft. 1978. *Computer Model of Two-Dimensional*

Solute Transport and Dispersion in Ground Water. Automated Data Processing and Computations. Techniques of Water Resources, Investigations of the USGS, Washington, D.C.

Lai, C. H., G. S. Bodvarsson, and P. A. Witherspoon. 1986. Second-order upwind differencing method for nonisothermal chemical transport in porous media. *Num. Heat Transfer* **9**: 453–471.

Lewis, B. D., and F. J. Goldstein. 1982. *Evaluation of a Predictive Ground Water Solute-Transport Model at the Idaho National Engineering Laboratory.* U.S. Geological Survey Water Resources Investigation 82-55.

Mangold, D. C., and C. F. Tsang. 1991. A summary of subsurface hydrological and hydrochemical models. *Rev. Geophys.* **29**: 51–80.

Mazor, E. 1991. *Applied Chemical and Isotopic Groundwater Hydrology.* Halsted Press, New York.

McWhorter, D. B., and D. K. Sunada. 1977. *Ground-Water Hydrology and Hydraulics.* Water Resources Publications, Fort Collins, Colo.

Mercer, J. M., L. R. Silka, and C. R. Faust. 1983. Modeling ground water flow at Love Canal, New York. *J. Environ. Eng. ASCE* **109**: 924–942.

Miller, C. W., and L. V. Benson. 1983. Simulation of solute transport in a chemically reactive heterogeneous system: Model development and application. *Water Resources Res.* **19**: 381–391.

Mills, A. L., G. M. Hornberger, J. S. Herman, J. E. Saiers, and D. E. Fontes. In press. Bacterial Transport in Heterogeneous Porous Media, in *Proceedings of the International Symposium on Deep Subsurface Microbiology,* C. B. Fliermans and T. C. Hazen (Eds.). U.S. Department of Energy, Washington, D.C.

Molz, F. J., M. A. Widdowson, and L. D. Benefield. 1986. Simulation of microbial growth dynamics coupled to nutrient and oxygen transport in porous media. *Water Resources Res.* **22**: 1207–1216.

Narasimhan, T. N., A. F. White, and T. Tokunaga. 1986. Groundwater contamination from an inactive uranium mill tailings pile: 2. Application of a dynamic mixing model. *Water Resources Res.* **22**: 1820–1834.

National Research Council Committee on Groundwater Modeling Assessment, Water Science and Technology Board, Commission of Physical Sciences, Mathematics and Resources, National Research Council. 1990. *Ground Water Models: Scientific and Regulatory Applications.* National Research Council, Washington, D.C.

Naymick, T. G. 1987. Mathematical modeling of solute transport in the subsurface. *Crit. Rev. Environ. Control* **17**: 229–251.

Noorishad, J., and M. Mehran. 1982. An upstream finite element method for solution of transient transport equations in fractured porous media. *Water Resources Res.* **18**: 588–596.

Parkhurst, D. L., D. C. Thorstenson, and L. M. Plummer. 1980. *PHREEQE— A Computer Program for Geochemical Calculations.* U.S. Geological Survey Water Resources Investigation 80-96.

Pickens, J. F., and G. E. Grisak. 1981. Modeling of scale-dependent dispersion in hydrogeologic systems. *Water Resources Res.* **17**: 1701–1711.

Pinder, G. F., and L. M. Abriola. 1986. On the simulation of nonaqueous phase organic compounds in the subsurface. *Water Resources Res.* **22**: 109s–119s.

Robertson, J. B. 1974. *Digital Modeling of Radioactive and Chemical Waste Transport in the Snake River Plain Aquifer at the National Reactor Testing Station, Idaho.* U.S. Geological Survey Open-File Report IDO-22054.

Runchal, A. K. 1985. *PORFLOW: A General Purpose Model for Fluid Flow, Heat Transfer and Mass Transport in Anisotropic Inhomogeneous, Equivalent Porous Media.* Technical Note TN-011. Analysis and Computer Research, Inc., Los Angeles, Calif.

Scholz, C. 1990. *The Mechanics of Earthquakes and Faulting.* Cambridge University Press, New York.

Sposito, G., and J. Coves. 1988. *SOILCHEM: A Computer Program for the Calculation of Chemical Speciation in Soils.* Report of the Kearny Foundation, University of California, Riverside and Berkeley.

Tracy, J. V. 1982. *User's Guide and Documentation for Absorption and Decay Modification to the USGS Solute Transport Model.* Report NUREG/CR-2502, U.S. Nuclear Regulatory Commission, Washington, D.C.

Travis, B. J. 1984. *TRACR3D: A Model of Flow and Transport in Porous/Fractured Media.* Report LA-9667-MS, Los Alamos National Laboratory, Los Alamos, N.M.

Van Genuchten, M. T. 1978. *Mass Transport in Saturated-Unsaturated Media: One-Dimensional Solutions.* Water Resources Program, Department of Civil Engineering, Princeton University, Princeton, N.J.

Van Genuchten, M. T. 1986. *A Numerical Model for Water and Solute Movement In and Below the Root Zone: Model Description and User Manual.* Draft Report, U.S. Salinity Laboratory, U.S. Department of Agriculture, Riverside, Calif.

Verdier, M. 1986. *Wireline Logging Tool Catalog,* Gulf Publishing Company, Houston, Tex.

Voss, C. I. 1984. *Saturated-Unsaturated Transport (SUTRA).* U.S. Geological Survey Water Resources Investigation 84-4369.

Wang, J. F., and M. P. Anderson. 1982. *Introduction to Ground Water Modeling.* Freeman, San Francisco, Calif.

4

Laboratory Biotreatability Studies

George J. Skladany
ENVIROGEN, Inc., Lawrenceville, N.J.

Katherine H. Baker
Environmental Microbiology Associates, Inc., Harrisburg, Pa.

The biotreatability laboratory serves a variety of purposes when dealing with pollution prevention and hazardous waste-site remediation projects. Both fundamental and complex questions regarding the suitability of using large-scale biological processes are frequently asked by those directly responsible for the project, as well as supporting hazardous waste professionals and regulatory personnel. These questions include (1) Are the contaminants biodegradable? (2) Are indigenous microorganisms suitable for the task? (3) How fast will the chemicals be destroyed? (4) What final contaminant concentrations are attainable? (5) Are any hazardous by-products produced? (6) What are the optimal growth conditions for the appropriate microorganisms? and (7) What bioreactor design (above ground or in situ) is optimal for the application? In many cases, biotreatability studies can be used to provide answers to these and other questions.

While laboratory biotreatability studies are used to answer widely diverse questions, a commonality can be found in scientifically sound and statistically valid tests. This chapter is designed to provide information about the key issues that must be addressed in any technically valid treatability study. These issues include the purpose of the study, the experimental design to be employed, supporting analytical methodologies chosen, the use of standard or customized experimental protocols, statistical considerations, and data evaluation. These items form the basis for all treatability work, ranging from simple to very complex studies.

An understanding of the benefits and limitations of laboratory biotreatability studies will be of value to all who work in the pollution-control field. In order to aid readers with different technical backgrounds and interests, this chapter has been organized into two major sections. The first section provides an overview of issues germane to planning and conducting proper biotreatability studies. This information is designed to familiarize environmental professionals with the complexities of laboratory biotreatability studies and their advantages and disadvantages. Information is covered in a conceptual and concise manner. The second section provides detailed information about analytical and microbiological methodologies as well as standardized testing protocols. This information may be most useful to scientists and engineers directly responsible for preparing, conducting, and supervising biotreatability tests. It is hoped that this dual organizational approach will allow readers to easily gain conceptual and detailed information as needed.

Overview of Biotreatability Studies

Choosing the appropriate treatment technologies for pollution prevention and waste-site remediation projects is frequently difficult because reliable compound treatability data and past project experiences are limited. To this lack of information is added the "case by case" nature of pollution-control work. That is, the same contaminant in two different environments or waste streams may respond differently to identical treatment efforts. In order to overcome these potential problems, laboratory and pilot-scale treatability studies play an increasingly important role in evaluating whether biological treatment technologies should be used in full-scale projects.

Laboratory biotreatability studies have two major purposes: (1) to determine if a bioremediation process will succeed in treating the contaminants of interest at a specific location and (2) to provide data to be used in properly planning, sizing, and optimizing the bioremediation process or equipment. If these goals are accomplished, site remediation usually can proceed in an efficient and cost-effective manner.

It is important to realize that biotreatability testing is used as a tool, frequently an inexact tool, to provide answers to difficult and complex questions. In most cases, actual field operating conditions are impossible to duplicate accurately in the laboratory. Thus the results of even carefully planned and executed studies must be interpreted with caution. This should not imply that all laboratory results are necessarily suspect; environmental professionals using results from a laboratory study must understand why the study was performed, how it was performed, and what limitations should be placed on interpreting the results obtained. The next subsection describes some of the uses of the biotreatability laboratory in the pollution control field.

Biotreatability Study Applications

The laboratory can serve several functions during the life of a biological treatment project. Five of these complementary functions can last from days to years and are described in greater detail below.

Microbial Site Characterization.   One of the first functions of treatability work is to help characterize the microbiological conditions of the site or waste stream. With either situation, it is important to know early in the life of a project whether biological processes will have a chance at treating the contaminants successfully. For example, if in situ bioremediation is being considered for use at a hazardous waste site, the microbial site characterization should define the degree of subsurface microbial activity already present, as well as delineate the major factors that control that activity. These factors include subsurface temperature, pH, inorganic and organic nutrients, percent moisture, dissolved oxygen, and others. Microbiological and analytical techniques will be discussed in greater detail in the second section of this chapter.

Microorganism Numbers.   Soil and groundwater microbial activity can be determined in a number of ways, including plate count analysis of microbial populations. Suitable sterile dilutions of field samples are pipetted onto agar dishes, and the dishes are incubated until visible microbial colonies appear. Ideally, each visible colony is the result of a single microorganism multiplying repeatedly under favorable growth conditions. Correcting for the dilution factor used, the minimum number of viable microorganisms present in a liquid or soil sample can be calculated.

Both mixtures of readily biodegradable substrates and defined chemical substrate plates can be used to evaluate the metabolic capabilities of indigenous microorganisms. Results are reported as colony forming units per milliliter or gram (CFU/ml or CFU/g) and provide a minimum estimate of viable microoganisms present in a measured sample. Microorganism numbers from soil samples are typically corrected for the moisture content of the soil and are reported as colony forming units per gram of dry-weight soil.

Soil that has favorable microbial growth conditions, metabolically suitable microorganisms, and a ready supply of biodegradable substrate will typically exhibit viable microorganism numbers of 10,000 CFU/g of dry-weight soil or greater. Soil that is unfavorable to microbial life may have very low viable plate count numbers (<100 CFU/g of dry-weight soil). Ground waters typically exhibit proportionally lower microbial counts under favorable and adverse conditions.

pH. The subsurface pH may affect the availability and mobility of contaminants either through sorption, desorption, or permanent bonding. In addition, extremes of soil pH can be inhibitory or toxic to microbial life. Several factors influence soil pH, including the types and quantities of organic and inorganic materials present, the soil-to-solution ratio, and the salt (electrolyte) content of the subsurface soils and ground water. Many environmentally important microorganisms grow only within a narrow optimal pH range of 6 to 8. There are exceptions to this rule. For example, many fungi grow best under acidic (low pH) conditions.

Temperature. Temperature is an important field factor to consider, especially when dealing with in situ projects. Both chemical and biological reaction rates are controlled to a large extent by temperature. In general, a 10°C increase in temperature will double the reaction rate. In addition, microorganisms have optimal metabolic temperatures, and typical organisms are unable to live in extremes of either hot or cold. It may not be practical to work with above-ground microbial soil piles in cold-weather seasons or climates. Likewise, the temperature of very hot or cold industrial process streams would have to be modified appropriately prior to biological treatment.

Inorganic Nutrients. The most important inorganic nutrients needed by microorganisms are nitrogen (typically in the form of ammonia or nitrate) and phosphorus (in the form of orthophosphate). Both nitrogen and phosphorus are found in soils in various forms. Some of these forms are freely available to microorganisms, while others are bound to soil constituents and are not freely available. Chemical tests can be used either to quantitate the total mass of each element present or to measure the specific compounds containing each element. Techniques are also available to measure freely available versus total mass concentrations present.

Soils Characterization. For soil remediation projects, it also may be important to characterize the soils themselves as to composition, particle size distribution, percent moisture, percent organics, and cation exchange capacity.

There are no set rules that specify the minimum or maximum number of samples needed to microbially characterize a site. The design of the specific sampling program will vary with the experience and knowledge of the project manager and technical support staff consulted. However, at a minimum, samples should be obtained from regions exhibiting low, average, and high concentrations of contaminants. In addition, a background sample from a clean region of the site should serve as a "control" sample. Keep in mind that microbial site characterization can be conducted simultaneously with site chemical and hydrogeologic characterizations. Doing this work concurrently saves the additional expense of mobilizing a sampling crew to return to the site at a later date.

Determining Biodegradation Potential. Most commonly, treatability testing is used to establish the potential of microorganisms to detoxify the contaminants present. Tests can be conducted using either contaminated site soils and liquids or suitably spiked materials. Indigenous or adapted microbial cultures can be used for these tests. Testing is also used to establish the expected duration of the bioremediation project, as well as to estimate the final attainable posttreatment contaminant concentrations.

The degree of sophistication needed to design and perform these tests depends on the specific needs of the project. For example, a company may wish to demonstrate to a regulatory agency that biological processes can treat site contaminants effectively to acceptable cleanup levels. In this case, a more complete testing program, with a suitable data quality assurance and quality control (QA/QC) program, most likely would be required. In other situations, a simple "yes or no" answer is required, and more formal programs and elaborate testing protocols are not needed.

The difficulty of establishing the biodegradability of contaminants depends on the specific chemical compound(s) under consideration, their physical and chemical properties, and the matrix in which they are present. Compounds readily biodegradable under one set of conditions may persist under different field conditions.

Unfortunately, there is no single biotreatability protocol that is applicable to every situation. Mostly academic protocols have appeared in the technical literature over a number of years describing aerobic and anaerobic biotreatability methodologies. The U.S. Environmental Protection Agency (EPA) has developed an initial aerobic bioscreening protocol for use with CERCLA (Superfund) projects, and efforts are being made to finalize testing procedures for evaluating the biodegradability of petroleum products. These protocols are discussed briefly later in this section and in greater detail in the second section of this chapter.

Bioprocess Optimization. Much of the laboratory work associated with bioremediation projects centers around the optimization of the treatment process. At many sites, especially those conducting in situ work, microbial metabolic activities are limited by the supply of readily available nutrients such as nitrogen, phosphorus, or oxygen. Laboratory testing can be used to evaluate the need or effectiveness of supplemental nutrient formulations and their mode of delivery. Laboratory optimization tests also may be concerned with the introduction or enrichment of metabolically capable or superior microorganisms.

In certain circumstances, biotreatability testing can be used to estimate process design data, including biodegradation rates. One must be careful not to scale up processes inappropriately from laboratory treatability tests. Many full-scale biological processes cannot be duplicated accurately on a laboratory scale. While numerical values for critical process variables can be calculated, their actual real-world usefulness may be suspect.

Testing of Novel Bioprocess Designs. One of the major advantages of properly conducted biotreatability work is the opportunity to evaluate novel bioprocess designs. For example, new reactor designs can be tested initially on a small scale in the laboratory before a commitment is made to larger and more expensive field pilot testing or full-scale implementation. Improvements in solid-phase and in situ treatment processes also can be evaluated on a more manageable and cost-effective scale.

Monitoring of Ongoing Biotreatment Projects.   Lastly, laboratory testing is also used to monitor the progress of ongoing bioremediation projects. This can be accomplished by quantitating increases in selected microbial populations or by examining changes in the metabolic capabilities and efficiencies of the microbial community.

Experimental Design and Data Analysis

Biotreatability testing can be used to provide information under very different experimental conditions. It is important to keep in mind that different protocols are needed to answer the specific questions generated by each application or site; there is no single "all-purpose" biotreatability protocol that is appropriate to evaluate every bioprocess design or field situation.

Major Biotreatability Study Questions.   Some of the most important questions that need to be considered carefully in experimental design are listed and discussed below.

1. *Will biodegradation of the contaminants occur at the site using indigenous microorganisms and environmental conditions?* In many situations, the site may have already begun to remediate itself. Successful in situ remediation may merely require the addition of nutrients or changes in the in situ conditions that are limiting the rate of reaction. Limitations are frequently caused by low oxygen and nutrient concentrations.

2. *Is the environment (soils and ground water) or process stream inhibitory or toxic to microbial life?* Many chemical compounds have adverse effects on microorganisms, and these effects can occur at either relatively low or high concentrations depending on the specific chemicals involved. Site soils or water can be tested to determine if microbial inhibition is occurring. If an inhibiting condition is found, then efforts can be made to identify the exact cause of the inhibition and remedy the situation.

3. *Can biodegradation of the contaminants be improved by changing the microbial populations or subsurface conditions?* Laboratory tests can demonstrate the inherent metabolic activity of indigenous organisms against the site contaminants. Tests also can be used to determine if environmental conditions can be easily optimized. If indigenous metabolic capabilities are known to be inadequate, the usefulness of a bioaugmentation program can be tested before field implementation.

4. *How fast will biodegradation occur at the site under the present (or modified) conditions?* Biological treatment depends on the type and number of organisms present, as well as on the type and concentration of specific organics present. Determining biodegradation rates will help to provide an estimate of the time needed to complete a project. Cleanup time directly influences the cost of the field remediation effort.

5. *Will biological processes meet the regulatory standards set for the site?* That is, what is the lowest contaminant concentration attainable following biological treatment? The effects of augmenting with nutrients and bacteria can be evaluated and optimized in the laboratory before pilot or full-scale work begins in the field.

6. *Will the biodegradation process produce any daughter products which are themselves hazardous?* If so, will these compounds remain in the environment or be further modified by biological, physical, or chemical activity? Potentially hazardous daughter products also can be produced during physical and chemical treatment processes. It may be important from a regulatory perspective to conduct toxicity tests on the original matrix material as well as materials produced during the biological treatment process to document that hazardous daughter products are either not being formed or, if they are generated, are transitory in nature.

7. *Is it important that data from the experiments performed be statistically valid or within a preestablished confidence interval? And how do these restrictions affect the sampling and analysis portion of the treatability study?* For example, if analytical data with a known certainty are desired, then a statistically valid protocol must be developed and implemented. Frequently, statistical considerations are required for evaluating chemical concentrations before and after biological treatment, especially when dealing with difficult-to-analyze matrices. This may result in the need to perform multiple analyses at each time point for both control and test samples. The added expense of providing statistically valid data may be the only way to compensate for variability in sample collection and analysis.

8. *Will the physical, chemical, or biological properties of the target compounds (such as volatility, sorbability, or biodegradability) affect the test? Will it be necessary to construct an accurate mass balance around the treatment system?* Laboratory studies frequently evaluate the biodegradability of volatile chemicals or compounds that sorb to glassware. It is important to understand how the properties of the contaminants under consideration will affect the test apparatus used, as well as

sampling methods and sample stability. The need to perform a complete mass balance on a soils bioremediation test would require the trapping and analysis of headspace or continuously generated off-gases, analysis of the soils, and solvent extraction and contaminant quantitation of the tubing and glassware used. Mixtures of contaminants with different physical, chemical, and biological properties rapidly increase both the complexity and the associated costs of laboratory treatability work. However, if these factors are not considered and accounted for, the study may generate data that will not stand up to close scientific and engineering scrutiny.

9. *Will the study be pragmatic in nature and attempt to portray actual or anticipated field conditions, or will it be more esoterically oriented?* Pragmatic studies may be designed to simulate suboptimal or actual field conditions. For example, shallow metal pans may be used to hold soil to be treated in a landfarm or solid-phase bioremediation mode. The soils would be mixed periodically by hand, representing field tilling. More sophisticated testing would be designed to evaluate biodegradation under optimal laboratory conditions. Temperature, moisture level, and nutrient concentrations would be monitored and adjusted periodically to optimal levels in the laboratory. While testing under these conditions does not reflect what would occur under actual site conditions, the question being answered is whether the contaminants are amenable to biological treatment at all. If the optimal-condition tests show favorable results, additional testing can be completed to evaluate the effectiveness of biological treatment under more realistic conditions.

10. *What physical configuration will the experimental setup take?* If a single container is used to house all the material (soil, liquid, or sludge) under investigation, provisions will have to be made for obtaining representative samples throughout the study. As an alternative setup arrangement, the material under evaluation can be divided into individual bottles or reactors. Samples can be obtained by sacrificing whole bottles at the appropriate time. It is easier to test liquids with either arrangement; work with soils and sludges is usually more complex and difficult.

11. *Will the study reflect regulatory limitations imposed on the field project?* In some cases, biological processes may require extended periods of time to treat the contaminants of interest. If there are severe field work time constraints on the project, treatability testing should take this site limitation into account. Bioprocess optimization may be

required to provide the shortest treatment time frame to meet site regulatory requirements.

12. *What time and cost constraints have been imposed on the treatability work?* Treatability studies have to be properly designed, conducted, and interpreted so as to generate useful data. Unrealistic limitations of duration for the study or financial limitations on the project may compromise proposed biotreatability studies. Many compounds, such as the higher number multiringed polycyclic aromatic hydrocarbons (PAHs), have biological half-lives of 6 to 12 months. A treatability study with an imposed 4-week duration period would fail to properly evaluate whether these compounds could be treated successfully. Likewise, lack of money to pay for supporting analytical testing may compromise the integrity of the study or result in the generation of data of dubious quality.

The usefulness of a feasibility study will be determined to a large extent by the validity of the experimental design. Experimental design must include considerations of the physical and chemical characteristics of the system, microbiological factors, and statistical requirements for appropriate data analysis. While many professionals are familiar with the basic physical, chemical, and biological requirements of microcosm studies, statistical considerations are often overlooked until the end of the project.

Statistical Considerations.   While it is beyond the scope of this chapter to thoroughly review experimental design and statistical analysis, some problems unique to environmental studies will be highlighted here. The reader is directed to standard statistical textbooks and books on experimental design for additional details (Lentner and Bishop, 1986; Fannin et al., 1981; Green, 1979; Box et al., 1978; Kleinbaum and Kupper, 1978; Odeh and Fox, 1975; Sokal and Rohlf, 1969).

Statistical considerations should be evaluated prior to the start of a feasibility study. These considerations will allow for optimal allocation of resources in the design of the study. Perhaps the most important experimental design factor is the question of the number of replicates which need to be evaluated. In order to make comparisons between different treatment options, it is necessary to have replicate determinations of the relevant parameters measured. Not only should experimental treatments be replicated, but perhaps more important, controls also must be replicated so as to allow for accurate comparisons. For example, assume that one purpose of the testing program is to determine if the addition of inorganic nutrients to an aquifer will enhance the degradation of the contaminants present compared with biodegradation occurring under controlled, nonsupplemented conditions. Also assume, for reasons of cost, that the test is limited to a total of four microcosms. Setting up a study in which three microcosms receive the treatment while one (the control) does not will preclude this comparison. Without a measure of the variance found in the control microcosm, it is impossible to determine whether a modified microcosm is significantly different from the control.

Frequently, the decision on the number of replicates and types of replicates (replicate microcosms or replicate subsamples from the same microcosm) is made based on the cost of performing the supporting chemical analyses. There is an obvious tradeoff between statistical considerations and cost. However, individuals responsible for designing and approving feasibility studies should always be aware that scrimping on the number of replicates may prove to be a false economy, leading to ambiguous data that fail to answer the questions originally asked.

There are two additional factors which must be considered when evaluating environmental data:

1. Environmental data such as contaminant concentration and microbial numbers rarely are normally distributed. For example, microbial numbers, based on plate counts, usually fit a Poisson distribution. In this case, the variance of the counts increases as the mean counts increase. Thus the use of parametric statistics such as "t tests" for comparing data points is invalid. Data transformation (e.g., log transformation, square root transformation, or arcsine transformation) often can be applied to such data to normalize it. If such transformations are not possible, the use of nonparametric, distribution-free techniques is appropriate.

2. Environmental data are frequently "censored." In statistical terms, a data set is censored if it contains missing points. This includes data points which are below or above the limits of detection of the analytical procedure. Arbitrarily assigning a value to these points, such as recording all data points below detectable limits as "zero," will result in errors in both the calculation of summary statistics (mean and standard deviation) and in the statistical comparison of data sets (Hensel, 1990). The use of robust methods involving extrapolation of missing data values or of distribution-free methods (nonparametric) allows for analysis of censored data without the introduction of these errors.

With these considerations in mind, it is time to examine how laboratory treatability studies can be used to answer the questions posed at the beginning of this section.

Biodegradation Studies Using Present Site Conditions. If it is essential to demonstrate that contaminant biodegradation will occur under present site conditions, appropriate field samples (soils, ground water, or process wastes) can be obtained and incubated in the laboratory for given periods of time. Periodically, samples can be obtained for direct or indirect contaminant analysis. For the most meaningful results, soil or water samples should be incubated under conditions (pH, oxygen concentration, temperature, etc.) similar to those occurring at the site. While laboratory conditions are more controlled and optimal than actual field conditions, these experiments are a good first test in evaluating bioremediation potential.

If desired, higher concentrations of contaminants can be added to site samples at the start of the experiment. This may provide more conclusive biotreatability data from the experiment, especially if contaminant concentrations at the start of the experiment are very low. The test allows confirmation that the target compounds can be degraded by biological techniques while in the specific soil or liquid matrix.

Biodegradation Studies Using Modified Site Conditions. In many cases it may be important to know if the biodegradation rate can be changed by modifying present site conditions. If low-cost environmental modifications can reduce the time needed for site remediation, a substantial cost savings may be realized. For example, a microbial inoculum already adapted to the contaminants of interest can be added to the site soils or water. The concentration of specific inorganic nutrients can be increased or temperature and oxygen concentrations changed. Additional carbon sources can be added to determine if the cometabolism of contaminants can be increased. As before, biodegradation of contaminants is monitored using either direct or indirect chemical tests.

In any of these tests, it is important to perform both *positive* and *negative* biodegradation controls to serve as a check on the results. Negative controls can consist of sterilized or poisoned soil and liquids. Negative controls are used to measure loss of the target compound due to abiotic means, such as volatilization or adsorption. Negative biological controls may be difficult actually to implement. Heat sterilization may volatilize contaminants or cause chemical reactions to occur. Adding specific chemical poisons, such as mercuric chloride or sodium azide, may not be effective against all the different microbial populations present and could cause problems with the final disposal of treatability materials once the study is concluded.

Positive controls can be composed of either clean soil or water with all needed nutrients. Positive controls are used to demonstrate that test conditions are indeed suitable for microbial growth. Testing a variety of microbial growth factors in the laboratory may provide an understanding of contaminant behavior at low cost. Testing may be especially valuable if conditions at the site periodically change.

Microbial Removal Rates. One of the most difficult questions to answer properly concerns the rate at which contaminants at a site will disappear. It should be evident by now that many factors influence the biodegradation of a contaminant in a given environment. With noninhibitory or nontoxic compounds as food sources, the rate of bacterial growth in general will be proportional to the concentration of contaminants present, up to some maximum value. Beyond this maximum value, the growth rate will remain constant regardless of the substrate concentration available. With inhibitory or toxic substrates, the bacterial growth rate will be proportional to the contaminant concentration up to a given concentration, beyond which the growth rate will decline. Simply put, the faster the microorganisms grow, the faster the contaminant will be removed, as long as the contaminant concentrations remain below inhibitory or toxic levels.

It is possible to determine the mathematical growth constants for a microbial population growing on a given food source in the laboratory. This can be done using liquid batch or continuous culture techniques (Grady et al., 1989; D'Adamo et al., 1984). However, there may be little correlation between what is observed in the laboratory and what actually happens in the field (Hickman and Novak, 1989; Larson, 1984; Subba-Roa et al., 1982; Boethling and Alexander, 1979). Removal rates from soils can be tested in the laboratory but generally require frequent monitoring of contaminant concentrations with time. For example, daily monitoring of contaminant concentration and bacterial numbers present in soil samples can allow a calculation to be made based on the mass of substrate removed per organism per time. Again, kinetic values generated in the laboratory may not truly reflect in situ values, and this frequent testing may greatly increase the cost of a study.

Optimization of In Situ Parameters. Using all the preliminary assessment and treatability data collected to date, it may be desirable to optimize those factors having the greatest influence on the biodegradation process. These include pH, temperature, nitrogen and phosphorus concentrations, and trace nutrient concentrations. Optimization can be done in the laboratory or on pilot-scale studies at the site.

Inherent Problems with Biotreatability Studies.   No laboratory study can ever exactly duplicate all the conditions found at a site. However, by paying close attention to experimental design and analytical methodologies, biotreatability studies can successfully generate much useful data. There are four major areas of difficulty in conducting biotreatability studies.

Quantitation of Contaminants.   Contaminants can be removed during a study by three major processes: biodegradation, volatilization, and sorption. It is imperative that these processes be considered during the planning stages of any study so as to minimize any inadvertent losses. Work with volatile compounds, especially at low concentrations, is exceedingly difficult. Quantitation of materials in a soil matrix also can present analytical problems. Unfortunately, the only way to surmount most of these problems is to plan a carefully controlled study and to choose the appropriate type and number of analytical tests. This frequently increases the cost of the study. Without these data, however, the study may fail to provide defensible data and conclusions.

Starting Concentrations of the Contaminants.   In order to help minimize contaminant concentration variability associated with site matrix materials, many studies frequently spike a clean soil or liquid sample with known concentrations of contaminants. Actual site soils and ground waters used in experiments may have either very high or extremely low concentrations of contaminants. Which is better to use, spiked samples or site matrix material? What concentrations of contaminants should be used in treatability studies? Unfortunately, there is no single "correct" answer to these questions. Experience and professional judgment must be used after weighing the advantages and disadvantages of each choice.

In general, studies should be designed to reflect conditions that exist (or will exist) at the project site. Extremely high or low contaminant concentrations will probably present difficulties to both the microorganisms present and the analytical staff that will have to analyze these samples. Keep in mind that concentration extremes may call for a combination of biological and physicochemical remediation technologies and that a single biotreatability study at one concentration may be inadequate to supply all the answers needed to make educated decisions about a project.

Microbial Inoculum. Only organisms genetically capable of producing the proper enzymes are able to effect chemical changes in a contaminant, and only under appropriate environmental conditions. The success of any biotreatability study therefore starts with suitable microorganisms. These organisms may be indigenous to a given site or may have to be added to site soils and liquids.

Where are the most appropriate organisms for any study found? Metabolically capable bacteria may be developed in the laboratory prior to the start of any biodegradation tests. By carefully manipulating laboratory growth conditions, experiments can be designed to select and enrich for desired microbial populations. These organisms can then be used to demonstrate that the target chemicals are biodegradable.

Bacteria suitable for laboratory experiments do not necessarily have to be initially selected in the laboratory. The same selection and enrichment process is constantly occurring in nature, especially at hazardous waste sites. Organisms which are able to degrade the contaminants or tolerate their presence under such conditions should increase in number. Thus site soils may already contain the most suitable organisms for use in the study.

It is also important to remember that the biodegradation of xenobiotic compounds is frequently the action of a diverse spectrum of mixed microbial populations able to catalyze a variety of reactions. Activated sludge, especially from a wastewater plant at a facility producing the chemical(s) of interest, should not be overlooked as a source of metabolically diverse microbial inoculum.

Laboratory Versus Site Conditions. Properly designed studies minimize the differences between laboratory and site conditions. However, even the best study is usually far from perfect. Most laboratory studies are unable to simulate the constantly changing conditions found at a hazardous waste site. In many cases, variations occurring in important site conditions are unknown. Therefore, the results of any study should be interpreted carefully as a whole, considering the individual strengths and weaknesses of both the experimental plan and the analytical methods used. Thought also should be given to conducting pilot-scale studies in addition to laboratory studies. In some cases, field pilot tests can even be entirely substituted for laboratory tests.

Biological treatability studies can supply data that are essential in evaluating the suitability of using different treatment processes. However, properly designed studies and analytical techniques must be used to ensure that the results obtained are scientifically valid and practically useful. Frequently, laboratory studies are designed to answer the wrong questions or are doomed to failure as a result of time or financial constraints. Many of these failures can be attributed to a lack of understanding of proper experimental design. It is evident that careful and exact experimental design is critical to the success of any laboratory endeavor.

Microbiological Methodologies

If one explores biotreatability studies from a historical perspective, the methodologies and purposes of such work become clear. Initially, biotreatability studies evolved along two parallel tracks. Basic researchers developed techniques to study microbial ecology and its effects on current pollutants of concern, including the biodegradation of natural products, hydrocarbons, and herbicides/pesticides. For the most part, the testing protocols and equipment developed were not designed to be widely applied in commercial testing laboratories. At the same time, applied research was concerned with developing tests that would aid in the design, sizing, and operation of biological wastewater plants for the treatment of municipal and industrial wastes.

During the period from the 1950s to the 1970s, gross parameters of pollution, such as biochemical oxygen demand (BOD), total organic carbon (TOC), and chemical oxygen demand (COD), were of interest. The regulatory community was not concerned with controlling specific chemicals being discharged into the environment. During the 1970s, concern shifted in many cases from regulating general pollution parameters to specific chemicals. Many aqueous-based treatability studies were developed to examine the effect that specific compounds would have on the activated sludge process. During the 1980s, regulatory focus shifted to the regulation of hazardous wastes, and renewed emphasis was placed on biotreatability studies to predict, for aqueous streams, what effect specific compounds would have on current treatment processes (USEPA, 1986b).

With the passage of the RCRA and CERCLA (Superfund) laws, regulatory attention also was focused on the treatment of solids and sludges in addition to aqueous streams. Physicochemical processes initially were thought to hold great hope for remediating hazardous waste sites, but as these technologies were implemented, both technical and financial problems arose. Around the mid-1980s, biological treatment processes were proposed as an efficient and cost-effective means of remediating contaminated soils, sludges, ground water, and industrial process wastes. Unfortunately, almost all the applied biotechnology protocols developed evaluated biodegradation occurring within the activated sludge process. As the 1980s drew to a close, researchers continued to demonstrate on the bench level that an increasing number of xenobiotic chemicals could be biodegraded in soils and sludges. Unfortunately, few standardized protocols were available to study the biodegradation of hazardous materials present in soils, sludges, and vapors.

As the 1990s opened, the desire for biotreatability protocols devoted specifically to the treatment of hazardous wastes in soil, water, and vapor matrices became more acute. The information derived from biotreatability testing is used by environmental professionals to decide whether biological processes are appropriate for use at a given waste site, as well as how to optimize their implementation. With this testing need now recognized, work is progressing along a number of fronts to develop suitable protocols for five major situations: (1) aerobic and anaerobic soils biotreatment, (2) aerobic and anaerobic subsurface biotreatment, (3) aerobic and anaerobic sediments biotreatment, (4) aerobic and anaerobic water biotreatment, and (5) aerobic and anaerobic vapor biotreatment. It will take many years before all these standardized testing protocols become available.

This subsection will briefly describe what specific protocols have been widely publicized and accepted. Their applicability covers only a limited number of situations, and therefore, customized biotreatability protocols will continue to serve a need for the foreseeable future.

Responding to the increased need for accepted biotreatability protocols to meet a variety of purposes, the U.S. EPA in 1988 published a series of testing procedures in the *Code of Federal Regulations* (USEPA, 1988a). These procedures are briefly described below.

Section 796.3100. Aerobic Aquatic Biodegradation (pp. 563–568). The guideline was designed to develop data on the rate and extent of aerobic biodegradation that might occur when chemical substances are released to aquatic environments. The method consists of a 2-week inoculum period during which soil and sewage microorganisms are provided the opportunity to adapt to the test compound. This inoculum is added to a specially equipped Ehrlenmeyer flask containing a defined medium with the test substance. A reservoir holding barium hydroxide solution is suspended in the test flask. After inoculation, the test flasks are sparged with CO_2-free air, sealed, and incubated, with shaking, in the dark. Periodically, samples of the test mixture containing water-soluble test substances are analyzed for dissolved organic carbon (DOC), and the $Ba(OH)_2$ from the reservoirs is titrated to measure the amount of CO_2 evolved. Differences in the extent of DOC disappearance and CO_2 evolution between control flasks containing no test substance and flasks containing the test substance are used to estimate the degree of ultimate biodegradation.

Section 796.3140. Anaerobic Biodegradability of Organic Chemicals (pp. 568–572). This procedure was developed to screen for the anaerobic biodegradability of organic chemicals. The test provides evidence that the target compound will be biodegradable in sewage treatment plant anaerobic digesters and in many natural anaerobic environments such as swamps, flooded soils, and surface water sediments. A chemically defined anaerobic medium containing resazurin as an oxidation-reduction indicator and 10 percent (v/v) primary anaerobic digestor sludge from a waste treatment plant is dispensed in 100-ml portions into 160-ml capacity serum bottles. Selected bottles are supplemented with a test substance at a concentration equivalent to 50 mg/liter as organic carbon. Gas production is measured with a pressure transducer. The extent of biodegradation is determined by comparing gas production from blank control bottles and bottles containing the test substance.

Section 796.3180. Ready Biodegradability: Modified AFNOR Test (pp. 572–576). This method allows for the evaluation in aqueous medium of so-called total biodegradability under experimental conditions which are easy to establish. These conditions, however, may not necessarily be the optimal test conditions that would result in the maximum total biodegradation of the target compound. The method is applicable for use with a single chemical or mixtures of chemicals, provided that the compound is soluble at the concentration specified in the test, that the chemical is nonvolatile or at least has a negligible vapor pressure under the conditions of the test, and that the chemical does not cause microbial toxicity or inhibition at the concentration used in the test. The organic carbon remaining in solution after 3, 7, 14, 28 (and 42) days is measured and the corresponding level of biodegradation calculated. The biodegradability is evaluated on the basis of this level.

Section 796.3200. Ready Biodegradability: Closed Bottle Test (pp. 576–583). This method is used to measure the biodegradability of organic chemicals at a starting concentration of 2 to 10 mg/liter in an aqueous medium. The test is most favorably used with water-soluble compounds. However, volatile compounds and low-solubility chemicals also can be tested in principle with this test. A predetermined amount of the compound is dissolved in an inorganic medium (mineral nutrient solution), providing a usual concentration of 2 mg active substance per liter. The solution is inoculated with a small number of microorganisms from a mixed population and kept in closed bottles in the dark in a constant-temperature bath at 20 ± 1°C. The degradation is followed by oxygen analyses over a 28-day period. A control with inoculum but without test material is run parallel for the determination of oxygen blanks.

Section 796.3220. Ready Biodegradability: Modified MITI Test (I) (pp. 583–594). This protocol was based on testing procedures developed in Japan. The procedure is most appropriate for use with nonvolatile and water-soluble (at least 100 mg/liter) chemicals and measures both the biochemical oxygen demand (BOD) and the concentration of residual chemicals remaining.

Section 796.3240. Ready Biodegradability: Modified OECD Screening Test (pp. 594–600). This procedure was designed to measure the ultimate biodegradability of water-soluble, nonvolatile organic compounds in an aerobic aqueous medium at a starting test concentration of 5 to 40 mg/liter of dissolved organic carbon (DOC). The solution is inoculated with a small number of microorganisms from a mixed population and aerated at 20 to 25°C in the dark or at least in diffused light only. The degradation is followed by DOC analysis over a 28-day period. The procedure is checked by means of a standard. A control with inoculation but without either test material or standard is run in parallel for the determination of DOC blanks.

Section 796.3260. Ready Biodegradability: Modified Sturm Test (pp. 600–606). This procedure was developed to screen biodegradable compounds serving as the sole source of carbon and energy and present in concentrations of 5 to 20 mg/liter. Volatile compounds cannot be tested with this method. The rate of biodegradation and the extent of catabolism occurring are determined based on the amount of carbon dioxide evolved during the test. A chemically defined liquid medium, essentially free of other organic carbon sources, is spiked with the test material and inoculated with sewage microorganisms. The CO_2 released is trapped as $BaCO_3$. After reference to suitable blank controls, the total amount of CO_2 produced by the test compound is determined for the test period and calculated as the percentage of total CO_2 that the test material could have theoretically produced based on carbon composition.

Section 796.3300. Simulation Test for Aerobic Sewage Treatment: Coupled Units Test (pp. 606–612). This test is used for the determination of ultimate biodegradability of organic compounds under conditions which simulate treatment in an activated sludge plant. The compounds to be tested in the activated sludge plant model must be water-soluble and nonvolatile and present at concentrations greater than 12 mg/liter of dissolved organic carbon (DOC). Two model activated sludge plants are operated in parallel, whereby the parallelism is enhanced and ensured by a transinoculation procedure. The test material is added to the influent (synthetic sewage) of one unit while the other is fed only with the synthetic sewage. The DOC (or COD) concentrations are measured in both effluents. The DOC (or COD) differences of these effluent values are due to nondegraded or only partially degraded test material.

Section 796.3340. Inherent Biodegradability: Modified SCAS Test (pp. 612–615).   This method is an adaptation of the Soap and Detergent Association's semicontinuous activated sludge (SCAS) procedure for assessing the biodegradability of alkylbenzene sulfonate. The method exposes the target compound to high concentrations of microorganisms for potentially extended periods of time. The viability of the microbial culture is maintained by daily addition of a settled sewage feed. The procedure is applicable to water-soluble, nonvolatile organic chemicals that are not inhibitory to bacteria at the concentrations tested.

Section 796.3360. Inherent Biodegradability: Modified Zahn-Wellens Test (pp. 616–622).   The static test described is a simple, reproducible method for evaluating the ultimate biodegradability of organic substances in water by microorganisms. The test is limited to the use of water-soluble, nonvolatile organic compounds. Starting concentrations range from 50 to 400 mg/liter of dissolved organic carbon (DOC). The high starting concentrations used help in providing analytical reliability; compounds toxic or inhibitory at high concentrations may delay or stop the degradation process. Activated sludge, mineral nutrients, and the test material as the sole carbon source in an aqueous solution are placed together in a 1- to 4-liter glass vessel equipped with an agitator and an aerator. The mixture is agitated and aerated at 22°C (± 3°) under diffuse illumination or in a dark room for up to 28 days. The degradation process is monitored by determination of the DOC (or COD) values in the filtered solution at daily or other appropriate regular time intervals. The ratio of eliminated DOC (or COD) after each interval to the value at the start is expressed as a percentage of biodegradation and serves as the measure for the rate of degradation at this time. The result is plotted versus time to give the biodegradation curve.

Section 796.3400. Inherent Biodegradability in Soil (pp. 622–631). This procedure is based on the soil biometer flask developed by Bartha and Pramer in 1965. A radiolabeled compound of concern is mixed into standardized soils chosen to represent the type of soil present at a given site. As the compound is metabolized, radioactive carbon dioxide is produced which is absorbed in an alkaline solution. The amount of radioactive carbon in the alkaline solution is measured using liquid scintillation counting. The protocol recommends that a mass balance be performed with each test. Volatilizing compounds, as well as soil extractable and nonextractable materials, are also examined. The mass balance allows for differentiation between biotic and abiotic removal processes.

The procedure can be modified for use under anaerobic conditions. In this case, the soil within the flask is flooded with water, and the head space of the flask is purged with nitrogen gas. Biodegradation is followed by measuring the amount of methane gas produced. In addition, the water and remaining soil can be tested for residual radioactive materials.

The procedure also can be modified for use with nonradiolabeled materials. Carbon dioxide produced during aerobic metabolism is still absorbed within the alkaline solution, but titration of the solution with an acid (rather than scintillation testing) can be performed so as to calculate the amount of carbon volatilized as carbon dioxide. It is also possible to test liquids or soil slurries within the same test apparatus.

An additional testing procedure for anaerobic soils and sediment appeared in the *Federal Register* (US EPA, 1988b). The procedure, *Section 795.54. Anaerobic Microbiological Transformation Rate Data for Chemicals in the Subsurface Environment,* uses serum bottles for reaction vessels. The test consists of 20 percent (w/v) soil/sediment slurries incubated under strict anaerobic conditions. Methanogenic, sulfate-reducing, and denitrifying conditions can be investigated with the technique. Direct compound quantitation is used to determine biotic and abiotic losses.

Biotreatment processes actively under consideration by the U.S. EPA for field implementation at hazardous waste sites include in situ treatment of the vadose and saturated zones, solid-phase bioremediation (controlled landfarming operations), slurry-phase bioremediation, above-ground engineered soil piles, composting, and bioventing. In order to assist environmental professionals specifically dealing with bioremediation projects at Superfund sites, the U.S. EPA has developed a three-tiered approach to the use of biotreatability studies. Tier one, remedy screening, is designed to provide a quick and relatively inexpensive indication of whether biodegradation is a potentially viable remediation technology. The laboratory study is designed to establish that contaminant loss is actually due to biodegradation, not abiotic processes, as well as to provide information needed for the next level of testing. It is expected that tier one testing would be performed during the site remedial investigation/feasibility study

(RI/FS) phases of Superfund project work. Remedy screening establishes technology feasibility.

In 1991, the U.S. EPA released a publication entitled, *Guide for Conducting Treatability Studies Under CERCLA: Aerobic Biodegradation Remedy Screening. Interim Guidance* (U.S. EPA, 1991). This first in a series manual was intended to answer questions specific to tier one activities. The manual relates that the appropriate level of biotreatability study needed for a specific project depends on the technical literature information available, expert technical judgment, and site-specific factors. The technical literature should be used initially to determine the physicochemical properties of the compound as well as whether it has already been treated successfully using biological processes. If satisfactory answers to biotreatability questions cannot be answered from the technical literature and experience, then a laboratory screening study is in order. Prior to conducting the study, however, the bioprocess technology limitations should be reviewed. These limitations include the specific site conditions expected to be encountered, time factors for the project, and financial concerns.

A number of different experimental approaches can be used to conduct the remedy screening test. Laboratory reactor systems can consist of shake flasks, soil pans, or slurry reactors. One large control and one large treatment reactor can be used, each containing enough site material for repeated periodic subsampling. As an alternative experimental design, multiple reactor vessels, each suitable for complete sample sacrifice per time point, can be used.

The test should include controls to allow for the measurement of abiotic contaminant losses. Inhibited controls can be established by using formaldehyde, mercuric chloride, or sodium azide to inhibit microbial activity. Complete sterilization of the control is not needed, provided that the biological activity is sufficiently inhibited that a statistically significant difference between the test and control sample means can be determined.

A statistical experimental design should be employed in conducting the treatability study so as to support any conclusions and decisions made from the data. The concentration of specific compounds should be monitored over the duration of the test. In general, at least three or four sampling points should be used, including the time-zero (T_0) analysis. Normally, triplicate subsamples should be taken at each time point. Triplicate samples provide a measure of the overall precision of the measurements made. Surrogate spikes also should be added to the matrix samples to ensure consistent analytical performance.

Normally, the average contaminant concentration should be reduced by at least 20 percent during a 6- to 8-week study, compared with an inhibited control, to conclude that aerobic biological treatment is a potentially suitable treatment technology for use at the site. The 20 percent contaminant reduction "success" level is arbitrary; it was designed to maximize the chances of technology success at the remedy screening tier. The duration of the study was chosen to provide a consistent endpoint for comparisons of different studies. A longer testing period may be needed depending on the contaminants and site-specific parameters encountered.

Keep in mind that the goal of the remedy screening treatability testing program is not to document that the biological treatment process can meet the required cleanup goals. Rather, the goal is to show that bioremediation is a possible treatment option for the site-specific waste material under consideration.

The guidance manual also provides information complementary to actually conducting the laboratory study. Major sections of the manual are entitled Introduction, Technology Description and Preliminary Screening, The Use of Treatability Studies in Remedy Evaluation, Remedy Screening Treatability Study Work Plan (with subsections covering test goals, experimental design, equipment and materials, sampling and analysis, data analysis and interpretation, reports, schedule, management and staffing, and budget), Sampling and Analysis Plan, Treatability Data Interpretation, and References.

Lastly, a new series of standard oil spill bioremediation product testing protocols is expected to become available in mid-1993. The protocols are being developed by the National Environmental Technology Applications Corporation (NETAC) in conjunction with the U.S. EPA's Office of Research and Development Laboratories and advisory Bioremediation Action Committee. The testing procedures will standardize procedures for identifying the effectiveness and safety of different commercially available bioremediation products proposed for use on oil spills. A much more complete description of these and other biotreatability protocols is presented later in this chapter.

Quantitating and Monitoring Biodegradation

Quantitating contaminants present in soils and ground water is one of the most difficult challenges facing environmental chemists. Frequently, extraction, cleanup, and concentration steps are needed before final analysis can be performed (Schulten and Schnitzer, 1991; Wahle et al., 1990; Cochran et al., 1988; Otson and Chan, 1987). The ability of some contaminants to bind to soils also makes their quantitation difficult (Ortego et al., 1991). Gas chromatography, gas chromatography with mass spectrometry, and high-performance liquid chromatography are some of the analytical methods most widely used to quantitate environmental contaminants. Standard procedures for organic and inorganic contaminant analysis are summarized in several publications (Winberry et al., 1990; American Public Health Association, 1989; USEPA, 1979, 1986a).

In the field, organic compounds may be quantified using portable gas chromatographs. Many of these techniques have only recently been developed, and improvements and innovations are constantly being made (Denahan et al., 1990; Kerfoot et al., 1990; Robbins et al., 1989, 1990). In the past few years, portable gas chromatographs with mass spectrometers have become more common, substantially increasing field analytical capabilities.

The progress of biotreatability studies can be followed in several ways, and a number of analytical methodologies are suitable for this purpose. The remainder of this section will briefly discuss specific and nonspecific testing strategies employed in the bioremediation field.

Contaminant-Specific Monitoring Methods. The preferred method of monitoring biodegradation typically involves the quantitation of specific target compounds using gas chromatography, gas chromatography with mass spectrometry, or other appropriate methods. Radiolabeling of target compounds and quantitating radioactive daughter products (such as radioactive carbon dioxide from a ^{14}C-labeled parent compound) are also very effective. Direct compound quantitation helps to reduce ambiguous results and conclusions from a study. In many cases, specific compound quantitation has a major disadvantage of being much more expensive than the use of indirect quantitation methods.

Nonspecific Monitoring Methods. Simple microbial enumeration methods can be used to indirectly monitor bioremediation efforts. Microscopic counting (with or without staining) and plate count enumerations can provide an indication of increases or decreases in the total or specific microbial population numbers. These changes can be correlated with periodic chemical quantitation of the target compounds.

Other indirect methods of tracking bioremediation progress include biochemical oxygen demand (BOD), chemical oxygen demand (COD), and total organic carbon (TOC). Modified "oil and grease" analyses can be used to quantitate petroleum products, with some method limitations (Nyer and Skladany, 1989). Ion-specific and pH electrodes may be used to monitor changes in the microbial growth environment as a result of biological activity. Liquids may have to be filtered to remove bacterial cells before some of these tests are performed to avoid interferences. Soil bioremediation projects may not be amenable to monitoring with some of these methods.

Special Considerations. Samples taken during biotreatability testing may remain biologically active unless steps are taken to cease this activity. The unwanted biological activity may interfere with the purpose of the testing program, such as the elucidation of biokinetic constants. The problem of sample integrity may become severe when dealing with very low concentrations of readily biodegradable materials.

Three commonly used methods to stop unwanted biological activity are rapidly cooling the sample, chemically poisoning the sample, and immediately extracting and/or analyzing the sample. The proper method to use depends on the needs of the study as well as the equipment and supplies readily available.

Economic Considerations

The costs associated with a biotreatability study are a direct function of the specific questions that need to be addressed and the degree of statistical certainty required of the analytical results. Assuming that these two factors are known and that a general experimental plan has been formulated, the following items significantly affect the cost of performing biotreatability work:

Controls. At a minimum, positive and negative biological controls should be performed. The types and numbers of additional controls needed vary based on the specific needs of the study. Controls also may be expanded to consider physical and chemical processes that affect the removal or transformation of target compounds during the testing procedures. In some cases, a total *mass balance* of the target compounds is required, and provisions must be made for distinguishing between abiotic and biotic compound losses.

Number of Test Variables to Evaluate. In some cases, it may be desirable and necessary to evaluate permutations of significant test factors; in other situations, a choice is made to concentrate only on the factors considered most critical to the success of the project. As the number of variables increases, the number of analytical samples and their costs also increase.

Number of Replicates.   The number of replicate microcosms or analytical samples required is based on the statistical level of certainty desired for the project. High levels of statistical certainty require large numbers of samples to be collected and analyzed, which may be prohibitively expensive. Professional judgment must be used to decide where statistically validity is required, as opposed to merely desired, in order to conduct a useful study.

Sampling Frequency.   Beginning (time zero) and end (time final) time points may not be adequate to discern trends in the biotreatment process or to calculate kinetic rates. Intermediate sampling times must be designed to provide the data needed to answer the study questions.

Depending on the specific characteristics of the study, it may be possible to sample many intermediate time points and store these samples for later testing. If the data from the stored samples are actually needed, the samples can then be analyzed. If the study questions can be answered without these data points, the samples are not analyzed.

Length of Study.   The duration of the study must be long enough for biological processes to act sufficiently. Compounds known to be biodegraded slowly, such as certain PAHs, may require study durations significantly longer than more readily biodegradable compounds, such as ketones or alcohols. Once again, a compromise is usually needed to satisfy both technical and financial considerations. It is wise also initially to prepare extra test materials in case the study time has to be extended. If the additional later time points are not needed, sampling and analytical costs are not incurred.

Supporting Chemical Analyses.   The types and numbers of chemical tests performed frequently are the major expense involved with treatability testing. Specific chemical tests, such as gas chromatography with mass spectrometry, typically cost several hundred dollars per sample. If a complex study is being performed with many replicate and control analyses included, analytical costs may make up 70 to 80 percent of the total project expenses. Study planners should investigate whether relatively inexpensive nonspecific chemical or biological testing can be used as a substitute for some of the specific analyses. In many cases, nonspecific testing can be used as a leading indicator for when subsequent sampling and specific chemical analysis should be performed.

Labor and Professional Services.   Many study costs can be negotiated on a cost-plus or fixed-price arrangement. In a cost-plus arrangement, the firm conducting the testing will bill the client at an hourly rate for personnel conducting the tests and also charge for the cost of supplies, reagents, equipment rental, and analytical expenses. Studies performed for a fixed price are just that; a "scope of work" is agreed to by both parties, and the testing laboratory receives a fixed sum to complete the work.

The choice of fixed-price or cost-plus payment arrangements is frequently made based on the complexity and flexibility required in a study. For example, a very complex study answering overlapping questions may be staged in a sequential, phased fashion. Frequent exchange of ongoing results between the testing laboratory and client allows for immediate testing modifications. This type of arrangement is best handled under a cost-plus contract because the signed contract will not have to be renegotiated and reapproved based on the revised "scope of work." For simple and straightforward treatability testing, fixed-price contracts are generally awarded.

Since biotreatability studies can cost anywhere from several thousand to several hundred thousand dollars, poorly designed and executed studies lose both money and the time spent during their completion. For companies and responsible individuals not familiar with biotreatability and analytical protocols, it may be money well spent to contract with appropriate technical specialists for assistance in preparing and evaluating bid documents. Outside assistance also can be used to verify the study results and conclusions made independently.

Benefits and Limitations of Biotreatability Studies

Every study design will have distinct advantages and disadvantages. For the study to be beneficial, the advantages must overcome major weaknesses. If this is so, the benefits of treatability work include the ability to

1. Unambiguously document that the compounds of concern are indeed susceptible to biological treatment and evaluate whether by-products of the process are nonhazardous.

2. Test treatment process designs on a small and cost-effective scale.

3. Test novel process designs without constructing more expensive pilot-scale equipment for testing.

4. Collect suitability and process operational data prior to full-scale implementation.

5. Determine the suitability of the technology to meet cleanup standards under either optimal or expected site conditions.

Many other project benefits are possible with treatability testing. There are also limitations to treatability testing, including the following:

1. Poorly conceived or executed studies may yield unusable results.
2. The laboratory typically provides optimal treatment conditions, and many laboratory studies therefore produce overly optimistic results.
3. Laboratory testing may be very expensive and require long periods of time to complete.
4. It is usually not possible to scale up laboratory processes to full-scale field applications directly.

While limitations to laboratory treatability work need to be considered carefully, properly designed and conducted experiments can overcome many of these problems. Decision makers and site-remediation specialists should remember that laboratory treatability testing is one of the many tools available to assist in developing an acceptable abatement program. In certain circumstances, laboratory testing may be superfluous and unnecessary. In other cases, treatability testing may be the only way to evaluate rigorously the suitability of using biological processes before expensive field work is begun.

Microbiological Protocols and Analytical Methodologies

This section will provide details of methodologies used in laboratory treatability studies. Initially, tests used to characterize the microbial environment will be discussed. This will be followed by a subsection dealing with biological treatability studies, including microbial growth inhibition testing, aerobic and anaerobic biodegradation studies, toxicity testing, and analytical methods used to quantitate biodegradation. Problems dealing with volatile and nonsoluble contaminants will be addressed throughout.

Microbiological Site Characterization

During the evaluation and planning stages of potential biological treatment projects, it is critical to gain a detailed understanding of the site conditions that will have a direct impact on the survival and function of the microorganisms employed. For in situ soil and ground water remediation projects, the subsurface environment must serve as the reaction tank (or reactor) in which all physical, chemical, and biological reactions occur. The composition and heterogeneity of the soils and liquids in the "reactor" will influence and control the types and speed of reactions that can take place. Therefore, an understanding of the makeup of the subsurface environment is important if an in situ project is to succeed.

For above-ground treatment of soils, ground water, and industrial process waters, the surrounding microbial environment may exist in a fabricated bioreactor vessel or engineered soil pile. Once again, it is important to understand the conditions present in the microbial environment so as to control and optimize the biochemical reactions taking place.

Sampling and Analysis. Laboratory analyses play a major role in elucidating the composition and characteristics of soils and liquids. This preliminary assessment, or microbial site characterization process, establishes the boundary conditions under which a biological treatment project must function. Soils, sludges, and aqueous samples can be characterized in a number of ways, and an objective evaluation as to the suitability of using biological treatment techniques really cannot be accurately made without at least some of this information.

It is critical to note that many different analytical methodologies are available to characterize soils and liquids. Some of these methods have been developed and compiled by the U.S. EPA (1979, 1986a). Other protocols are developed by suitable professional societies, such as the American Society for Testing Materials (ASTM, 1988) and the American Society of Agronomy (Page et al., 1982). In other cases, testing laboratories or company laboratories may have developed or modified standardized protocols to meet their individual needs. Depending on the eventual use of the study, the choice of what analytical methodologies to use may be mandated by regulatory agencies or left to best professional judgment. Individuals planning, approving, conducting, and evaluating laboratory treatability tests should be knowledgeable about the limitations and proper applications of these different protocols.

Proper field sampling procedures and equipment should be used during collection to minimize contaminating or effecting changes in the samples. Sampling strategies must be designed so that representative samples of the area can be obtained (Karlen and Fenton, 1991; Oliver and Webster, 1991; Splitstone, 1989; Mason, 1983; Ackler, 1974).

Care must be taken that the samples are not contaminated or altered chemically during collection (Maitre et al., 1991). In some instances, special aseptic techniques must be employed to prevent contamination of the sample with nonindigenous microbes. Several recent papers and reviews discuss aseptic sampling techniques (Russell et al., 1992; Leach and Ross, 1991; Beeman and Suflita, 1989; Phelps et al., 1989; Wilson et al., 1989; McNabb and Mallard, 1984).

The initial site assessment should identify and quantitate both the organic and the inorganic contaminants present. In addition, general tests performed on site soils usually include temperature, pH, inorganic nutrients (particularly phosphorus and nitrogen), particle size composition (sieve analysis), organic content, moisture content, cation exchange capacity (CEC), and bacterial numbers.

Ground water can be analyzed for a variety of organic and inorganic constituents, including temperature, pH, inorganic nutrients, iron, and dissolved oxygen. In general, ground water is easier to analyze than soil because it is more homogeneous and contains fewer interfering substances. Characterization of the ground water may give totally different results than those obtained from the analysis of site soils. These differences reflect the solubility and physical and chemical properties of the contaminants and other compounds present in the subsurface environment. Materials trapped in interstitial pore spaces may be present in much higher concentrations than their water concentrations would indicate. Since bacteria rely on ground water to transport themselves and nutrients through the subsurface, accurate characterization of ground water constituents is very important.

A comprehensive review of all the soil and water characterization tests possible is outside of the scope of this chapter and will not be attempted. The most pertinent of these characterization tests are discussed below.

Temperature. Temperature is an important factor to consider, especially when dealing with in situ projects. Both chemical and biological reaction rates are controlled to a large extent by temperature. In general, a 10°C increase in temperature will double the reaction rate. Correspondingly, a drop in soil temperature, such as during the fall or winter seasons, will lower reaction rates. Keep in mind that microorganisms have optimal growth temperatures and that typical organisms are unable to live in extremes of either hot or cold.

The subsurface temperature at a site can be measured directly using a temperature probe. However, in many cases it is more convenient to use average soil temperature graphs prepared by the U.S. Geological

Survey. These graphs are available for each month of the calendar year for different regions of the country.

It is important to know the depth to which contamination extends beneath the surface, because the subsurface temperature remains relatively constant below the frost line. Sites with shallow contamination may have to consider seasonal changes in soil and water temperature, while projects with contamination below the frost line may encounter relatively constant subsurface temperatures year round. Ground water temperature remains relatively constant throughout the year. However, these temperatures change significantly as one moves from the southern United States to the northern United States. If possible, water temperature should be measured in place using a probe or thermocouple. Water samples that are removed from a deep well before testing may change temperature rapidly, and the resulting data may not be accurate.

pH. pH is a measure of the positively charged hydrogen ions present in a sample. One way of measuring soil pH is by resuspending air-dried soil in distilled water. The suspension can then be tested colorimetrically, with litmus paper, or with a pH-selective electrode. Direct soil pH methods are also available, but the water addition method tends to yield more consistent results (Page et al., 1982).

The subsurface pH may affect the availability and mobility of contaminants either through sorption, desorption, or permanent bonding. In addition, soil pH itself can be inhibitory or toxic to microbial life. Several factors influence soil pH, including the types and quantities of organic and inorganic materials present, the soil solution ratio, and the salt (electrolyte) content of the subsurface soil or ground water. These factors should be considered when interpreting soil pH data.

Ground water pH can be measured easily using either a probe or litmus paper. The pH of a sample can change with exposure to the atmosphere. In addition, biological activity also can change the pH of a sample. If possible, ground water pH should be measured immediately in the field as samples are collected. If this cannot be done, samples should be brought back to the laboratory and checked as soon as possible.

Inorganic Nutrients. The most important inorganic nutrients needed by microorganisms are nitrogen (typically in the form of ammonia) and phosphorus (typically in the form of orthophosphate). Both nitrogen and phosphorus are found in soils and ground water in various forms. Some of these forms are freely available to microorganisms, while others are bound to soil constituents and aquifer solids and are not freely available. Chemical tests can be used to quantitate either the total mass of each element present or the specific compounds containing each element. Single techniques also can be used to measure freely available versus total mass concentrations present (Flowers and Bremner, 1991; Glenn and Abeles, 1991; Roth et al., 1991; Baker, 1990; Dahnke and Johnson, 1990; Fixen and Grove, 1990; Irving and McLaughlin, 1990; Wolf and Baker, 1990; Page et al., 1982).

To quantitate freely available ammonia or orthophosphate easily, a given mass of soil can be mixed with ammonia-free or phosphate-free water and the supernatant of this mixture tested. However, this method may underestimate the true amount of an element available to microorganisms because of the presence of soil-bound nutrients. Finding low measurable concentrations of ammonia or orthophosphate at a site does not necessarily preclude the ability of the soil to support high concentrations of bacteria.

Soil composition and pH should be considered when interpreting nutrient test results (Sharpley, 1991). Under certain conditions, for example, available phosphate may tightly complex to soil particles and not be available to organisms. Adding supplemental phosphate under these conditions may not stimulate microbial growth if the additional phosphate also becomes complexed and is unavailable for microbial use.

Other inorganic microbial nutrients needed in low concentrations (such as iron or copper) also can be quantified by specific chemical analysis if needed. Since these elements are typically required in very low concentrations (1 mg/liter or less), they usually do not limit biological growth and metabolism.

Particle Size Composition.　Soil is a mixture of differently sized organic and inorganic materials. Soils can be described by the dry-weight fraction of material able to pass through differently sized sieves. The size composition of the subsurface material influences the amount of open space between the soil particles (the porosity). Porosity is important in transporting gases and liquids through the subsurface environment (Page et al., 1982).

Soils composed primarily of very small particles, such as clay or fine sand, can be compacted to high densities. This retards the passage of liquids or air through the soil and may hinder effective in situ treatment. Soils consisting of large particles, such as coarse sand, gravel, or stone, allow rapid passage of air and liquids. The heterogeneous nature of the subsurface environment is one of the most important factors that must be considered during in situ projects.

Organic Content.　Both ground water and soils typically contain naturally occurring, nonhazardous organic matter. Organic materials in the soil consist of living plant and animal matter, decaying plant and animal matter, and biological/chemical transformation products of these materials. Rapidly degradable materials (such as carbohydrates, fats, and proteins) are known as *nonhumic substances*. Other materials resistant to rapid biodegradation (such as lignin and complex polymers) are called *humic substances*. Humic substances range in molecular weight from several hundred to several thousand and are also generally hydrophilic, acidic, and partly aromatic in nature (Tate, 1987).

Humic substances are further classified as *humic acids, fulvic acids,* or *humin*. Humic acids are soluble in weakly basic solutions but precipitate on acidification of the base solution. Fulvic acids are soluble in both dilute basic and acid solutions. Humin is the humic fraction that cannot be extracted from the soil using either acid or base (Aiken et al., 1985; Tate, 1987). All humic substances are able to form water-soluble and water-insoluble complexes with metal ions and hydrous oxides. They also can interact with clay liners. The presence and concentration of humic materials in the subsurface environment can increase or decrease the solubility of contaminants present.

The amount of humic substances present in a soil sample can be quantified by dilute base extraction and purification. A more general method of determining the organic content of soils is through the use of total carbon or organic carbon tests. Total organic carbon (TOC) tests convert all forms of carbon present in soils to carbon dioxide through dry or wet chemical processes. The carbon dioxide is then quantified by one of several methods. Organic carbon tests can be conducted in two ways. With the first, inorganic carbon is preferentially removed from a sample prior to total organic analysis. The second method quantifies total organic carbon and inorganic carbon (carbonates) separately. Organic carbon then becomes the difference of total carbon less inorganic carbon (Page et al., 1982).

As previously mentioned, soils with high organic content may act to reversibly or irreversibly bind the contaminants of interest (Ong and Lion, 1991; Tell and Uchrin, 1991; Bollag and Bollag, 1990). This factor may have direct bearing on the suitability of using in situ remediation. Another factor to consider is that higher-carbon-content soils may serve as an oxygen sink. The total oxygen demand in the subsurface includes both the organic material already present in the soil and the contaminants of concern. This situation could add significant operating costs to a project if, for example, supplemental hydrogen peroxide is used to supply oxygen.

The amount of organic material present in a liquid sample is usually determined by one of three tests: biochemical oxygen demand (BOD), chemical oxygen demand (COD), or total organic carbon (TOC). Each of these tests has strengths and weaknesses (American Public Health Association, 1989).

BOD is an indirect measurement of the readily biodegradable material present in a sample. The amount of oxygen consumed during the microbial metabolism of a sample solution is used as a measure of the concentration of biodegradable material present. A microbial inoculum is added to the ground water or industrial process water for BOD tests. The metabolic capabilities of the inoculum used in the experiment may influence the results significantly. Organisms already adapted to the carbon sources present (such as contaminating chemicals) will consume oxygen as the chemicals are biodegraded. An inoculum not acclimated to the organics present will show little response in the test and will provide artificially low BOD values because oxygen will not be consumed.

The BOD test is more suited for determining the impact of readily biodegradable wastes on receiving waters, such as discharges of municipal sewage into streams or lakes. It is not a particularly good method for determining the general quantity of specific toxic or inhibitory chemicals present in a sample. In addition, the fact that 5 days are required to complete the standard BOD assay also hampers the practical usefulness of the test.

To overcome the variability inherent in using different microbial inocula, the COD test chemically oxidizes organics present in a sample to carbon dioxide and water. This test does not differentiate between biodegradable and nonbiodegradable material. The COD test has the advantage of requiring only several hours to perform, but not all organic compounds are equally oxidized. Most notably, aromatic ring compounds and pyridine are poorly oxidized by the test. In addition, high chloride concentrations may interfere and produce artificially high COD values. Chloride concentration interferences can be corrected for in many situations.

Lastly, TOC initially acidifies and purges the inorganic carbon from a sample and then oxidizes the organic matter remaining to carbon dioxide. The carbon dioxide is usually measured using an infrared detector. TOC testing requires little time per sample and can be used on both liquid and soil samples.

There is no direct conversion factor between BOD, COD, and TOC for all compounds. The oxygen demand of many biological treatment system designs is based on the total amount of biodegradable carbon

(BOD) present. The BOD concentration is usually very different from merely summing the concentrations of the specific organic contaminants present. For this reason, it is important to include at least one test for total organic content along with the specific organic tests performed when characterizing a site.

Moisture Content. Water is necessary for microbial life. Water also serves as a transport medium for bacteria, nutrients, and contaminants in the subsurface. Soil moisture content is easily determined by noting the weight change in a sample before and after it has been oven dried to a consistent weight.

For vadose zone bioremediation projects, either too much or too little water can cause problems from a microbial perspective. Too much water completely fills the soil pore spaces, limiting the direct transport of oxygen to the bacteria. Oxygen diffusion through the water to the bacteria may be slow or ineffective. Too little water in the environment stops the transport of needed materials to the bacteria and likewise the transport of inhibitory or waste products away from the cells. Soil microbiologists studying soil metabolism typically rehydrate soil to 50 to 60 percent of its holding capacity in their tests.

Moisture content is not as important a factor in the saturated zone. In these areas, attention is focused on the movement of water through the soil as opposed to soil moisture content. Alternative strategies are available to provide oxygen or suitable electron acceptors to the microorganisms.

Cation Exchange Capacity (CEC). Cation exchange capacity (CEC) is a measure of the readily exchangeable positively charged ions neutralizing negatively charged ions found in the soil. CEC is determined by using different combinations of soil pretreatments, followed by the controlled exchange and quantitation of "target" cations. CEC values are usually expressed as milliequivalents per 100 grams of soil (Page et al., 1982).

Negative charges in soils are formed by a variety of means. Some of the charges are permanent; others are dependent on the soil pH. In either case, soils with high CECs may exhibit increased binding of positively charged contaminants. Heavy metals (lead, copper, zinc, cadmium, etc.) will be removed by the cation exchange mechanisms in the soil. The total amount of heavy metals immobilized by the soils will be directly related to the CEC.

Dissolved Oxygen. Dissolved oxygen measurements can be made quickly in the field using a meter or prepared reagent kits. Wet chemical techniques are also available for use in the laboratory (American Public Health Association, 1989).

Dissolved oxygen concentrations can be difficult to measure accurately in the field. Any mixing of the sample within or outside the well can introduce additional oxygen into the water, resulting in elevated high oxygen concentration values. Mixing during pumping also can serve to reaerate a sample. In many cases, however, it is important to know only approximately how much oxygen is present in a sample to determine if the subsurface environment is either aerobic or anoxic.

Bacterial Numbers. The number of microorganisms (primarily bacteria and fungi) present in soils and ground water can be determined either microscopically through direct counting procedures or by traditional plate count methods. In either test, a microbial suspension is made using a known mass of soil and sterile extraction liquid. Cell recoveries sometimes differ with the composition of the extraction liquid (Jensen, 1968). Mechanical shaking, blending, sonication, or centrifugation to remove bacteria bound to soil particles also can be used to more accurately quantitate the total number of bacteria present in a soil sample (Hopkins et al., 1991a, 1991b).

For direct microbial counts, organisms are typically stained using a fluorescent dye (acridine orange or DAPI) and then observed with the use of a fluorescence microscope. Special microscope slides that hold a known volume of liquid are used. The number of organisms present in subsections of these slides is counted, and this number is multiplied by the appropriate dilution factor to arrive at the total number of organisms per volume or mass originally present (Hobbie et al., 1977).

The fluorescence microscopy technique is quite sensitive, but it requires specialized and expensive equipment. Staining techniques provide a number of total organisms present, be they viable, dormant, or dead. Recently, direct microscopic procedures (combined acridine orange–INT staining) have been developed which address this limitation (King and Parker, 1988). However, use of these procedures is not widespread. Acridine orange staining procedures should be interpreted with caution because the total number of organisms present is frequently much larger than the number of viable organisms.

A simpler but not as sensitive method of quantitating the number of microorganisms present is to use traditional agar plating techniques. Agar plating techniques quantitate viable organisms present by growing them on agar dishes containing different nutrients (Difco Laboratories, 1984). The extraction liquid is serially diluted into sterile dilution buffer, and aliquots from the dilution tubes are spread across the surface of the agar. The plates are incubated at an appropriate temperature until visible microbial colonies appear. Theoretically, each colony is the result of growth from a single organism. The colonies are counted, and the total number of colonies obtained is multiplied by the dilution factor used to provide a total cell number. Results are typically reported as colony forming units per gram of dry-weight soil (CFU/g). Bacterial concentrations in liquids are commonly expressed as colony forming units per milliliter (CFU/ml).

Typical plate count values for total soil organisms are between 10,000 and 10,000,000 CFU/g (Atlas and Bartha, 1987). Average concentrations for ground water are usually lower, ranging from 10,000 to 100,000 CFU/ml (Hazen et al., 1991; Balkwill and Ghiorse, 1985; Ghiorse and Balkwill, 1983). Numbers much lower or higher than this should serve as a signal that the environmental conditions (such as carbon source, pH, nutrients, etc.) are either extremely unfavorable or favorable to microbial growth.

In addition, the colors and shapes (the morphology) of the colonies present also can provide an indication of stresses in the subsurface environment. Most healthy soils and aquifers support diverse microbial populations, evidenced by a variety of microbial colony shapes, colors, and sizes. Highly stressed or toxic environments, on the other hand, typically show colonies with little diversity (Diltz, et al., 1992; Atlas, 1984; Wassel and Mills, 1983).

The plating method of microbial quantitation has the drawback that no single type of agar and nutrients will support the growth of all types of microorganisms. For example, subsurface organisms may not grow on agar plates containing high concentrations of organic carbon, such as those typically used for medical or wastewater purposes. Rather, these organisms may grow only when incubated on plates containing low organic carbon concentrations, such as those present in R2A medium or plates made from dilutions of more concentrated media. For these reasons, plate count numbers should be interpreted as the minimum rather than the actual number of viable organisms present in a sample.

General microbiological media such as trypticase soy agar, nutrient agar, R2A, brain-heart agar, and soil extract agar, to name a few, have all been used to determine total heterotrophic microorganisms in soil samples (American Public Health Association, 1989). Viable counts on the same sample using these different media can vary considerably. Therefore, caution should be used in comparing plate counts on different samples made using different media. In general, meaningful comparisons can only be made between samples plated on the same medium and incubated under the same conditions.

At times, the difference between direct count and agar plate numbers may be significant. The difficulty in quantitating subsurface soil organisms on agar plates was one of the driving factors behind the development of the acridine orange test. Direct count methods can be used to quantitate both viable and nonviable organisms; plate counts can only quantitate viable organisms. It is not unusual to find that only 1 to 10 percent of the total cells present are viable (King and Parker, 1988; Ghiorse and Balkwill, 1983).

One of the advantages of the plate count technique is that the media used can be tailored to select for specific physiological groups of microorganisms. This technique allows for determining the numbers of microorganisms capable of using a specific contaminant of interest in a sample. Specialized media consisting of a mineral salts solution with the contaminant as the sole source of carbon and energy are used for these types of enumerations. For example, microorganisms degrading volatile organics, such as benzene, toluene, ethylbenzene, and xylenes (BTEX), can be enumerated by plating onto a mineral salts medium and incubating in an atmosphere saturated with BTEX vapors.

Lastly, attempts may be made to identify and classify soil organisms to the genus and species level. In some cases, this identification may be impossible because subsurface organisms do not always respond "properly" in tests designed primarily to identify medically significant bacteria. In addition, successful identification may not be practically useful because microbial metabolic capabilities, rather than the specific species of microorganisms present, are most important to the practicing pollution-control field. Much of the biodegradation taking place is due to the combined action of many different types of bacteria, and it is likely that the overall metabolic capabilities of different microbial communities will vary from site to site (Mueller et al., 1989; Fogel et al., 1986; Sylvestre et al., 1985; Sleat and Robinson, 1984).

While bacterial enumeration methods can provide useful information about the types and numbers of indigenous microbial populations present at a site, interpretation of these numbers should be made with caution. These numbers are most properly used in a relative rather than an absolute sense. Professional judgment must be used in drawing significance and conclusions from such numbers.

Standard Biotreatability Protocols

Biodegradation studies are commonly designed to determine whether a specific contaminant is susceptible to microbial metabolism. However, the innate biodegradability of a compound frequently can be established initially by reviewing the scientific literature, lists of known biodegradable chemicals (Pitter and Chudoba, 1990; Dragun, 1988; Verschueren, 1983; Kobayashi and Rittmann, 1982; Tabak et al., 1981), or any of several computer databases (such as EPA's ATTIC) containing information on biodegradation. A brief "paper study" is usually sufficient to suggest that biological treatment is a remediation technique that should be considered further.

Keep in mind that the published literature is not without limitations. In many cases, details about the tests performed are unavailable, and it may not be possible to evaluate whether conclusions drawn from the study are valid. False-positive answers can result from compound volatilization or adsorption taking place during the biodegradation test. False-negative answers can be the result of inadequate culture adaptation before the test or of inadequate test conditions.

Once the need for biotreatability testing has been established, decisions regarding what tests to be performed must be resolved. Earlier sections of this chapter conveyed the idea that each treatability study is designed to answer specific questions. A separate experimental protocol may be needed to answer each question. While some "standard" protocols exist, there is no universally applicable, generic treatability protocol designed to answer all questions. This lack of a generic protocol offers both advantages and disadvantages to environmental professionals.

The chief advantage lies in the flexibility available to knowledgeable professionals to design a scientifically sound, statistically valid experimental plan tailored to their specific needs. Attention must be placed on including the proper experimental controls, obtaining representative samples in proper number and at appropriate times during the testing program, using correct analytical methods, and carefully analyzing and interpreting the data generated. The major disadvantage to this flexibility is that individuals without a good understanding of the critical factors involved in a proper study will most likely design and carry out a flawed experimental plan. This will result in the generation of suspect or worthless data.

Standardized testing protocols are meant to overcome limitations in personal experience and professional judgment by specifying, step by step, how an experimental plan should be conducted. These protocols have the advantage of ensuring that biotreatability testing meets minimum technical standards regardless of the background of those performing the tests. Because the study design flexibility is limited, the types of questions that can be answered properly are also limited. This disadvantage will undoubtedly lead to the generation of technically sound answers to the wrong questions.

The remainder of this subsection discusses protocols and methodologies for microbial inhibition testing, aerobic and anaerobic biodegradation testing, and toxicity testing.

Microbial Growth-Inhibition Tests. In many cases, site microbial characterization data are incomplete or unavailable. Under these conditions, it is advantageous to conduct a microbial growth-inhibition test to determine whether substances detrimental to microbial growth and survival are present in the site ground water or soils. These tests can quickly supply pertinent information without actually identifying and quantitating the specific harmful materials present.

There are four basic tests for inhibition that can be performed inexpensively and in relatively short periods of time. These inhibition tests should not be confused with toxicity tests, which are typically used to investigate whether the bioremediation process itself produces any daughter products that are toxic to living organisms. Toxicity testing is discussed later in this chapter.

Perhaps the simplest test used to investigate microbial growth inhibition is the oxygen uptake test. This test has been used extensively to test for chemical inhibition of activated sludge populations used in sanitary and industrial wastewater treatment (American Public Health Association, 1989; Volskay and Grady, 1988; King and Dutka, 1986). An aerated liquid microbial culture, such as a sample from the aeration basin of a wastewater treatment plant, is spiked with the chemical(s) to be evaluated. The spiked liquid is held in a suitable container such as a BOD bottle, and a dissolved oxygen probe is used to measure the decrease in liquid dissolved oxygen concentration over time. If the compound being tested is inhibitory to the microorganisms, oxygen uptake will be very slow or negligible. Noninhibitory compounds will have no impact on oxygen uptake.

The test can be conducted with zero headspace present in the bottles, minimizing volatilization problems. However, the inhibitory effect must be sufficiently great as to cause a major change in the respiration rate of the culture. Low concentrations of test compounds may not produce this effect and would be unsuitable for testing with this method. More sophisticated versions of respirometric testing are described later in the chapter.

The microbial plate count assay also has been used to monitor inhibition. In reality, this test does not accurately determine a specific growth-inhibition factor but rather can be used as an indirect measure of acute inhibition. Inhibiting conditions in the soil or ground water would substantially reduce the concentration of viable microorganisms present, resulting in a low plate count. Thus a very low plate count number or limited microbial diversity should serve as warning signs that something may be inhibitory to microorganisms in the surrounding environment.

Caution should be used in interpreting the test results because false-negative conclusions can occur. For example, organisms that are dormant in the field may grow freely once removed, through dilution, from high concentrations of inhibitory materials. Under these conditions, a high colony count can develop once the organisms are plated onto the agar dishes.

A third type of growth-inhibition test monitors changes in microbial numbers present in liquid samples. The liquid sample to be tested is diluted to a series of different concentrations. For instance, various dilutions (1:10, 1:100, 1:1000, etc.) and full-strength samples can be made. To each of these tubes is added small quantities of organic nutrients, inorganic nutrients, and a suitable microbial inoculum. "Control" tubes for the experiment consist of contaminant-free water with the supplemental aliquots. As the bacterial numbers increase with time, the test solution becomes turbid. The differences in turbidity are monitored using a spectrophotometer, and the amount of growth occurring in sample dilution tubes is compared with that obtained in the control tubes. Inhibition at a given dilution is said to occur if growth in that tube is less than that obtained in the control tube. This simple test can be performed in as little as 48 h and is applicable primarily to liquid samples (Lui, 1985; Alsop et al., 1980).

The fourth test for microbial toxicity monitors if the cells present are producing enzymes that can affect an indicator molecule such as rezasurin or tetrazolium salts (Lui and Thomson, 1983; Lui, 1981). Living cells produce enzymes that can catalyze changes in these chemicals; the test solution changes color as the molecules are modified. Once again, dilutions of the liquid to be evaluated are made. Reagent chemicals and a mixed microbial inoculum are added to each tube. If the microorganisms remain alive, there is a color change in the solution due to the action of cellular enzymes on the indicator molecules. If the bacterial cells are killed, no color change takes place.

These four tests are designed primarily to test liquid solutions for the presence of materials inhibitory to the growth of microorganisms. It also may be possible to perform these tests with extracts made from soil samples.

Verification of acute microbial inhibition (including cell death) can be demonstrated by inoculating a soil or liquid sample with exogenous microorganisms and following the changes in cell numbers with time. Either turbidometric or plate counts can be used to monitor changes in microbial numbers.

Aerobic and Anaerobic Biotreatability Protocols. Many different types of biotreatability protocols have been developed over the past 30 years by microbiologists, soil scientists, agronomists, biochemists, as well as by sanitary and environmental engineers. These researchers in academic, industrial, and government laboratories had a dual purpose: (1) to gain an improved understanding of the basic science involved in biodegradation and (2), in response to ever-expanding state and federal environmental laws and regulations, to predict and quantify the potential adverse impacts of chemicals released into the environment.

To accomplish the first goal, researchers have developed a wide variety of complementary testing techniques to learn about applied microbial ecology and biochemistry. These efforts frequently centered on the biodegradation of chemical contaminants of current concern and included petroleum hydrocarbons in water or soil, herbicides, and pesticides.

The second goal was concerned with developing testing protocols needed for regulatory purposes, such as the biochemical oxygen demand (BOD) and growth-inhibition tests used to size and monitor municipal and industrial biological treatment systems. As regulatory requirements continued to evolve, various toxicity testing protocols also were developed to monitor the impact of chemicals on receiving bodies of water. In recent years, the use of bioremediation to treat hazardous waste sites has generated the need for additional standardized testing protocols. These later protocols would be used to demonstrate whether biological processes were suitable for use at waste-site remediation projects.

This subsection will initially discuss biotreatability protocols developed for a variety of purposes. These protocols originated with individual basic and applied research laboratories. While it is impossible within the confines of this chapter to discuss all the specialized biotreatability testing protocols developed over the years, a representative chronological selection will be presented to provide the reader with an appreciation for the continual evolution of such tests. Keep in mind that individual protocols may have been developed strictly for a single purpose and may have to be modified for less specific current needs. Following this information, the more limited number of biodegradation protocols used for hazardous waste remediation purposes will be presented.

One of the first biodegradation methodology papers was published in 1965 by Bartha and Pramer. They described an experimental apparatus consisting of a 250-ml Erlenmeyer flask fused to a 50-ml tube with a round bottom. Soil spiked with an organic contaminant of interest, in this case the pesticide 2,4-dichlorophenoxyacetic acid (2,4-D), was placed in the bottom of the Erlenmeyer flask. As the indigenous soil microorganisms respired, the carbon dioxide generated was absorbed in a potassium hydroxide solution contained within the 50-ml tube. The alkaline solution can be periodically removed, and the amount of carbon dioxide absorbed can be quantitated by titration of the liquid with an acid. Both positive and negative control flasks, containing glucose and azide, respectively, were used to establish limits for the test.

Pramer and Bartha (1972) later recommended that some standardization of handling procedures for use with soil samples be implemented. They remarked that currently accepted practices, such as air drying, prolonged storage, and freezing and thawing of soils, will drastically alter the biochemical activity of soils by inactivating extracellular enzymes and modifying the composition and density of microbial populations present.

In 1976, Young and Baumann reported the results of a comprehensive study designed to determine the limitations of using electrolytic respirometry to measure the biochemical oxygen demand (BOD) and oxygen uptake rates of wastewater samples. The electrolytic respirometer consisted of a lower reaction vessel containing the liquid sample, as well as a mixing device such as a magnetic stirrer. An adaptor unit held an alkaline solution in the headspace above the test liquid to absorb carbon dioxide produced through microbial metabolism. As oxygen was consumed and the resulting carbon dioxide absorbed, the partial pressure within the test flask was reduced. An electrolysis cell located above the adaptor served as a manometer to detect pressure changes and automatically generated oxygen as needed to maintain a constant pressure within the flask. The authors concluded that the electrolytic respirometer held several advantages over standard BOD testing procedures.

Also in 1976, Goswami and Koch described a simple apparatus for measuring the biodegradation of ^{14}C-labeled pesticides in soil. The radioactive carbon dioxide generated by metabolism of the labeled target compound was trapped in a phenethylamine-containing scintillation "cocktail" solution. The amount of compound destruction was then quantitated by direct scintillation counting. Use of radiolabeled compounds provided nearly certain evidence that target compound destruction was occurring as a result of biological activity. The only exception would occur if the labeled compound could undergo a chemical reaction with the soil constituents present to produce carbon dioxide. The method, however, had limitations if volatile target compounds were used. For example, some of the unreacted target chemical removed from the soil sample by volatilization could be trapped within the cocktail, which would be interpreted incorrectly as artificially high rates of biodegradation.

To help overcome the problems of dealing with volatile target compounds, Marinucci and Bartha (1979) described a new test apparatus consisting of a 125-ml micro-Fernbach flask. The flask was sealed with a Teflon-lined screw cap through which two 16-gauge syringe needles were inserted and secured with epoxy cement. After the flask was loaded with an appropriate soil mixture and any needed additives, the radiolabeled volatile test compound was added by a microliter syringe fed through one of the 16-gauge needles. Connections to the flask were then sealed, and the flask was incubated under suitable test conditions until a sample was to be taken. To collect a sample, air was flushed through the test flask for a known period of time. The volatilized radiolabeled parent compound as well as the radiolabeled carbon dioxide generated by microbial respiration were passed initially through two liquid traps containing either a toluene- or xylene-based universal scintillation cocktail (to remove the parent compound), followed by two traps containing phenethylamine (to remove the carbon dioxide). The authors claimed that the two phenethylamine traps could effectively capture greater than 99 percent of the radiolabeled carbon dioxide.

Many soil scientists and agronomists also were involved with systematically exploring the respiration rate of soils. Van Cleve et al. (1979) compared four different methods (infrared gas analysis, gas chromotography, potassium hydroxide absorption of carbon dioxide, and Gilson respirometry) to measure soil respiration via carbon dioxide production using uniformly prepared birch-forest-floor organic matter. Respiration estimates were made at a single temperature and also at variable temperatures at constant moisture contents. Not surprisingly, the different analytical methods provided different estimates of respiration rates. Gas chromatography provided the lowest rate, followed by Gilson respirometry, and the infrared gas analysis and potassium hydroxide methods provided maximum rates.

Bacterial growth inhibition tests were developed to answer pragmatic questions such as the effect that a chemical or liquid waste stream would have on a receiving industrial or municipal wastewater treatment system. Many of the early toxicity tests were based on measuring changes in test culture respiration rates, but a new test procedure described by Alsop, Waggy, and Conway (1980) used a different approach. They developed a growth-inhibition test based on bacterial growth, as opposed to respiration. The test would offer the advantages of being more sensitive and less costly to perform because simpler equipment was needed. The material to be evaluated for toxicity or growth inhibition was diluted into a series of 250-ml bottles. To each bottle was added 1 ml of a stock seed suspension of microorganisms, 20 ml of standard BOD dilution water, 4 ml of stock BOD buffer, 10 ml of a nutrient broth–sodium acetate solution, and 4 ml of the diluted liquid sample to be tested. The bottles were incubated for 16 h at room temperature on a shaker platform. At the end of the incubation period, the optical density of each solution was determined at 530 nm. Controls for the experiment include azide-poisoned bottles, seeded and unseeded bottles, and seeded bottles without the test solution. The degree of microbial inhibition was determined by comparing the turbidity in each diluted test bottle with the turbidity of the control bottles.

Continuing with the desire to investigate the biodegradability of environmentally significant chemicals, Tabak et al. (1981) published a study describing the acclimation and degradability rate of 96 organic priority pollutants included in the U.S. EPA consent decree list. The 96 compounds were grouped into the following classes of compounds: phenols, phthalate esters, naphthalenes, monocyclic aromatics, polycyclic aromatics, polychlorinated biphenyls, halogenated ethers, nitrogenous organics, halogenated aliphatics, and organochlorine insecticides. A modified static culture flask screening procedure was used. Standard BOD dilution water was augmented with 5 mg/liter of yeast extract, and the target chemicals were added to final concentrations of 5 and 10 mg/liter. The testing procedure involved an initial 7-day static incubation in the dark at 25°C, followed by three weekly subcultures. Settled domestic wastewater served as the source of microbial inoculum for all tests. Volatile chemicals were incubated in stoppered BOD bottles with headspace, while nonvolatile compounds were incubated in Erlenmeyer flasks stopped with sterile cotton plugs. Gas chromatography, dissolved organic carbon (DOC), and total organic carbon (TOC) were used to monitor the extent of biodegradation occurring, and attempts were made to correct for abiotic volatility losses. The authors point out that failure of specific compounds to undergo biodegradation under the present test conditions does not inherently classify that chemical as nonbiodegradable. Rather, the microbial inoculum or test conditions may have been inadequate to completely determine the compound's true biodegradation potential.

In 1981, Lui et al. commented that many testing procedures being developed to document biodegradation failed to account for abiotic removal processes (such as water hydrolysis, volatilization, and photolysis), the need for a cometabolite, and biodegradation occurring under anoxic conditions. To overcome these problems, they designed a glass cyclone fermenter apparatus comprised of three main components: the cyclone column, the circulating pump, and the recirculation condenser. Exhaust gases released from the apparatus could be trapped to provide a more accurate mass balance when working with volatile target chemicals. The microbial inoculum used in the experiments typically consisted of soil/sediment microorganisms mixed with activated sludge. The cyclone reactors were operated under both aerobic and anoxic conditions, more representative of the microbial growth environments encountered in the real world. The authors state that "[t]here is a great tendency to underestimate the complexity of biological processes involved in the biodegradation of an organic compound, and consequently, when laboratory data alone are used to predict a substance's behavior in the natural environment, errors may occur."

As a better understanding was being gained about the wide distribution of xenobiotic compounds in the environment, at least some attention was focused on isolating subsurface microorganisms and characterizing their metabolic capabilities. It was unknown whether these microorganisms could in fact metabolize the various pollutants released over decades of time into the vadose and saturated zones. Ghiorse and Balkwell (1983) described different microbial enumeration techniques and morphologic characteristics for subsurface microorganisms, including acridine orange fluorescent cell counts, plate count agars, transmission electron microscopy, and cultivation methods.

While more numerous sophisticated biotreatability studies were being performed in the laboratory, the relevance of these results to those which occur under natural environmental conditions continued to be questioned. Spain et al. (1984) attempted to compare the action of p-nitrophenol–acclimated cultures in a small freshwater pond and in different laboratory systems. For the field test, p-nitrophenol was initially added to a landlocked pond. Shake flasks with water, shake flasks with water and sediment, ecocores, and two differently sized microcosms were tested using materials from the pond. The p-nitrophenol was rapidly biodegraded in the pond after a 6-day acclimation period. When the pond was respiked with p-nitrophenol, biodegradation of the chemical began immediately. In the laboratory tests, microbial acclimation and degradation of p-nitrophenol were quickest in samples that contained pond sediments. In general, the laboratory testing procedures could be used to predict adequately the fate of p-nitrophenol under real-world conditions.

The majority of biodegradation research focused on aerobic biological processes. However, liquids, soils, and sediments found in the natural world frequently are anoxic or anaerobic. It was probable that these specific microbial environments would harbor microorganisms possessing useful metabolic capabilities different from those exhibited by aerobic organisms. In 1984, Shelton and Tiedje published a general method for determining anaerobic biodegradation potential in the laboratory. The method tested whether specific organic compounds could be metabolized to methane and carbon dioxide under appropriate conditions, using a digested sewage sludge as a general inoculum. Biodegradation was determined by measuring the increase in gas pressure within sealed test vials with a pressure transducer, as well as by quantitating methane production by gas chromatography. A chemical was judged to be anaerobically biodegradable if greater than 75 percent of the theoretical gas production possible was observed. If only 30 to 75 percent of the theoretical gas production was observed, the chemical was said to be partially degradable. Over 100 chemicals with different chemical properties were tested with this method; 46 were found to be anaerobically degraded. As expected, the authors also reported that the source of the microbial inoculum also affected which substrates could be biodegraded.

Ward (1985) published data showing that the biodegradation rates for a specific chemical compound could vary widely when comparing the action of microbial activity in surface versus subsurface soils of septic tank tile fields. For example, the half-lives for the aerobic biodegradation of glucose and glutamic acid ranged from 2 to 5 h in surface soils compared with 8 to 25 h in subsurface soils. Under anaerobic (denitrifying) conditions, the microbial populations present in surface and subsurface soils could again biodegrade a wide range of compounds. In general, however, subsurface microbial populations required longer periods of time to metabolize the target substrate than did surface microbial populations.

The increased desire to study the metabolic capabilities of subsurface microorganisms led to the development of drilling, coring, and sampling techniques that allowed subsurface soil samples to be collected without inadvertently contaminating them with typical surface-dwelling microorganisms. Bengtsson (1985) described a core barrel apparatus used for sampling saturated soil, as well as an extruding device to obtain aseptic soil cores. The aseptic cores could be produced within a sealed glovebox environment, allowing for the maintenance of subsurface anoxic or anaerobic conditions. The same publication described a laboratory microcosm system designed for ground water research. The microcosm consisted of a sealed soil column through which a contaminated water could be passed. If a volatile chemical was to be tested, it would be dissolved in water held within a collapsible balloon. Water pumped from the balloon does not create any headspace or corresponding compound volatilization, allowing the soil column to receive a constant concentration of the volatile target compound. Sampling ports prior to, along the length, and after the column allow appropriate samples to be drawn for mass balance purposes.

The metabolic activity occurring within lake or river sediments presents a major challenge to study in the laboratory. Sediments may be anoxic or anaerobic, and nutrient fluxes are frequently seen across the sediment-water interface. Ideally, a laboratory apparatus designed to represent natural conditions should maintain anoxic/anaerobic conditions, mix the sediments to maintain a homogeneous microbial population, continually add the substrate to desired concentrations, and prevent sediment dilution. Smith and Klug described such a laboratory system in 1987. The flowthrough reactor system consisted of a 1-liter reagent bottle with a modified bottom consisting of porous polyethylene sheets. Feed from a reservoir of stock water was fed continuously through the stoppered opening of the flask. An equal volume of water was removed from the vessel after it had passed through the sediment layer and the supporting polyethylene sheets. The experimental apparatus is suitable for studying the fate of naturally occurring organic compounds or xenobiotic compounds within aerobic or anaerobic sediments.

Responding to the increased need for accepted biotreatability protocols to meet a variety of purposes, the U.S. EPA in 1988 published a series of testing procedures in the *Code of Federal Regulations* (US EPA, 1988a). These procedures are briefly described below.

Section 796.3100. Aerobic Aquatic Biodegradation (pp. 563–568). This guideline was designed to develop data on the rate and extent of aerobic biodegradation that might occur when chemical substances are released to aquatic environments.

Principle of the Test Method

This Guideline method is based on the method described by William Gledhill (1975) under paragraph (d)(1) of this section. The method consists of a two-week inoculum period during which soil and sewage microorganisms are provided the opportunity to adapt to the test compound. This inoculum is added to a specially equipped Ehrlenmeyer flask containing a defined medium with test substance. A reservoir holding barium hydroxide solution is suspended in the test flask. After inoculation, the test flasks are sparged with CO_2-free air, sealed, and incubated, with shaking in the dark. Periodically, samples of the test mixture containing water-soluble test substances are analyzed for dissolved organic carbon (DOC) and the $Ba(OH)_2$ from the reservoirs is titrated to measure the amount of CO_2 evolved. Differences in the extent of DOC disappearance and CO_2 evolution between control flasks containing no test substance and flasks containing test substance are used to estimate the degree of ultimate biodegradation.

Section 796.3140. Anaerobic Biodegradability of Organic Chemicals (pp. 568–572). This procedure was developed to screen for the anaerobic biodegradability of organic chemicals. The test provides evidence that the target compound will be biodegradable in sewage treatment plant anaerobic digesters and in many natural anaerobic environments such as swamps, flooded soils, and surface water sediments.

Principle of the Test Method

(i) This section is based on anaerobic biodegradability methods referenced in paragraph (d) of this section. (ii) A chemically defined anaerobic medium, containing resazurin as an oxidation/reduction indicator and 10 percent (v/v) primary anaerobic digestor sludge from a waste treatment plant, is dispensed in 100 ml portions into 160 ml capacity serum bottles. Selected bottles are supplemented with test substance at a concentration equivalent to 50 mg/liter as organic carbon. Gas production is measured with a pressure transducer. The extent of biodegradation is determined by comparing gas production from blank control bottles and bottles containing the test substance. (iii) The average cumulative gas production (CH_4 + CO_2), in ml, is reported for blank controls, solvents controls, test substances and any reference compounds. Also reported is the percent of theoretical anaerobic biodegradation at test completion or 56 days (whichever comes first) and the standard deviation between replicate bottles.

Section 796.3180. Ready Biodegradability: Modified AFNOR Test (pp. 572–576). This procedure is based on "Norme Experimentale AFNOR T 90-302." The method allows the evaluation in aqueous medium of so-called total biodegradability under experimental conditions which are easy to establish. These conditions, however, may not necessarily be the optimal test conditions that would result in the maximum total biodegradation of the target compound. The method is applicable for use with a single chemical or mixtures of chemicals, provided that (1) the compound(s) is soluble at the concentration specified in the test, (2) the chemical(s) is nonvolatile or at least has a negligible vapor pressure under the conditions of the test, and (3) the chemical(s) does not cause microbial toxicity or inhibition at the concentration used in the test.

Principle of the Test Method

The biodegradation of organic products dissolved in water by chemico-organotrophic microorganisms using the products as the sole source of carbon and energy is observed. These products are studied at a concentration such that the initial content of organic carbon is 40 mg/liter. The organic carbon remaining in solution after 3, 7, 14, 28 (and 42) days is measured and the corresponding level of biodegradation calculated. The biodegradability is evaluated on the basis of this level.

Section 796.3200. Ready Biodegradability: Closed Bottle Test (pp. 576–583). This method is used to measure the biodegradability of organic chemicals at a starting concentration of 2 to 10 mg/liter in an aqueous medium. The test is most favorably used with water-soluble compounds. However, volatile compounds and low-solubility chemicals also can be tested in principle with this test.

Principle of the Test Method

A predetermined amount of the compound is dissolved in an inorganic medium (mineral nutrient solution), providing a usual concentration of 2 mg active substance per liter (AS/liter). The solution is inoculated with a small number of microorganisms from a mixed population and kept in closed bottles in the dark in a constant temperature bath at 20 ± 1°C. The degradation is followed by oxygen analyses over a 28-day period. A control with inoculum, but without test material, is run parallel for the determination of oxygen blanks.

Section 796.3220. Ready Biodegradability: Modified MITI Test (I) (pp. 583–594). This protocol is based on testing procedures developed in Japan. The test was found suitable by the OECD Expert Group on Degradation/Accumulation for determining the ready biodegradability of organic chemicals under aerobic conditions. It was also tested in the OECD Laboratory Intercomparison Test Program of 1978–1980. The procedure is most appropriate for use with nonvolatile and water-soluble (at least 100 mg/liter) chemicals and measures both the biochemical oxygen demand and the concentration of residual chemicals remaining.

Principle of the Test Method

This test method is based on the following conditions: (A) test chemicals as sole organic carbon sources and (B) no adaptation of microorganisms to test chemicals. An automated closed system oxygen consumption measuring apparatus (BOD meter) is used. Chemicals to be tested are inoculated with microorganisms in the testing vessels. During the test period, biochemical oxygen demand is measured continuously by the BOD meter. Biodegradability is calculated on the basis of BOD and supplemental chemical analysis is undertaken, such as measuring dissolved organic carbon concentrations, concentration of residual chemicals, etc.

Section 796.3240. Ready Biodegradability: Modified OECD Screening Test (pp. 594–600). This procedure was designed to measure the ultimate biodegradability of water-soluble, nonvolatile organic compounds in an aerobic aqueous medium at a starting test concentration of 5 to 40 mg/liter of dissolved organic carbon (DOC). The test was found suitable by the OECD Expert Group on Degradation/Accumulation for determining the ready biodegradability of organic chemicals under aerobic conditions. It was also tested in the OECD Laboratory Intercomparison Test Program of 1978–1980.

Principle of the Test Method

(A) A predetermined amount of compound is dissolved in an inorganic medium (mineral nutrient solution, fortified with a trace element and essential vitamin solution), providing a concentration corresponding to 5 to 40 mg DOC/liter. The solution is inoculated with a small number of microorganisms from a mixed population and aerated at 20 to 25°C in the dark or at least in diffuse light only. The degradation is followed by DOC analysis over a 28-day period. The procedure is checked by means of a standard. (B) A control with inoculation, but without either test material or standard, is run in parallel for the determination of DOC blanks.

Section 796.3260. Ready Biodegradability: Modified Sturm Test (pp. 600–606). The procedure was developed to screen biodegradable compounds serving as the sole source of carbon and energy and present in concentrations of 5 to 20 mg/liter. Volatile compounds cannot be tested with this method. The rate of biodegradation and the extent of catabolism occurring are determined based on the amount of carbon dioxide evolved during the test.

Principle of the Test Method

(A) A chemically defined liquid medium, essentially free of other organic carbon sources, is spiked with the test material and inoculated with sewage microorganisms. The CO_2 released is trapped as $BaCO_3$. (B) After reference to suitable blank controls, the total amount of CO_2 produced by the test compound is determined for the test period and calculated as the percentage of total CO_2 that the test material could have theoretically produced based on carbon composition.

Section 796.3300. Simulation Test—Aerobic Sewage Treatment: Coupled Units Test (pp. 606–612). This test is recommended by the OECD Expert Group on Degradation/Accumulation as a test for the determination of ultimate biodegradability of organic compounds under conditions which simulate treatment in an activated sludge plant. The compounds to be tested in the activated sludge plant model must be water-soluble and nonvolatile and present at concentrations greater than 12 mg/liter of dissolved organic carbon (DOC).

Principle of the Test Method

Two OECD Confirmatory Test Units, i.e., model activated sludge plants, are operated in parallel whereby the parallelism is enhanced and assured by a transinoculation procedure. The test material is added to the influent (synthetic sewage) of one unit while the other is fed only with the synthetic sewage. The DOC (or COD) concentrations are measured in both effluents. The DOC (or COD) difference of these effluent values is due to non- or only partially-degraded test material. The raw data obtained require a statistical procedure; i.e., the tolerance limits of the mean must be calculated in addition to the mean.

Section 796.3340. Inherent Biodegradability: Modified SCAS Test (pp. 612–615). The method is an adaptation of the Soap and Detergent Association semicontinuous activated sludge (SCAS) procedure for assessing the biodegradability of alkylbenzene sulfonate. The method exposes the target compound to high concentrations of microorganisms for potentially extended periods of time. The viability of the microbial culture is maintained by daily addition of a settled sewage feed. The procedure is applicable to water-soluble, nonvolatile organic chemicals that are not inhibitory to bacteria at the concentrations tested.

Principle of the Test Method

(A) Activated sludge from a sewage treatment plant is placed in an aeration (SCAS) unit. The test compound and settled domestic sewage are added, and the mixture is aerated for 23 hours. The aeration is then stopped, the sludge is allowed to settle, and the supernatant liquor is removed. The sludge remaining in the aeration chamber is then mixed with a further aliquot of test compound and sewage and the cycle is repeated. (B) Biodegradation is established by determination of the dissolved organic carbon content of the supernatant liquor. This value is compared with that found for the liquor obtained from a control tube dosed with settled sewage only.

Section 796.3360. Inherent Biodegradability: Modified Zahn-Wellens Test (pp. 616–622). The static test described is a simple, reproducible method for evaluating the ultimate biodegradability of organic substances in water by microorganisms. The test is limited to the use of water-soluble, nonvolatile organic compounds. Starting concentrations range from 50 to 400 mg per liter of dissolved organic carbon (DOC). The high starting concentrations used provide greater analytical reliability; compounds toxic or inhibitory at high concentrations may delay or stop the degradation process.

Principle of the Test Method

Activated sludge, mineral nutrients and the test material as the sole carbon source in an aqueous solution are placed together in a 1-4 litre glass vessel equipped with an agitator and an aerator. The mixture is agitated and aerated at 22°C (± 3°) under diffuse illumination or in a dark room for up to 28 days. The degradation process is monitored by determination of the DOC (or COD) values in the filtered solution at daily or other appropriate regular time intervals. The ratio of eliminated DOC (or COD) after each interval to the value at the start is expressed as percentage of biodegradation and serves as the measure for the rate of degradation at this time. The result is plotted versus time to give the biodegradation curve.

Section 796.3400. Inherent Biodegradability in Soil (pp. 622–631). This procedure is based on the soil biometer flask developed by Bartha and Pramer in 1965. A radiolabeled compound of concern is mixed into standardized soils chosen to represent the type of soil present at a given site. As the compound is metabolized, radioactive carbon dioxide is produced which is absorbed in an alkaline solution. The amount of radioactive carbon in the alkaline solution is measured using liquid scintillation counting. The protocol recommends that a mass balance be performed with each test. Volatilizing compounds, as well as soil extractable and nonextractable materials, are examined. The mass balance allows for differentiation between biotic and abiotic removal processes.

The procedure can be modified for use under anaerobic conditions. In this case, the soil within the flask is flooded with water, and the headspace of the flask is purged with nitrogen gas. Biodegradation is followed by measuring the amount of methane gas produced. In addition, the water and remaining soil can be tested for residual radioactive materials.

The procedure also can be modified for use with nonradiolabeled materials. Carbon dioxide produced during aerobic metabolisms is still absorbed within the alkaline solution, but titration of the solution with an acid (rather than scintillation testing) can be performed to calculate the amount of carbon volatilized as carbon dioxide. It is also possible to test liquids or soil slurries within the same test apparatus.

Principle of the Test Method

(A) Basic test (1) A small sample of soil is treated with the ^{14}C-labeled test chemical in a biometer flask apparatus. Release of

$^{14}CO_2$ from the test chemical is measured by means of alkali absorption and liquid scintillation counting. (2) Optional experiments include the following tests. (B) Evaporation test. When testing chemicals of a vapor pressure higher than 0.0133 Pa, a polyurethane foam plug is placed into the biometer flask apparatus to absorb the labeled volatile metabolites for liquid scintillation counting. (C) Residue test. At the point of 50 percent mineralization the test soil may be extracted. The extractable portion of the compound, and its metabolites remaining in the soil, may be determined by liquid scintillation counting. Furthermore, data on the bound residue part may be obtained by measuring the $^{14}CO_2$ released after combustion of the soil.

An additional testing procedure for anaerobic soils and sediment appeared in the *Federal Register* (USEPA, 1988b). The procedure, *Section 795.54. Anaerobic Microbiological Transformation Rate Data for Chemicals in the Subsurface Environment,* uses serum bottles for reaction vessels. The test consists of 20 percent (w/v) soil/sediment slurries incubated under strict anaerobic conditions. Methanotrophic, sulfate-reducing, and denitrifying conditions can be investigated with the technique. Direct compound quantitation is used to determine biotic and abiotic losses.

While in many cases it is acceptable simply to know whether a specific compound is susceptible to biodegradation, in other situations it is more important to know the rate at which the compound can be treated. Being able to determine the intrinsic biodegradation kinetics for a given biomass has typically been very difficult. In 1989, Grady et al. published a paper describing the use of electrolytic respirometry to determine biodegradation kinetics for environmentally significant chemicals. The biodegradation kinetics for single organic compounds based on oxygen consumption in batch cultures were calculated from these tests. The automatic data-collection system allowed the kinetic constants to be calculated in a much less labor-intensive manner than with other procedures. The kinetic values established in this new manner also were comparable with those obtained using more classical techniques.

As the knowledge base regarding the biodegradation of different chemicals continued to expand, an attempt was made to develop a predictive model for use in evaluating both proposed and existing commercial chemicals. Boethling and Sabljic described such a model in 1989. Initially, 22 experts in the microbial degradation of xenobiotic chemicals were asked to predict the biodegradability of 50 chemicals representing the wide range of chemicals encountered in premanufacturing notices. From the group's recommendations, a number of generalizations regarding the effects of chemical structure on biodegradability were derived. Negative influences on biodegradability included molecular mass, alkyl branching, halogenation, and nitrogen heterocycles. Positive influences on biodegradation included hydrolyzable groups, hydroxyl or carboxylic acid groups, and linear alkyl chains. The model predicted ultimate aerobic biodegradability for receiving waters. Based on environmentally relevant experimental data, the model correctly classified 36 of 40 chemicals (90 percent) in two test prediction sets.

While much of the biotreatability database consists of information on the treatment of chemicals in a liquid matrix, soils remediation projects at hazardous waste sites (including RCRA, CERCLA, and state-led cleanups) were becoming increasingly important to the waste-treatment field. Unfortunately, the knowledge base for the bioremediation of soils is much less complete than that for water-based biodegradation systems.

Mueller et al. (1991a) published a study describing bench-scale biotreatability studies designed to evaluate the potential of using a solid-phase bioremediation process to treat pentachlorophenol- and creosote-contaminated soils and sediment from a Superfund site. The effects of tilling the soil and adding inorganic nutrients were evaluated. The indigenous microflora were used, and contaminant removal was monitored by using gas chromatography. Specially designed "landfarming chambers" allowed for the capture and analysis of contaminants lost through abiotic processes. The authors reported an increase in the number of PAH-degrading microorganisms present over the life of the project and that soils receiving inorganic nutrients enhanced the removal of certain creosote constituents. However, the high-molecular-weight PAHs and other relatively nonbiodegradable chemicals present were not effectively treated under the landfarming conditions employed.

In a follow-up publication, Mueller et al. (1991b) evaluated the use of bench-scale slurry-phase biotreatment of the pentachlorophenol- and creosote-contaminated soils. Soil slurries prepared from site soils and sediments were incubated for 30 days. The 1.5-liter bioreactors were operated in the batch mode at 28.5°C with continuous mixing. The slurry was kept at a pH of 7.0 and at a dissolved oxygen concentration of 90 percent. Samples were periodically removed from the slurry for analysis using gas chromatography. Pentachlorophenol and

42 individual creosote constituents were quantitated. The bioslurry-phase treatment system showed considerable improvement in contaminant biodegradation over solid-phase bioremediation. Except for pentachlorophenol, benzo[*b*]fluoranthene, benzo[*k*]fluoranthene, and indeno[1,2,3-*cd*]pyrene, the pH-adjusted soil slurry system biodegraded more than 50 percent of the targeted organics within 3 to 5 days. The data suggested that slurry-phase bioremediation strategies can be utilized effectively to remediate creosote-contaminated soil.

Biotreatment processes actively under consideration by the U.S. EPA for field implementation at hazardous waste sites include in situ treatment of the vadose and saturated zones, solid-phase bioremediation (controlled landfarming operations), slurry-phase bioremediation, above-ground engineered soil piles, composting, and bioventing. To assist environmental professionals specifically dealing with bioremediation projects at Superfund sites, the U.S. EPA has developed a three-tiered approach to the use of biotreatability studies.

Tier one, remedy screening, is designed to provide a quick and relatively inexpensive indication of whether biodegradation is a potentially viable remediation technology. The laboratory study is designed to establish that contaminant loss is actually due to biodegradation, not abiotic processes, as well as to provide information needed for the next level of testing. It is expected that tier one testing would be performed during the site remedial investigation/feasibility study (RI/FS) phases of Superfund project work. Remedy screening establishes technology feasibility.

Tier two, remedy selection, consists of more involved bench-scale testing designed to establish the design and operating parameters for optimization of technology performance. This work is also expected to be completed during the remedial investigation/feasibility study (RI/FS) phases of Superfund project work. Remedy selection is used to develop performance and cost data.

Tier three, remedy design, is used to scale up and test the selected biotreatment technology on a pilot scale in the field. This work is expected to occur during the remedial design/remedial action (RD/RA) phases of Superfund project work. The goal of remedy design is to develop scale-up, design, and detailed cost data for the biotreatment process.

In 1991, the U.S. EPA released a publication entitled *Guide for Conducting Treatability Studies Under CERCLA: Aerobic Biodegradation Remedy Screening. Interim Guidance* (USEPA, 1991). This first in a series manual was intended to answer questions specific to tier one activities. A brief summary of some of the information provided in the guidance manual follows.

The manual relates that the appropriate level of biotreatability study needed for a specific project depends on the technical literature information available, expert technical judgment, and site-specific factors. The technical literature should be used to determine the physicochemical properties of the compound as well as whether it has already been treated successfully using biological processes. If satisfactory answers to biotreatability questions cannot be answered from the technical literature and experience, then a laboratory screening study is in order. Prior to conducting the study, however, the bioprocess technology limitations should be reviewed. These limitations include the specific site conditions expected to be encountered, time factors for the project, and financial concerns.

Carefully planned treatability studies are necessary to ensure that the data generated are useful for validating the performance of the technology. The goals of the treatability study should be clearly defined before any laboratory work is performed. The two major goals of the remedy screening evaluation are to

1. Provide an indication that contaminant reductions are due to biodegradation and not to abiotic processes such as volatilization, photodecomposition, and adsorption

2. Produce the data and design information required for the next level of treatability testing, assuming that the screening evaluation was successful

Normally, the average contaminant concentration should be reduced by at least 20 percent during a 6- to 8-week study, compared with an inhibited control, to conclude that aerobic biological treatment is a potentially suitable treatment technology for use at the site. The 20 percent contaminant reduction "success" level is arbitrary; it was designed to maximize the chances of technology success at the remedy screening tier. The duration of the study was chosen to provide a consistent endpoint so as to be able to compare different studies. A longer testing period may be needed depending on the contaminants and site-specific parameters encountered.

A number of different experimental approaches can be used to conduct the remedy screening test. Laboratory reactor systems can consist of shake flasks, soil pans, or slurry reactors. One large control and one large treatment reactor can be used, each containing enough site material for repeated periodic subsampling. As an alternative experimental design, multiple reactor vessels, each suitable for complete sample sacrifice per time point, can be used.

The test should include controls to allow for the measurement of abiotic contaminant losses. Inhibited controls can be established by using formaldehyde, mercuric chloride, or sodium azide to inhibit microbial activity. Complete sterilization of the control is not needed, provided that the biological activity is inhibited sufficiently so that a statistically significant difference between the test and control sample means can be determined.

A statistical experimental design should be employed in conducting the treatability study in order to support any conclusions and decisions made from the data. The concentration of specific compounds should be monitored over the duration of the test. In general, at least three or four sampling points should be used, including the time-zero (T_0) analysis. Normally, triplicate subsamples should be taken at each time point. Triplicate samples provide a measure of the overall precision of the measurements made. Surrogate spikes also should be added to the matrix samples to ensure consistent analytical performance.

The mean contaminant concentrations in the test and control reactors are then compared to see whether a statistically significant change in concentration has occurred. In general, the experiment should be designed so that a difference of 20 to 50 percent in contaminant removal can be observed between the the treatment reactor and the inhibited control. Keep in mind that the goal of the remedy screening treatability testing program is not to document that the biological treatment process can meet the required cleanup goals. Rather, the goal is to show that bioremediation is a possible treatment option for the site-specific waste material under consideration.

The guidance manual also provides information complementary to the actual conduct of the laboratory study. Major sections of the manual are entitled Introduction, Technology Description and Preliminary Screening, The Use of Treatability Studies in Remedy Evaluation, Remedy Screening Treatability Study Work Plan (with subsections covering test goals, experimental design, equipment and materials, sampling and analysis, data analysis and interpretation, reports, schedule, management and staffing, and budget), Sampling and Analysis Plan, Treatability Data Interpretation, and References.

Lastly, a new series of standard oil spill bioremediation product testing protocols is expected to become available in mid-1993. The protocols are being developed by the National Environmental

Technology Applications Corporation (NETAC) in conjunction with the U.S. EPA's Office of Research and Development Laboratories and advisory Bioremediation Action Committee. The testing procedures are being developed to satisfy a number of critical oil spill bioremediation product information needs and will standardize procedures for identifying the effectiveness and safety of different commercially available bioremediation products.

Toxicity Testing. Concerns over environmental contamination with organic compounds are frequently based on the human and environmental health risk associated with these chemicals. The goal of treatment is to reduce the toxicity, carcinogenicity, and mutagenicity of compounds and the contaminated matrix. Adverse health effects can be associated with both untreated hazardous materials and materials that have undergone treatment using physical, chemical, or biological means. It is possible for some treatment technologies to produce hazardous by-products which remain a concern to regulatory agencies and the general public.

Frequently, the potential toxicity and other adverse health effects of the starting target compound can be determined by a review of the technical literature. The end result of biologically mediated transformations and interactions between chemicals is not as readily apparent. With this in mind, the inclusion of some form of toxicity testing in a bioremediation feasibility study should be considered to demonstrate that any microbial by-products produced are indeed nontoxic. In these instances, direct toxicity testing may provide the needed data.

Toxicity can be measured using any of a number of bioassays. In these procedures, an appropriate test organism is exposed to various concentrations of a suspected toxicant for a predetermined length of time. The influence of the material on survival, growth, and reproduction or any number of other indicators of physiological status of the organisms are then determined. Based on these data, an LC_{50} (dose corresponding to 50 percent mortality), IC_{50} (dose corresponding to 50 percent inhibition), or EC_{50} (dose corresponding to 50 percent of the response measured) concentration can be calculated. Several standard bioassays have been developed by the U.S. EPA and are routinely used for assaying toxicity as a part of the National Pollution Discharge Elimination Standards (NPDES) discharge permits. These include chronic and acute tests with daphnia and fathead minnows. Numerous other tests have been described in the literature (Dutka et al., 1991; Knobeloch et al., 1990; Swartz et al., 1990; Devillers et al., 1989; Hao et al., 1987; Middaugh et al., 1987; Dutka and Bitton, 1986).

In addition, several toxicity tests are commercially available. The Microtox™ test measures the inhibition of the marine luminescent bacterium *Photobacterium phosphoreum*. The ATP-TOX system screens microbial toxicity by determining the concentration of ATP present (Xu and Dutka, 1987).

There are several factors that complicate the application of bioassays to the determination of toxicity reduction as a result of bioremediation. First, bioremediation frequently is conducted in a soil or water matrix. The standardized toxicity procedures for the most part are designed for aquatic toxicity testing, and therefore, some type of extraction or solubilization procedure must be employed to remove the compounds of interest from the soil matrix for testing. Bennett et al. (1990) have recently reported on a procedure for the preparation of oil products for use in toxicity testing which addresses the solubility problems associated with conventional assays.

A more fundamental concern is the proper choice of a bioassay organism for use in the testing program. Ideally, a bioassay organism should be an organism which is indigenous to the type of environment being tested. In addition, the organism should be among the most sensitive organisms typical of that environment to prevent false-negative conclusions. No single bioassay procedure meets these requirements in all cases. It is necessary, therefore, to consider carefully the advantages and limitations of a bioassay system before it is applied to the evaluation of bioremediation efforts.

Several researchers have recently reported on the use of bioassays to evaluate and monitor bioremediation. Dasappa and Loehr (1991) conducted laboratory studies on the bioremediation of sandy soils to which phenolic compounds had been added. They monitored bioremediation progress by determining the amount of the added contaminant remaining with time. In addition, they evaluated the toxicity of water extracts of the soils using the Microtox™ system. They found a correlation between removal of the target contaminants and a reduction of the soils toxicity as measured in the bioassay system. Wang and Bartha (1990) evaluated the bioremediation of soils contaminated with jet fuel, heating oil, and diesel oil. They assessed toxicity using the Microtox™ system as well as seed germination and plant growth assays. Their results indicated that there was a good correlation between the disappearance of the contaminants and a reduction in soils toxicity. Kuehl et al. (1990) reported on the use of *Cerodaphnia dubia* acute assays in evaluating toxicity reduction procedures for an aqueous leachate from a creosote-contaminated site. Finally,

Middaugh et al. (1991) used a combined toxicity-teratogenicity test on silverside minnows (*Menidia beryllina*) as well as Microtox™ tests to evaluate the toxicity of ground water contaminated with pentachlorophenol and creosote.

In addition to tests for direct toxicity, there are several tests which can be used to measure carcinogenicity and mutagenicity. Probably the simplest of these tests is the Ames test, which measures the ability of a test compound (or extract) to induce revertant mutations in a culture of *Salmonella* (Ames et al., 1975). Other tests for mutagenicity include the SOS Chromotest, which measures the activity of the SOS DNA repair system of *Escherchia coli* (Fish et al., 1985), and the Mutatox test, which measures the induction of mutations in a dark mutant of the luminescent organism *Photobacterium phosphoreum* (Dutka et al., 1991).

Claxton et al. (1991) used the Ames test to evaluate changes in mutagenicity of samples from Prince William Sound following the *Exxon Valdez* spill. They found that mutagenicity existed in samples from both untreated (control) and bioremediated beaches and that mutagenic contaminants were rapidly lost from the oil-contaminated beaches under both the control and treated conditions. They concluded that the loss of mutagenic contaminants was the result of both natural biodegradation and active bioremediation efforts. April et al. (1990) used the Ames test to monitor mutagenicity on four types of waste (API separator sludges, slop oil emulsion solids, creosote sludges, and PCP-creosote mixed sludge) in model soil remediation systems. Both slop oil emulsified solids and creosote sludge/soil mixtures were initially mutagenic. After approximately 1 year of treatment in the soil system, however, no mutagenic potential was detected in either of those samples. API separator sludge and PCP-creosote mixed sludge samples were not found to be mutagenic at any time.

Quantitating Biodegradation

In general, the best biotreatability data come from tests that directly monitor changes in the specific contaminant concentration. These tests may be able to monitor several compounds simultaneously but are correspondingly more complex and expensive.

Biotreatability data from tests using indirect contaminant monitoring methods usually are easier to perform, are less expensive, and may be limited to liquid samples. They provide the most information when only a single contaminant is present as a source of carbon and energy for the organisms. Depending on what information is needed from the test, indirect monitoring methods may provide little or no useful data when used with mixtures of contaminants. Bacterial plate count numbers frequently are useful to cross-check and independently verify the decrease in contaminant concentration observed with expensive specific analytical tests such as gas chromatography with mass spectrometry. Proper application of these simple tests can decrease the overall cost of a laboratory treatability study by reducing the number of analytical samples needed. If done properly, this results in only a marginal decrease in overall data quality.

Many monitoring methods are suitable for use with either soil or water samples. A brief summary of the most common analytical methods follows.

Direct Quantitation of Contaminants. The U.S. EPA and other professional societies have developed specific sampling and testing methodologies for analyzing liquids, soils, sludges, and vapors (Winberry et al., 1990; American Public Health Association, 1989; American Society for Testing Materials, 1988; USEPA, 1979, 1986a). Using these (or customized) methods, the concentrations of specific contaminants are followed as a function of time. Compounds biodegradable under the test conditions will show a decrease in concentration, while compounds resistant to biodegradation will show little or no decrease in concentration. This method works best when investigating the biodegradability of nonvolatile contaminants. Volatile compounds can be used in these tests as long as appropriate positive and negative controls are also included. These controls may include suitable sorbent traps to capture volatile compounds. The traps can then be analyzed to gain information about abiotic contaminant loss due to volatility. Gas chromatography, gas chromatography with mass spectrometry, and high-performance liquid chromatography are commonly used to quantitate specific compounds.

While these analytical methods provide definitive data regarding reductions in target concentrations, the cost is usually expensive. This is especially true if daily or more frequent monitoring results are needed. Consideration also must be given to the size of the sample needed for each timepoint.

Direct compound quantitation also may have recovery problems if one is working with a complex liquid or soil matrix. For example, nonreversible bonding of some chemicals to soil constituents may make accurate compound quantitation difficult if not impossible to perform accurately (Ong and Lion, 1991; Tell and Uchrin, 1991; Bollag and Bollag, 1990; Sawhney and Brown, 1989). The detection limits needed for the analysis are also important. Very low detection limits may only be attained when large sample volumes of material are extracted and concentrated. It may be impractical to set up treatability studies using large volumes of liquid or soil.

Radiolabeling of Contaminants. This technique unambiguously demonstrates that biodegradation of the contaminants is taking place. A quantity of the target chemical to be tested is "tagged" with radioactive atoms, usually carbon in the form of isotope ^{14}C. These compounds can be either purchased commercially or synthesized as needed. The radioactive compound is mixed in with site soils or water. When the radioactive compound is aerobically metabolized by the microorganisms, part of the carbon atoms are used to form carbon dioxide, while the remaining carbon atoms typically form biomass. Carbon dioxide formed from the radioactive carbon atoms is trapped in an alkaline solution, and the amount of radioactivity present in the alkali solution is then quantified (Loos et al., 1980; Marinucci and Bartha, 1979; Bartha and Pramer, 1965). Since only the target compound contained the radioactive carbon atoms, radioactive carbon dioxide could only be the result of mineralization of the target compound. In addition, the remaining radioactivity present in the soil or water can be quantified to provide an overall mass balance for the target compound.

Radiolabeling also can be employed during anaerobic biotreatability tests. Under anaerobic conditions, metabolic by-products include methane and carbon dioxide.

The radiolabeling technique is especially valuable in demonstrating compound biodegradability in soils, because a suitable mass balance can be performed without having to extract the remaining chemical mass from the soils. In addition, it may be possible to identify and quantitate intermediate products of biodegradation when used in conjunction with mass spectrometry.

There are several major drawbacks to the common use of this method. Many compounds of environmental significance must be custom synthesized using the radioactive atoms, usually a costly procedure. Generally, only one labeled compound at a time can be studied. In addition, specialized equipment, training, and licenses are needed to work with radioactive materials. The specialized equipment and materials may make studies of this kind very expensive. Despite these drawbacks, use of radiolabeled compounds may be the only technique to convincingly demonstrate the biodegradability of certain chemicals.

Indirect Monitoring of Biodegradation. In contrast to specifically quantitating the targets of interest, it may be possible to indirectly follow the biodegradation of chemical compounds. Several of the most commonly employed methods are discussed below.

Respirometry. Under aerobic metabolism, oxygen is consumed while carbon dioxide is produced. Indirect biodegradation monitoring in the laboratory can quantitate changes in either oxygen or carbon dioxide concentrations present in a test vessel (Baker et al., 1988; King and Dutka, 1986; Flathman et al., 1985; Anderson, 1982). While this technique can be used with mixtures of contaminants, it is impossible to state with certainty which of the contaminants is (are) being metabolized at any point. Other biodegradable material present in the test sample also may contribute to positive results with the test.

With a closed-vessel setup, carbon dioxide from the atmosphere can be removed from air prior to its entering the test biodegradation flask; only the carbon dioxide evolved due to biological action will be present within the flask. The carbon dioxide produced can be scrubbed into an alkaline solution and later quantitated.

Several devices exist to monitor respiring cultures. The equipment is based on measuring changes in the partial pressure of gases in the headspace of a sealed biodegradation flask. Effluent carbon dioxide produced is again removed through the use of an alkaline solution. As the oxygen is depleted and carbon dioxide is precipitated in the alkaline solution, the partial pressure within the flask decreases. The Warburg respirometer measures the decrease in pressure within the flask and is the apparatus most suitable for solutions containing low concentrations of organics.

The electrolytic respirometer monitors the decrease in partial pressure within sealed vessels and electrically hydrolyzes water to hydrogen and oxygen to resupply the flasks with oxygen. The amount of oxygen produced within the cell is electronically monitored (Young and Baumann, 1976). A variation of the water hydrolysis electrolytic respirometer hydrolyzes copper sulfate held in an adjacent flask to generate oxygen without producing hydrogen gas (Venosa et al., 1992; Grady et al., 1989).

With each of these systems, biodegradation is indirectly monitored by following the change in air pressure occurring within the sealed flask. These systems have been used traditionally to monitor biodegradation taking place in liquids or soil slurries and have only occasionally been used to monitor soils alone. Data from soil slurry work may be useful from a technical perspective but may not represent the same rate of microbial activity that will occur in the field under nonslurry treatment conditions.

Similar indirect analytical techniques can be used to monitor anaerobic respiration. Under anaerobic conditions, the major metabolic products produced are carbon dioxide and methane. If sealed tubes or bottles with a septum are used in the tests, two simple methods are available for monitoring. Gas samples can be removed from the bottles and the amount of methane and carbon dioxide quantified using gas chromatography with appropriate detectors. Pressure transducers also can be used to measure increases in pressure with the test vessels as methane and carbon dioxide are produced.

Quantitation of Side-Chain Materials. It may be possible to measure composition changes in the liquid test medium as biodegradation occurs. By-products of biodegradation or end products of microbial metabolism may affect changes to the growth medium. For example, a chloride ion–specific electrode can measure changes in chloride concentration in the test liquid as these atoms are removed from a compound such as methylene chloride (Flathman et al., 1985). Colorimetric testing, changes in solution absorbance at particular wavelengths, and pH have all been used to monitor biodegradation. Most of these methods are not directly applicable to soil degradation studies.

Changes in Nonspecific Test Parameters. It is also possible to monitor biodegradation by following concentration changes in nonspecific test parameters, such as soluble BOD, COD, TOC, or turbidity. As the contaminants are degraded by microorganisms, some of the soluble carbon is transformed into nonsoluble biomass. By filtering out the biomass prior to BOD, COD, or TOC testing, soluble organic concentration changes may become evident. The results should be interpreted with caution, since cell lysis products, metabolic by-products, and contaminant daughter products can contribute to soluble organic concentration levels. These methods are generally not sensitive enough to monitor changes with very low initial contaminant concentrations.

Turbidity measurement techniques rely on the fact that increases in cell numbers cause a corresponding increase in the turbidity of test solutions. A spectrophotometer can be used to measure the increase in solution absorbance at an appropriate wavelength.

All these methods, with the exception of BOD, have the advantages of being quick, inexpensive, and easy to perform. However, no conclusions about the biodegradation of specific compounds present in mixtures of chemicals can be made using nonspecific tests.

Bacterial Plate Counts. Lastly, it may be possible to monitor biodegradation by following increases in the number and types of bacteria present in the test solution. Biodegradable materials will act as food for the cells, which will then grow and multiply. Compounds resistant to biodegradation cannot act as food, and under these conditions, cell numbers would be expected to remain level or decline over time. Plate counts can be done using liquid, soil, or soil slurries as starting materials. The method is also relatively inexpensive.

Summary

Biological treatment processes are playing an increasingly important role in pollution prevention and hazardous waste site remediation projects. While each site presents different technical and regulatory challenges, the goal for environmental professionals is to find an efficient, cost-effective, and environmentally acceptable technology for use with each project.

The difficulty of establishing the biodegradability of contaminants depends on the specific chemical compound(s) under consideration, their physical and chemical properties, and the matrix in which they are present. Compounds readily biodegradable under one set of conditions may persist under different field conditions. Biological treatability studies can supply data that are essential in evaluating the suitability of using biotreatment processes.

Biotreatability studies are typically used both to show that the contaminants of concern are amenable to biological treatment and to provide the data needed to plan, design, and optimize field processes. The usefulness of a feasibility study will be determined to a large extent by the validity of its experimental design. Experimental design must consider the physical and chemical properties of the contaminants, microbiological factors, laboratory testing protocols, supporting analytical methodologies, the controls needed, and statistical requirements for appropriate data analysis. Frequently, laboratory studies are designed to answer the wrong questions or are doomed to failure due to time or financial constraints. Many of these failures can be attributed to a lack of understanding of proper experimental design. It is evident that careful and exact experimental design is critical to the success of any laboratory endeavor.

Unfortunately, there is no single standardized biotreatability protocol that is appropriate for all uses. Rather, a number of aerobic and anaerobic protocols have been developed by academic, private, and government laboratories over the past 25 years. Most of these protocols and testing equipment were designed to answer very specific questions and were not intended to be implemented on a wide scale by commercial testing laboratories. To fill the need for standardized biotreatability protocols, the U.S. EPA and other technical organizations are developing suitable laboratory methodologies for the aerobic and anaerobic biological treatment of soils, liquids, and sediments. To date, only an interim aerobic screening protocol for use on Superfund projects has been completed (USEPA, 1991). Until additional protocols become available, care should be used when adapting protocols designed for one purpose to a totally different application. The skill and experience of those involved with planning and conducting biotreatability studies may provide the key to their success.

Both the advantages and limitations of laboratory studies must be considered when evaluating data from treatability tests. Only properly planned, controlled, and executed studies will provide valid scientific data for technology evaluation or process optimization. Most important, very few laboratory studies will ever provide data as relevant as that obtained from properly conducted tests on site.

References

Ackler, W. L. 1974. *Basic Procedures for Soil Sampling and Core Drilling.* Acker Drill Company, Inc., Scranton, Pa.

Aiken, G. R., D. M. McKnight, R. L. Wershaw, and P. MacCarthy. 1985. *Humic Substances in Soil, Sediment, and Water.* John Wiley & Sons, New York, N.Y.

Alsop, G. M., G. T. Waggy, and R. A. Conway. 1980. Bacterial growth inhibition test. *Journal of the Water Pollution Control Federation.* 52:2452–2456.

American Public Health Association. 1989. *Standard Methods for the Examination of Water and Wastewater,* 17th ed. American Public Health Association, Washington, D.C.

American Society for Testing Materials. 1988. *Ground Water Contamination: Field Methods,* ASTM Special Technical Publication No. 963. American Society for Testing Materials, Philadelphia, Pa.

Ames, B. N., J. McCann, and E. Yamasaki. 1975. Methods for detecting carcinogens and mutagens with the *Salmonella*/mammalian-microsome mutagenicity test. *Mutation Research* 31:347–364.

Anderson, J. P. E. 1982. Soil Respiration, pp. 831–871. In A. L. Page, R. H. Miller, and D. R. Kinney (eds.), *Methods of Soil Analysis,* Part 2: *Chemical and Microbiological Properties,* 2d ed. American Society for Agronomy/Soil Science Society of America, Madison, Wisc.

April, W., R. C. Sims, J. L. Sims, and J. E. Matthews. 1990. Assessing detoxification and degradation of wood preserving and petroleum wastes in contaminated soil. *Waste Management and Research* 8:45–65.

Atlas, R. M. 1984. Use of microbial diversity measurements to assess environmental stress, pp. 540–545. In M. J. Klug and C. A. Reddy (eds.), *Current Perspectives in Microbial Ecology*. American Society for Microbiology, Washington, D.C.

Atlas, R. M., and R. Bartha. 1987. *Microbial Ecology: Fundamentals and Applications*. The Benjamin/Cummings Publishing Co., Inc., Menlo Park, Calif.

Baker, D. E. 1990. Baker soil test theory and application. *Communications in Soil Science and Plant Analysis* 21:13–16.

Baker, K. H., D. S. Herson, and D. A Bunisky. 1988. Bioremediation of soils contaminated with a mixture of hydrocarbon wastes: A case study, pp. 490–494. In *Superfund '88*. Hazardous Materials Control Research Institute, Silver Spring, Md.

Balkwell, D. L., and W. C. Ghiorse. 1985. Characterization of subsurface bacteria associated with two shallow aquifers in Oklahoma. *Applied and Environmental Microbiology*, 50:580–588.

Bartha, R., and D. Pramer. 1965. Features of a flask and method for measuring the persistence and biological effects of pesticides in soil. *Soil Science* 100:68–70.

Beeman, R. E., and J. M. Suflita. 1989. Evaluation of deep subsurface sampling procedures using serendipitous microbial contaminants as tracer organisms. *Geomicrobiology Journal* 7:223–233.

Bengtsson, G. 1985. Microcosm for ground water research, pp. 330–341. In C. H. Ward, W. Giger, and P. L. McCarty (eds.), *Ground Water Quality*. John Wiley & Sons, New York, N.Y.

Bennett, D., A. E. Girling, and A. Bounds. 1990. Ecotoxicology of oil products: Preparation and characterization of aqueous test media. *Chemosphere* 21:659–669.

Boethling, R. S., and M. Alexander. 1979. Effect of concentration of organic chemicals on their biodegradation by natural microbial communities. *Applied and Environmental Microbiology* 37:1211–1216.

Boethling, R. S., and A. Sabljic. 1989. Screening-level model for aerobic biodegradability based on a survey of expert knowledge. *Environmental Science & Technology* 23:672–679.

Bollag, J. M., and W. B. Bollag. 1990. A model for enzymatic binding of pollutants in the soil. *International Journal of Environmental Analytical Chemistry* 39:147–157.

Box, G. E. P., W. G. Hunter, and J. S. Hunter. 1978. *Statistics for Experimenters*. John Wiley & Sons, New York, N.Y.

Claxton, L. D., V. S. Houk, R. Williams, and F. Kramer. 1991. Effect of bioremediation on the mutagenicity of oil spilled in Prince William Sound, Alaska. *Chemosphere* 23:643–650.

Cochran, J. R., M. V. Yates, and J. M. Henson. 1988. A modified purge-and-trap/gas chromatography method for analysis of volatile halocarbons in microbiological degradation studies. *Journal of Microbiological Methods* 8:347–354.

D'Adamo, P. D., A. F. Rozich, and A. F. Gaudy, Jr. 1984. Analysis of growth data with inhibitory carbon sources. *Biotechnology and Bioengineering* 26:397–402.

Dahnke, W. C., and G. V. Johnson. 1990. Testing soils for available nitrogen, pp. 127–139, In *Soil Testing and Plant Analysis*, 3d ed. SSSA Book Series No. 3. Soil Science Society of America, Madison, Wisc.

Dasappa, S. M., and R. C. Loehr. 1991. Toxicity reduction in contaminated soil bioremediation processes. *Water Research* 25:1121–1130.

Denahan, S. A., B. J. Denahan, W. G. Elliott, W. A. Tucker, M. G. Winslaw, and S. R. Boyes. 1990. Relationships between chemical screening methodologies for petroleum contaminated soils: Theory and practice, pp. 93–109. In P. T. Kostecki and E. J. Calabrese (eds.), *Petroleum Contaminated Soils*, Vol. 3. Lewis Publishers, Chelsea, Mich.

Devillers, J., R. Steiman, and F. Seigle-Murandi. 1989. The usefulness of the agar-well diffusion method for assessing chemical toxicity to bacteria and fungi. *Chemosphere* 19:1693–1700.

Difco Laboratories. 1984. *Difco Manual.* Difco Laboratories, Detroit, Mich.

Diltz, M. S., C. E. Hepfer, E. Hartz, and K. H. Baker. 1992. Recovery of heterotrophic bacterial guilds from transient gasoline pollution. *Hazardous Waste and Hazardous Materials* 9:267–273.

Dragun, J. 1988. *The Soil Chemistry of Hazardous Materials.* Hazardous Materials Control Research Institute, Silver Spring, Md.

Dutka, B. J., K. K. Kwan, S. S. Rao, A. Jurkovic, R. McInnis, G. A. Palmateer, and B. Hawkins. 1991. Use of bioassays to evaluate river water and sediment quality. *Environmental Toxicology and Water Quality* 6:309–327.

Dutka, B. J., and G. Bitton (eds.). 1986. *Toxicity Testing Using Microorganisms,* Vols. I and II. CRC Press, Inc., Boca Raton, Fla.

Fannin, T. E., M. D. Marcus, and D. A. Anderson. 1981. Use of a fractional factorial design to evaluate interactions of environmental factors affecting biodegradation rates. *Applied and Environmental Microbiology* 42:936–943.

Fish, F., I. Lamper, A. Halachmi, G. Riesenfeld, and M. Hezberg. 1985. The SOS chromotest kit—A rapid method for the detection of genotoxicity. Second International Symposium of Toxicity Testing using Bacteria, Banff, Canada, May 6–10.

Fixen, P. E., and J. H. Grove. 1990. Testing soils for phosphorus, pp. 141–180. In *Soil Testing and Plant Analysis,* 3d ed. SSSA Book Series No. 3. Soil Science Society of America, Madison, Wisc.

Flathman, P. E., M. J. McCloskey, J. J. Vondrick, and D. M. Pimlett. 1985. In situ physical/biological treatment of methylene chloride (dichloromethane) contaminated ground water, pp. 571–577. In *Proceedings of the 5th National Symposium and Exposition on Aquifer Restoration and Ground Water Monitoring.* National Water Well Association, Worthington, Ohio.

Flowers, T. H., and J. M. Bremner. 1991. A rapid dichromate procedure for routine estimation of total nitrogen in soils. *Communications in Soil Science and Plant Analysis* **22**:1409–1416.

Fogel, M. M., A. R. Taddeo, and S. Fogel. 1986. Biodegradation of chlorinated ethanes by a methane-utilizing mixed culture. *Applied and Environmental Microbiology* **44**:720–724.

Ghiorse, W. C., and D. L. Balkwill. 1983. Enumeration and morphological characterization of bacteria indigenous to subsurface environments, pp. 213–224. In *Developments in Industrial Microbiology*, Vol. 24: *Proceedings of the Thirty-Ninth General Meeting of the Society for Industrial Microbiology*. Society for Industrial Microbiology, Arlington, Va.

Glenn, D. M., and F. B. Abeles. 1991. Use of a microtitre plate reader in soil phosphorus analysis. *Communications in Soil Science and Plant Analysis* **22**:1503–1506.

Goswami, K. P., and B. L. Koch. 1976. A simple apparatus for measuring degradation of ^{14}C-labeled pesticides in soil. *Soil Biology and Biochemistry* **8**:527–528.

Grady, C. P. L., Jr., J. S. Dang, D. M. Harvey, A. Jobbagy, and X.-L. Wang. 1989. Determination of biodegradation kinetics through use of electrolytic respirometry. *Water Science and Technology* **21**:957–968.

Grady, C. P. L., Jr. 1985. Biodegradation: Its measurement and microbial basis. *Biotechnology and Bioengineering* **27**:660–674.

Green, R. H. 1979. *Sampling Design and Statistical Methods for Environmental Biologists*. John Wiley & Sons, New York, N.Y.

Hao, X., B. J. Dutka, and K. K. Kwan. 1987. Genotoxicity studies on sediments using a modified SOS chromotest. *Toxicity Assessment* **2**:1630–1631.

Hazen, T. C., L. Jimenez, G. Lopez de Victoria, and C. B. Fliermans. 1991. Comparison of bacteria from deep subsurface sediment and adjacent ground water. *Microbial Ecology* **22**:293–304.

Hensel D. R. 1990. Less than obvious. *Environmental Science & Technology* **24**:1766–1774.

Hickman, G. T., and J. T. Novak. 1989. Relationship between subsurface biodegradation rates and microbial density. *Environmental Science & Technology* **23**:525–532.

Hobbie, J. E., R. J. Daley, and S. Jasper. 1977. Use of nucleopore filters for counting bacteria by fluorescence microscopy. *Applied and Environmental Microbiology* **33**:1225–1229.

Hopkins, D. W., S. J. MacNaughton, and A. G. O'Donnell. 1991a. A dispersion and differential centrifugation technique for representatively sampling microorganisms from soil. *Soil Biology and Biochemistry* **23**:217–225.

Hopkins, D. W., S. J. MacNaughton, and A. G. O'Donnell. 1991b. Evaluation of a dispersion and elutration technique for sampling microorganisms from soil. *Soil Biology and Biochemistry* **23**:227–233.

Irving, G. C. J., and M. J. McLaughlin. 1990. A rapid and simple field test for phosphorus in Olsen and Bray no. 1 extracts of soil. *Communications in Soil Science and Plant Analysis* **21**:2245–2255.

Jensen, V. 1968. The plate count technique, pp. 158–170. In T. R. G. Gray and D. Parkinson (eds.), *The Ecology of Soil Bacteria.* University of Toronto Press, Toronto, Canada.

Karlen, D. L., and T. E. Fenton. 1991. Soil map units: Basis for agrochemical-residue sampling, pp. 182–194. In R. G. Nash and A. R. Leslie (eds.), *Ground Water Residue Sampling Design.* ACS Symposium Series 465. American Chemical Society, Washington, D.C.

Kerfoot, H. B., S. L. Pierett, E. N. Amick, D. W. Bottrell, and J. D. Petty. 1990. Analytical performance of four portable gas chromatographs under field conditions. *Journal of the Air & Waste Management Association* **40**:1106–1113.

King, L. K., and B. C. Parker. 1988. A simple, rapid method for enumerating total viable and metabolically active bacteria in ground water. *Applied and Environmental Microbiology* **54**:1630–1631.

King, E. F., and B. J. Dutka. 1986. Respirometric techniques. In B. J. Dutka and G. Bitton (eds.), *Toxicity Testing Using Microorganisms,* Vol. I. CRC Press, Inc., Boca Raton, Fla.

Kleinbaum, D. G., and L. L. Kupper. 1978. *Applied Regression Analysis and Other Multivariable Methods.* Duxbury Press, North Scituate, Mass.

Knobeloch, L. M., G. A. Blondin, H. W. Read, and J. M. Harkin. 1990. Assessment of chemical toxicity using mammalian mitrochondrial electron transport particles. *Archives of Environmental Contamination & Toxicology* **19**:828–835.

Kobayashi, H., and B. E. Rittmann. 1982. Microbial removal of hazardous organic compounds. *Environmental Science & Technology* **16**:170A–183A.

Kuehl, D. W., G. T. Ankley, L. P. Burkhard, and D. Jensen. 1990. Bioassay directed characterization of the acute aquatic toxicity of a creosote leachate. *Hazardous Waste and Hazardous Materials* **7**:283–291.

Larson, R. J. 1984. Kinetic and ecological approaches for predicting biodegradation rates of xenobiotic organic chemicals in natural ecosystems, pp. 677–686. In M. J. Klug and C. A. Reddy (eds.), *Current Perspectives in Microbial Ecology.* American Society for Microbiology, Washington, D.C.

Leach, L. E., and R. R. Ross. 1991. Aseptic sampling of unconsolidated heaving soils in saturated zones, pp. 334–348. In R. G. Nash and A. R. Leslie (eds.), *Ground Water Residue Sampling Design.* ACS Symposium Series 465. American Chemical Society, Washington, D.C.

Lentner, M., and T. D. Bishop. 1986. *Experimental Design and Analysis.* Valley Book Company, Blacksburg, Va.

Loos, M. A., A. Kontson, and P. C. Kearney. 1980. Inexpensive soil flask for ^{14}C-pesticide degradation studies. *Soil Biology and Biochemistry* **12**:583–585.

Lui, D. 1985. Effect of bacterial cultures on microbial toxicity assessment. *Bulletin of Environmental Contamination & Toxicity* **34**:331–339.

Lui, D. 1981. A rapid biochemical test for measuring chemical toxicity. *Bulletin of Environmental Contamination & Toxicity* **26**:145–149.

Lui, D., and K. Thomson. 1983. Toxicity assessment of chlorobenzenes using bacteria. *Bulletin of Environmental Contamination & Toxicity* **31**:105–111.

Lui, D. W. M. J. Strachan, K. Thomson, and K. Kwasneiwska. 1981. Determination of the biodegradability of organic compounds. *Environmental Science and Technology* 15: 788-793.

Maitre, V., G. Bourrie, and P. Curmi. 1991. Contamination of collected soil water samples by the dissolution of the mineral constituents of porous P.T.F.E. cups. *Soil Science* 152:289–293.

Marinucci, A. C., and R. Bartha. 1979. Apparatus for monitoring the mineralization of volatile ^{14}C-labeled compounds. *Applied and Environmental Microbiology* 38:1020–1022.

Mason, B. J. 1983. *Preparation of Soil Sampling Protocol: Techniques and Strategies.* U.S. EPP Report No. EPA/600/4083–020, Washington, D. C.

McNabb, J. F., and G. E. Mallard. 1984. Microbiological sampling in the assessment of ground water pollution, pp. 235–260. In G. Bitton and C. P. Gerba (eds.), *Ground water Pollution Microbiology.* John Wiley & Sons, New York, N.Y.

Middaugh, D. P., J. G. Mueller, R. L. Thomas, S. E. Lantz, M. H. Hemmer, G. T. Brooks, and P. J. Chapman. 1991. Detoxification of pentachlorophenol and creosote contaminated ground water by physical extraction: chemical and biological assessment. *Archives of Environmental Contamination & Toxicology* 21:233–244.

Middaugh, D. P., M. J. Hemmer, and L. R. Goodman. 1987. *Methods for Spawning, Culturing and Conducting Toxicity Tests with Early Life Stages of Four Atherinid Fishes: The Inland Silverside,* Menidia beryllina, *Atlantic Silverside,* Menidia menidia, *Tidewater Silverside,* Medidia peninsulae, *and the California Grunion,* Leuresthes tennuis. USEPA Report No. EPA/600/8-87/004, Washington, D.C.

Mueller, J. G., S. E. Lantz, B. O. Blattmann, and P. J. Chapman. 1991a. Bench-scale evaluation of alternative biological treatment processes for the remediation of pentachlorophenol- and creosote-contaminated materials: Solid-phase bioremediation. *Environmental Science & Technology* 25:1045–1054.

Mueller, J. G., S. E. Lantz, B. O. Blattmann, and P. J. Chapman. 1991b. Bench-scale evaluation of alternative biological treatment processes for the remediation of pentachlorophenol- and creosote-contaminated materials: Slurry-phase bioremediation. *Environmental Science & Technology* 25:1055–1061.

Mueller, J. G., P. J. Chapman, and P. H. Pritchard. 1989. Action of a fluoranthene-utilizing bacterial community on polycyclic aromatic hydrocarbon components of creosote. *Applied and Environmental Microbiology* 55:3085–3090.

Nyer, E. K., and G. J. Skladany. 1989. Treatment technology: Relating the physical and chemical properties of petroleum hydrocarbons to soil and aquifer remediation. *Ground Water Monitoring Review* 9:54–60.

Odeh, R. E., and M. Fox. 1975. *Sample Size Choice.* Marcel Dekker, Inc., New York, N.Y.

Oliver, M. A., and R. Webster. 1991. How geostatistics can help you. *Soil Use and Management* 7:206–217.

Ong, S. K., and L. W. Lion. 1991. Trichloroethylene vapor sorption onto soil minerals. *Journal of the Soil Science Society of America* **55**:1559–1568.

Ortego, J. D., M. Kowalska, and D. L. Cocke. 1991. Interactions of montmorillonite with organic compounds: Adsorptive and catalytic properties. *Chemosphere* **22**:769–798.

Otson, R., and C. Chan. 1987. Sample handling and analysis for 51 volatile organics by an adapted purge and trap GC-MS technique. *International Journal of Environmental Analysis & Chemistry* **30**:275–287.

Page, A. L., R. H. Miller, and D. R. Kinney (eds.). 1982. *Methods of Soil Analysis*, part 2: *Chemical and Microbiological Properties*, 2d ed. American Society for Agronomy/Soil Science Society of America, Madison, Wisc.

Phelps, T. J., E. G. Raione, D. C. White, and C. B. Fliermans. 1989. Recovery of deep subsurface sediments for microbiological studies. *Journal of Microbiological Methods* **9**:267–279.

Pitter, P., and J. Chudoba. 1990. *Biodegradability of Organic Substances in the Aquatic Environment.* CRC Press, Inc., Boca Raton, Fla.

Pramer, D., and R. Bartha. 1972. Preparation and processing of soil samples for biodegradation studies. *Environmental Letters* **2**:217–224.

Robbins, G. A., B. G. Deyo, M. R. Temple, J. D. Stuart, and M. J. Lacy. 1990. Soil-gas surveying for subsurface gasoline contamination using total organic vapor detection instruments: II. Field experimentation. *Ground Water Monitoring Review* **10**:110–117.

Robbins, G. A., R. D. Bristol, and V. D. Roe. 1989. A field screening method for gasoline contamination using a polyethylene bag sampling system. *Ground Water Monitoring Review* **9**:87–97.

Roth, G. W., D. B. Beegle, R. H. Fox, J. D. Toth, and W. P. Piekeilek. 1991. Development of a quick test kit method to measure soil nitrate. *Communications in Soil Science & Plant Analysis* **22**:191–200.

Russell, B. F., T. J. Phelps, W. T. Griffin, and K. A. Sargent. 1992. Procedures for sampling deep subsurface microbial communities in unconsolidated sediments. *Ground Water Monitoring Review* **12**:96–104.

Sawhney, B. L., and K. Brown (eds.). 1989. *Reactions and Movement of Organic Chemicals in Soil,* SSSA Special Publication No. 22. Soil Science Society of America, Madison, Wisc.

Schulten, H.-R., and M. Schnitzer. 1991. Supercritical carbon dioxide extraction of long-chain aliphatics from two soils. *Soil Science Society of America Journal* **55**:1603–1611.

Sharpley, A. N. 1991. Effect of soil pH on cation and anion solubility. *Communications in Soil Science & Plant Analysis* **22**:827–841.

Shelton, D. R., and J. M. Tiedje. 1984. General method for determining anaerobic biodegradation potential. *Applied and Environmental Microbiology* **47**:850–857.

Sleat, R., and J. P. Robinson. 1984. The bacteriology of anaerobic degradation of aromatic compounds. *Journal of Applied Bacteriology* **57**:381–394.

Smith, R. L., and M. J. Klug. 1987. Flowthrough reactor flasks for study of microbial metabolism in sediments. *Applied and Environmental Microbiology* **53**:371–374.

Sokal, R. R., and F. J. Rohlf. 1969. *Biometry: The Principles and Practice of Statistics in Biological Research.* W. H. Freeman & Company, San Francisco, Calif.

Spain, J. C., P. A. Van Veld, C. A. Monti, P. H. Pritchard, and C. R. Cripe. 1984. Comparison of *p*-nitrophenol biodegradation in field and laboratory test systems. *Applied and Environmental Microbiology* **48**:944–950.

Splitstone, D. E. 1989. Ground water monitoring needs and requirements: 7. A statistician's view of ground water monitoring. *Hazardous Materials Control* **3(2)**:40–45.

Subba-Roa, R. V., H. E. Rubin, and M. Alexander. 1982. Kinetics and extent of mineralization of organic chemicals at trace levels in freshwater and sewage. *Applied and Environmental Microbiology* **43**:1139–1150.

Swartz, R. C., D. W. Schults, T. H. Dewitt, G. R. Ditsworth, and J. O. Lamberson. 1990. Toxicity of fluoranthene in sediment to marine amphipods: A test of the equilibrium partitioning approach to sediment quality criteria. *Environmental Toxicology and Chemistry* **9**:1071–1080.

Sylvestre, M., R. Masse, C. Ayotte, F. Messier, and J. Fauteaux. 1985. Total biodegradation of 4-chlorobiphenyl (4CB) by a two-membered bacterial culture. *Applied Microbiology and Biotechnology* **21**:192–195.

Tabak, H. H., S. A. Quave, C. I. Mashni, and E. F. Barth. 1981. Biodegradability studies with organic priority pollutant compounds. *Journal of the Water Pollution Control Federation* **53**:1503–1518.

Tate, R. L. 1987. *Soil Organic Matter: Biological and Ecological Effects.* John Wiley & Sons, New York, N.Y.

Tell, J. G., and C. G. Uchrin. 1991. Relative contributions of soil humic acid and humin to the adsorption of toluene onto an aquifer solid. *Bulletin of Environmental Contamination & Toxicology* **47**:547–554.

United States Environmental Protection Agency. 1991. *Guide for Conducting Treatability Studies under CERCLA: Aerobic Biodegradation Remedy Screening—Interim Guidance.* USEPA Publication No. EPA/540/2-91/013A, Washington, D.C.

United States Environmental Protection Agency. 1988a. *Code of Federal Regulations* (40 CFR Ch. 1, 7-1-88 Edition), pp. 563–631. U.S. Government Printing Office, Washington, D.C.

United States Environmental Protection Agency. 1988b. §795.54 Anaerobic microbiological transformation rate data for chemicals in the subsurface environment. *Federal Register* **53**:22320–22323.

United States Environmental Protection Agency. 1986a. *Test Methods for Evaluating Solid Waste,* 3d ed. USEPA Publication No. SW8-46, Washington, D.C.

United States Environmental Protection Agency. 1986b. *Protocol Development for the Prediction of the Fate of Organic Priority Pollutants in Biological Wastewater Treatment Systems.* USEPA Publication No. EPA/600/S2-85/141, Washington, D.C.

United States Environmental Protection Agency. 1982. *Pesticide Assessment Guidelines Subdivision N Chemistry: Environmental Fate.* Office of Pesticides and Toxic Substances, Washington, D.C.

United States Environmental Protection Agency. 1979. *Methods for Analysis of Water and Wastes.* USEPA Publication No. EPA/600/4-79/020, Washington, D.C.

Van Cleve, K., P. I. Coyne, E. Goodwin, C. Johnson and M. Kelley. 1979. A comparison of four methods for measuring respiration in organic material. *Soil Biology and Biochemistry* 11:237–246.

Venosa, A. D., J. R. Haines, and D.M. Allen. 1992. Efficacy of commercial inocula in enhancing biodegradation in closed laboratory bioreactors. *Journal of Industrial Microbiology* 10:13–23.

Verschueren, K. 1983. *Handbook of Data on Organic Chemicals,* 2d ed. Van Nostrand Reinhold Company, Inc., New York, N.Y.

Volskay, V. T., Jr., and C. P. L. Grady, Jr. 1988. Toxicity of selected RCRA compounds to activated sludge microorganisms. *Journal of the Water Pollution Control Federation* 60:1850–1856.

Wahle, U., W. Kordel, and W. Klein. 1990. Methodology for the exposure assessment of soil for organic chemicals: I. Extraction procedure of nonvolatile organic chemicals. *International Journal of Environmental Analytical Chemistry* 39:121–128.

Wang, X., and R. Bartha. 1990. Effects of bioremediation on residues, activity and toxicity in soil contaminated by fuel spills. *Soil Biology and Biochemistry* 22:501–504.

Ward, T. E. 1985. Characterizing the aerobic and anaerobic microbial activities in surface and subsurface soils. *Environmental Toxicology and Chemistry* 4:727–737.

Wassel, R. A., and A. L. Mills. 1983. Changes in water and sediment bacterial community structure in a lake receiving acid mine drainage. *Microbiol Ecology* 9:155–170.

Wilson, J. T., L. E. Leach, J. Michalowski, S. Vandergrift, and R. Callaway. 1989. In situ *Bioremediation of Spills from Underground Storage Tanks: New Approaches for Site Characterization, Project Design, and Evaluation of Performance.* USEPA Report No. EPA/600/2-89/042, Washington, D.C.

Winberry, W. T., N. T. Murphy, and R. M. Riggan. 1990. *Methods for Determination of Toxic Organic Compounds in Air-EPA Methods.* Noyes Data Corporation, Inc., Park Ridge, N.J.

Wolf, A. M., and D. E. Baker. 1990. Colorimetric method for phosphorus measurement in ammonium oxalate soil extracts. *Communications in Soil Science & Plant Analysis* 21:2257–2263.

Xu, H., and B. J. Dutka. 1987. ATP-TOX system—A new, rapid, sensitive bacterial toxicity screening system based on the determination of ATP. *Toxicity Assessment* 2:149–166.

Young, J. C., and E. R. Baumann. 1976. The electrolytic respirometer: Factors affecting oxygen uptake measurements. *Water Research* 10:1031–1040.

5

Bioengineering of Soils and Ground Waters

Marleen A. Troy

O.H.M. Remediation Services Corp.
Princeton, N.J.

The term *bioengineering* implies the alliance of a biological system and an engineered system for the remediation of undesirable chemical contamination. In most instances, it denotes the use of engineering techniques to optimize and/or maximize an already ongoing biological process. In this chapter, both the in situ and ex situ bioengineering of soils and ground waters will be discussed. These environments can become contaminated from leakage or spillage from storage tanks, from accidental discharge or release from truck or train tanker cars, from industrial discharge, from seepage from landfills, and as a consequence of agricultural activities. *In situ treatment* refers to the performance of in-place remediation of the contamination. The advantages of in situ bioengineered treatment are that it does not require excavation of the contaminated material and movement to another locale, it can allow above-ground site activities to occur while treatment is going on, and it minimizes exposure to the contaminants. *Ex situ treatment* refers to the extraction and/or excavation of the contaminated material for treatment aboveground in an engineered treatment system.

The objectives of the bioengineered remediation treatment process are analogous to those of conventional biological treatment operations. With conventional biological treatment systems, a treatment vessel is "engineered" to provide optimal conditions for the microorganisms to grow. As a result of their growth, the microorganisms will metabolize the compound(s) of interest, usually resulting in the production of innocuous end products. An example of this concept would be a wastewater treatment facility. For this process, the conditions in a treatment vessel (i.e., a large tank) are optimized (pH adjusted, aerated, provisions to control flow rates to provide adequate contact time) to promote biodegradation of the organic materials in the wastewater. For a bioengineered treatment system, instead of utilizing a manufactured container to accommodate the treatment process, the soil environment could be "bioengineered" to create an in-place treatment vessel and to provide optimal growth conditions for the indigenous microorganisms present. The effective application of this type of biological treatment can result in the complete breakdown of the contaminant(s) to innocuous end products in many instances.

The successful implementation of bioengineered remediation techniques will involve a multidisciplinary approach requiring input from individuals with expertise in microbiology, chemistry, geology, soil science, environmental engineering, and chemical engineering (18). In order to use bioengineering successfully for the remediation of environmental contamination problems, the first step is to obtain a thorough understanding of the matrix characteristics of the media to be treated and the properties (physical, chemical, microbiological) of the contaminant(s).

Matrix Characteristics

The overall goal of a bioengineered process is to provide optimal conditions for the indigenous bacteria such that they metabolize the contaminant(s) of interest in the most efficient manner possible. The following is a brief description of the important characteristics of the soil/subsurface and ground water environments that will have to be considered for proper bioengineering design.

Soil/Subsurface

Soil is composed of particles that are the result of the weathering (disintegration and decomposition) of rocks and the decay of vegetation (37).

As part of the definition of *soil,* air and water have to be included along with the solid material. *Soil* is a term that is used to describe a "natural resource" that resides at the earth's surface as a continuum of many different types of soils (17). For the proper design and installation of a biological remediation treatment system, it is important to understand the physical, chemical, and biological characteristics of a particular soil.

Soil properties can vary greatly from one region to another. They also can vary greatly within the same region spatially and with depth. The characteristics that define a soil type are properties such as cation exchange capacity, clay type, gradation, intrinsic permeability, liquid limit, organic matter content, particle size, pH, porosity, and soil texture. A brief definition for each of these terms is presented in Table 5-1. In order to describe the heterogeneities of soils uniformly, soil classification systems have been developed. The two classification systems in most common use are the Unified Soil Classification System (the USCS) and the United States Comprehensive Soil Classification System of the Soil Staff of the United States Department of Agriculture (the USDA). Detailed descriptions of these systems are provided in other sources (17,37). Understanding the heterogeneities of the soil matrix that is being considered for remediation is critical because these characteristics will influence the movement of the contaminant(s) as well as of water and air and are essential to successful bioengineering treatment system design.

Ground water

All water located beneath the land surface is referred to as *underground water* (27). The underground water can be found in two different zones, as illustrated in Figure 5-1. The *unsaturated zone,* found immediately below the ground surface, contains both water and air. In most areas, underneath the unsaturated zone is the *saturated zone.* This zone is characterized by all openings (pore spaces) being filled with water. According to the United States Geological Survey (27), water located in the saturated zone is the only water to which the term *ground water* is correctly applied. Replenishment or "recharge" of the saturated zone occurs through the percolation of water from the land surface through the unsaturated zone. The subzone located between the saturated and unsaturated zones is referred to as the *capillary fringe.* The water in this zone is held by surface tension. The water table is that level in the saturated zone at which the water is held by surface tension. There are many excellent references on ground water

Table 5-1. Glossary of Parameters Used to Classify Soils

Cation exchange capacity	Ion exchange occurs when ions (electrically charged particles) in solution displace ions associated with geologic materials. The term *cation exchange* refers to the exchange between cations balancing the surface charge on a soil surface and cations dissolved in water.
Clay type	Clays are a complex and loosely defined group of silicate minerals, essentially aluminum. Most clay minerals belong to the kaolin, smectite (montmorilonite), and illite groups, the micas and chlorites are close relatives.
Effective porosity	The ratio, usually expressed as a percentage, of the total volume of voids available for fluid transmission to the total volume of the porous medium.
Gradation	The procedure for systematically and arbitrarily dividing an essentially continuous range of particle sizes (of a soil, sediment, or rock) into a series of classes or grades for the purpose of standardization of terms for statistical analysis.
Intrinsic permeability	A measure of the relative ease with which a porous medium can transmit a liquid under a potential gradient. Intrinsic permeability is a property of the medium alone that is dependent on the shape and size of the openings through which the liquid moves.
Liquid limit	The water-content boundary between the semiliquid and the plastic states of soil.
Organic matter content	Pertaining to or relating to the amount (percentage) of carbon in substance. Organic compounds have hydrogen bonded to the carbon atom.
Particle size	The general dimensions, such as average diameter or volume, of the particles in a sediment or rock or of the grains of a particular mineral that make up a sediment or rock based on the premise that the particles are spheres or that measurements can be expressed as diameters of equivalent spheres. It is commonly measured by sieving, by calculating settling velocitiesN, or by determining areas of microscopic images.
pH	The negative $\log_{10}$ of the hydrogen-ion activity in solution; a measure of acidity or basicity.

Table 5-1. Glossary of Parameters Used to Classify Soils
(*Continued*)

Porosity	A measure of interstitial space contained in a rock (or soil) expressed as the percentage ratio of void space to the total (gross) volume of the rock.
Soil texture	The physical nature of the soil according to the relative proportion of sand, silt, and clay.

Note: see Refs. 7, 17, 41.

available for more detailed discussions. These include Freeze and Cherry (21), Heath (27), and Mercer et al. (41).

Microbial Activity in Soil and Ground Water

The subsurface soils have been reported to provide favorable habitats for the proliferation of microorganisms (5,6,21,23). However, the vertical distribution of microorganisms in a soil profile is reported to vary greatly as a function of soil type. The soil type also may determine whether or not contaminants are biodegraded as they pass through the vadose zone (22). There have been numerous studies documenting microbial degradation of various contaminants in soil (34,61,62,64).

There have been a number of investigations documenting that aquifers contain considerable microbial populations (6,22,26,29,30). There is also abundant information available documenting microbial degradation of chemical contamination in these environments (1–3,24,31,35,43,49,51,60,61,63).

The Properties of the Contaminant

The nature of the contaminant(s) will dictate whether biological processes should even be considered for bioremediation. Therefore, the first step in evaluating possible bioengineering applications is understanding the characteristics of the contaminants. The first place

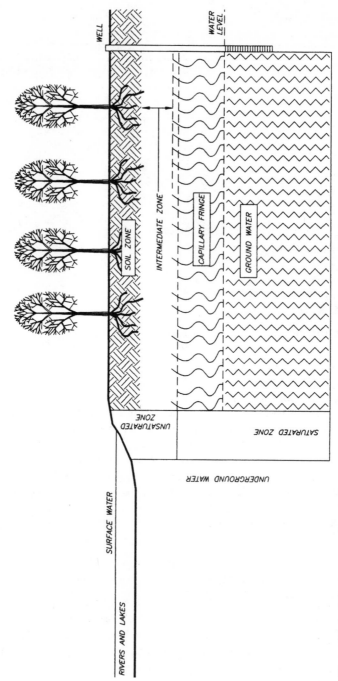

Figure 5-1. Schematic representation of undergroundwater. (*Drawing by Willie E. Jetter.*)

to gather information about a contaminant is from a material safety data sheet (MSDS) sheet. This is a document provided by the manufacturer of the compound that documents its physical, chemical, and toxicological properties. A literature review also should be performed to assess whether biodegradation of the particular compound(s) has been achieved or what difficulties have been encountered. The U.S. EPA has several databases that can be consulted for this purpose such as ATTIC (61) and VISITT (63).

The objectives of the search are to obtain as much information about the contaminants as possible so as to be able to predict the potential migration pathways and to assess whether or not the conditions at the site are or can be made suitable for microbial action on the contaminants. This involves understanding the physical, chemical, and microbiological properties of the contaminant.

Physicochemical Properties

Determination of the chemical properties of the contaminant is necessary to assess its fate and transport potential, as well as for evaluating potential for biodegradation. The following properties are used to characterize a contaminant and dictate its potential migration pathways once it is released into the environment (17,39,41,47):

- *Density.* The density of a substance is the ratio of its mass to its volume. This parameter is important in evaluating whether liquids and solids will float or sink in water or whether a gas is lighter or heavier than air.

- *Adsorption coefficient for soils.* The degree to which the contaminant distributes itself between the solid and solution phases of a water-saturated or unsaturated soil is a result of several physical and chemical properties of both the contaminant and the soil. The parameter K_{oc} is used to symbolize the tendency of the contaminant to be adsorbed. K_{oc} can be described as the ratio of the amount of chemical adsorbed per unit weight of organic carbon in the soil to the concentration of the contaminant in solution at equilibrium. The extent of adsorption not only will influence the contaminants mobility but also may affect volatilization, photolysis, hydrolysis, and biodegradation.

- *Solubility in water.* This is a measure of the degree to which the contaminant will dissolve in water. Solubility in water is a function of temperature. The presence of dissolved salts or minerals in water also can influence solubility. Naturally occurring dissolved organic material also can affect the solubility of the contaminants.

- *Solubility in various solvents.* This is a measure of the degree to which one liquid is soluble in another, solid solute in a liquid solute, and gases in liquids. These solubilities are also temperature-dependent. This parameter may be important to consider at sites where there is more than one contaminant present.

- *Octanol-water partition coefficient.* This is a measure of a contaminant's tendency to partition between the soil and the ground water. K_{ow}, the octanol-water partition coefficient, is defined as the ratio of a contaminant's concentration in the octanol phase to its concentration in the aqueous phase of a two-phase octanol-water system.

- *Volatilization from soil.* This is the process by which the contaminant evaporates in the vapor phase from the soil to the atmosphere. How long the contaminant may persist in the soil is a function of the contaminant's volatility. Factors such as soil characteristics, contaminant properties, and site conditions can affect the rate at which a chemical will volatilize from the soil.

- *Volatilization from water.* This is the process by which the contaminant evaporates in the vapor phase to the atmosphere from water. The tendency for a contaminant to volatilize from an aqueous phase to the atmosphere or in soil can be estimated from Henry's law. Henry's law contends that when a solution becomes very dilute, the vapor pressure of a contaminant should be proportional to its concentration, i.e., that the vapor pressure of the contaminant is equal to Henry's law constant multiplied by the concentration of the contaminant in water.

Values for these parameters can be obtained or estimated from sources such as Lyman et al. (39), Mercer et al. (41), and Verschueren (58).

Microbiological Properties

When evaluating the potential of using microbiological processes for the remediation of contaminated soils and ground waters, it is beneficial to develop a checklist of requirements for bacterial growth that will be the basis of the evaluation, as suggested by Parkin and Calabria (45). If the existing site matrix can be bioengineered to meet these criteria, the use of bioremediation will be a practical remedial option. This checklist includes carbon and energy source, electron acceptor (nature of the environment), nutrients, proper pH, temperature, absence of toxic materials, and adequate contact. The following is a brief summary of each of these criteria:

- *Carbon and energy source.* In order for the microorganisms to sustain themselves, biodegradable organic compounds must be present to serve as the carbon and energy source. The contaminant(s) in some cases can serve this purpose. However, this depends on the contaminant(s) concentration and the mechanism by which it is transformed, i.e., through primary or secondary utilization or through cometabolism.

- *Electron acceptor.* The electron acceptor is the substance that accepts electrons during an oxidation-reduction reaction (which is the process by which the contaminant is broken down into innocuous end products). Oxygen is the electron acceptor in aerobic environments. Nitrate is the electron acceptor in anoxic environments. Sulfate, carbon dioxide, or reduced organics act as electron acceptors in anaerobic environments.

- *Nutrients.* Nutrients are substances taken in by the microorganisms from their environments and used in metabolic reactions. Nutrients are substances such as nitrogen (which can be present as ammonium, nitrite, nitrate, or organic nitrogen), phosphorus (present as orthophosphate), and sulfur (required as sulfate for aerobic environments and sulfide for anaerobic environments). There may be some situations where trace metals (such as potassium, iron, molybdenum, zinc) may not be present in sufficient quantities and may need to be added.

- *Proper pH.* For most microbiological activities, pH will need to be kept in the range of 6.5 to 8.5.

- *Temperature.* Temperature is an important consideration because there are specific ranges in which the microbiological metabolic pathways and enzymes can operate. For bioengineered remediation systems, temperature can affect the required contact time and, for aerobic treatment, the solubility of oxygen.

- *Absence of toxic materials.* A toxic material is one that inhibits the rate at which microbiological activity occurs. Whether or not a material is toxic depends on concentration, the characteristics of the material, the ability of the microorganisms to adapt to its presence, and time of exposure.

■ *Adequate contact.*   In order for the biodegradation of a contaminant to occur, there must be sufficient interaction between the contaminated matrix and the microorganisms. This is a key parameter to be considered in the design of a bioengineered system. Adequate contact must be provided between the contaminant(s), the microorganisms performing the biodegradation process, and the added amendments (nutrients, electron donor, pH adjusters). For in situ treatment systems this will be greatly affected by the hydrogeologic characteristics of the site.

Examples of Bioengineered Process Technologies for Soil and Ground Water Treatment

Many factors influence the proper choice of a bioengineered remediation treatment system. These include the nature of the contaminant(s), the extent of contamination, site characteristics, cleanup goals, and economics. Therefore, it is not possible to have a single remedial approach that will be suitable for every site, every contaminant, and every remedial situation. As a result, there does not exist one biological treatment scheme that would be applicable to every remedial situation. Each site needs to be evaluated individually. The proper design, implementation, and operation of a bioengineered treatment process requires a multidisciplinary approach to characterize the hydrogeologic, geologic, chemical, and microbiological characteristics of the locale effectively (18). Only when the site characteristics have been thoroughly evaluated can a successful treatment process be designed. Biological treatment offers the distinct advantage of partial or complete destruction of the contaminants as opposed to simply transferring the pollution to another phase of the environment (35). Possible bioengineered treatment processes would include pump and treat, bioventing, land treatment, slurry reactors, and combined "treatment train" technologies.

Pump and Treat

Pump and treat treatment technologies consist of extracting contaminated ground water and treating it on the surface. This technology is generally considered when significant levels of ground water contamination exist (41). A possible use for this water after it has been extracted and surface treated (physically, chemically, or biologically) is reinjection to promote soil flushing as well as to stimulate in situ biological activities.

The technology was first demonstrated by Raymond (48) for the remediation of a pipeline leak in which 3186 barrels of high-octane gasoline leaked into an aquifer. This project indicated that by providing the appropriate conditions in the subsurface through the addition of oxygen and nutrients, the indigenous microorganisms could be stimulated to promote degradation of the hydrocarbon contaminants.

Aerobic biodegradation of the contaminant(s) has been the objective of many of the "bioengineered" pump and treat systems. Because of the local hydrogeologic conditions at a particular site, oxygen availability can become the critical factor that can limit the ability of aerobic microorganisms to degrade the contaminant(s). As a result, other mechanisms are being investigated to introduce oxygen to contaminated low dissolved oxygen areas. Hydrogen peroxide injection into the ground water is one method of overcoming the oxygen availability difficulties. Two moles of decomposed hydrogen peroxide can yield two moles of water and one mole of oxygen (2). This process can result in the introduction of oxygen into the ground water. This process has been studied by several investigators in the laboratory and the field (2,33,35,36,56,57).

The degradation of chlorinated aliphatic compounds trichloroethylene (TCE), cis-1,2-dichloroethylene (cis-DCE), trans-1,2-dichloroethylene (trans-DCE), and vinyl chloride (VC) was field demonstrated by a group from Stanford University at the Moffett Naval Air Station in Mountain View, California (50). Biological treatment of chlorinated aliphatic compounds requires a different approach because of the inability of indigenous microorganisms to utilize these compounds as primary growth substrates. However, it has been observed that these compounds can be broken down as secondary substrates by microorganisms which have available another primary substrate for growth (50). The Stanford study conclusively demonstrated that a site could be bioengineered to promote the in situ biotransformations of chlorinated aliphatic compounds by methanotrophic and heterotrophic microbial communities. The study demonstrated that it was feasible to encourage the development of the native population of methanotrophic bacteria by supplying appropriate quantities of oxygen and methane. It was observed that once the mixed methane-grown microbial communities were stimulated, the target chlorinated aliphatic compounds were degraded at rates that spanned from moderately rapid (half-lives of a few days) to very rapid (half-life of less than one day). Except for one instance, the chlorinated aliphatic compounds were metabolized to innocuous end products. The one exception was the production of a transitory intermediate product. This compound was identified as the epoxide of *trans*-DCE, *trans*-1,2-dichlorooxirane. However, the compound was observed to degrade via a hydrolysis mechanism with a half-life of approximately 4 days at 18°C.

Bioventing

Bioventing is a term used to describe the merger of soil vapor extraction technologies with bioremediation. The operation of a soil vapor extraction system will result in the stimulation of aerobic soil biodegradation activities, even though that is not the primary goal of the treatment technology. The objectives of bioventing are to engineer the vapor flow rates and nutrient balance in soils to maximize microbiological activity (46). For bioventing to be successful, adequate water must be present in the unsaturated zone to allow the required biodegradation enzyme transfers (38).

Limitations to bioventing can be physical, chemical, or operational (10). The physical limitations are a function of the degree to which air can be moved through the soil matrix. The air movement will be governed by the soil's permeability, degree of heterogeneity, and water saturation. The main chemical limitations relate to the degree of a contaminant's volatility, biodegradability, and solubility. Operational limitations such as proper application and competent operation and maintenance of the system are also keys to the success of the bioventing treatment system.

Bioventing is particularly attractive for the in situ remediation of petroleum hydrocarbons in unsaturated soils (28,32). The first reported study which documented biodegradation of gasoline being stimulated as a result of forced aeration into the vadose zone was performed by the Texas Research Institute, Inc. (28). Their study indicated that venting would be able to remove the gasoline physically as well as enhancing microbiological activity.

Bioventing was used in the spill of approximately 100,000 liters of JP-4 jet fuel at the Hill Air Force Base near Ogden, Utah (28). The contamination was primarily found in the upper 20 m at the site. Regional ground water was located at approximately 200 m, with water also occasionally found in discontinuous perched zones. Initially, vent wells were installed and operated to promote volatilization of the jet fuel. After a year of operation, the venting rate was reduced from approximately 2500 to 500 to 1000 m^3/h to provide aeration for bioremediation and to reduce volatilization. This combined operation was successful in removing a total mass of JP-4 as carbon of approximately 53,000 kg through ventilation and 42,000 kg through biodegradation.

Bioventing was investigated as a remedial alternative for the treatment of an unknown volume of JP-4 jet fuel that had spilled or leaked at an abandoned tank farm located on Tyndall Air Force Base in Florida. An in situ field demonstration was conducted for 7 months (42). Soils at the site were reported to be a fine- to medium-grained quartz sand. Depth to ground water was 0.5 to 1.0 m.

Biodegradation and volatilization rates were observed to be much higher than those observed at the Hill Air Force Base (42). These higher rates were believed to be a result of higher average levels of contamination, warmer temperatures, and ambient moisture conditions. Results of the investigation also indicated that nutrient and moisture additions did not increase biodegradation rates. Data from the airflow tests indicated that decreasing flow rates increased the percent of hydrocarbon removed by biodegradation and decreased the percent of hydrocarbon removed by volatilization relative to the total mass of hydrocarbon removed.

Biological Land Treatment

Biological land treatment is the remediation technology whereby contaminated soil is placed in an engineered treatment cell and managed to enhance microbiological activity and promote biodegradation. This type of bioremediation technology can be an attractive option when contaminated soils have to be excavated and there is enough room onsite for placement of the treatment cell.

Most biological land treatment operations are designed to promote aerobic microbiological activity. Therefore, oxygen transfer is often the most critical component of the operation. This factor becomes apparent in the construction of the treatment cell. To provide adequate aeration, the maximum depth to which contaminated soil can be spread in the treatment cell is 18 in. If large volumes of soil are to be treated, a great deal of area is required for treatment cell construction to guarantee the 18-in limitation. Depending on where the treatment cell is to be constructed, as well as on regulatory requirements, a synthetic liner and stormwater collection system may be necessary. Air emission standards also may require enclosing the treatment area or monitoring it during operation.

Operations of the landfarm generally include regular tilling of the soil using conventional farming equipment such as a tractor, bottom plow, and disk to aerate the soil. The soils in the treatment cell are also regularly monitored for pH, temperature, available nitrogen and phosphorus, moisture content, and bacterial count, as well as contaminant(s) concentration. Amendments (such as fertilizer, lime, water, etc.) are added as necessary to the soils to make the conditions optimal for microbial growth. Biological land treatment has been used successfully in the treatment of soils contaminated with petroleum hydrocarbons and polyaromatic hydrocarbons (53,56,65,69).

A variation on this biological land treatment technology is a process called *enhanced composting* (or *soil piles* or *soil mounds*). Instead of relying on physical methods such as tilling to promote aeration of the soil, this process employs the use of a forced ventilation system to aerate the soils. This type of system allows the use of deeper (>18 in) piles of soil and can incorporate bioventing techniques. Requirements for the enhancement of biodegradation in enhanced compost piles are the exchange of air to remove and treat volatile components as well as for providing the necessary oxygen for microbial growth, adequate moisture, adequate pH, and adequate nutrients (46). Studies have shown that enhanced composting can be an effective technology for the remediation of petroleum-contaminated soils (55,63,64,65).

Increasing attention is also being given to the use of compost piles (and/or soil mounds) for use as biofilters for the treatment of contaminated airstreams, particularly those resulting from soil vapor extraction systems (46). *Biofiltration* refers to the removal and oxidation of organic gases (volatile organic compounds, or VOCs) from contaminated air by beds of compost or soil (8). As the contaminated air flows through the biofilter, the VOCs sorb onto the organic surfaces of the pile. The sorbed contaminants are then oxidized by the microorganisms. However, biofiltration may not be a good candidate for the removal of some of the more highly halogenated compounds such as trichloroethylene (TCE), trichloroethane (TCA), and carbon tetrachloride (8). These contaminants biodegrade very slowly under aerobic conditions, which would result in long residence times (i.e., very large biofilters) or treatment of very low flow rates of air containing the contaminants. However, aerobic biodegradation of these contaminants can result in the innocuous end products CO_2, H_2O, and Cl^- instead of toxic intermediates such as vinyl chloride that are a result of anaerobic biodegradation (8).

Slurry Reactors

Slurry-phase bioremediation is a batch treatment technique in which contaminated soil is excavated and mixed with water to form a slurry that is mechanically aerated in a reactor vessel. Conditions within the reactor are optimized for microbial growth by providing adequate oxygen and nutrients and adjusting the pH. Other amendments, such as surfactants, dispersants, and materials supporting microbial growth and inducing degradation of the contaminants, are also used to improve the contaminated soil's handling characteristics or increase substrate availability for biodegradation (66). Temperature is also optimized for microbial growth. Seed microorganisms also may be required initially or during operation to maintain an optimal biomass concentration. After treatment has been completed, the slurry is dewatered and disposed of. Process water either may be treated onsite in a treatment system prior to discharge or recycled into the reactor vessel for use in further treatment batches. Depending on the characteristics of the contaminant(s) being treated, vapor emission treatment also may be required.

Slurry biodegradation has been reported to be effective in the treatment of highly contaminated soils ranging from 2500 to 250,000 ppm. Studies have indicated that this technology has the potential to bioremediate a wide range of organic contaminants such as pesticides, petroleum hydrocarbons, creosote, pentachlorophenol, PCBs, some halogenated organics, coal tars, refining wastes, wood-preserving wastes, and organic and chlorinated organic sludges (13,66,68). Slurry reactor treatment is not capable of treating inorganics and/or heavy metals. If these materials are present, pretreatment may be required to prevent the inhibition of biodegradation.

Combined "Treatment Train" Systems

Chemical contamination of soils and ground waters can occur in many combinations, i.e., volatiles and nonvolatiles, organics and inorganics, etc. Depending on the type of accidental discharge that occurred or the industrial activities that took place over many years at a site, it is very likely that several different categories of chemical compounds may be present in a matrix simultaneously. The presence of some types of chemical compounds, such as heavy metals, may preclude the use of biological treatment; however, in many situations, the use of biological treatment in conjunction with other physicochemical treatment methods, such as air stripping, chemical oxidation, carbon adsorption, vapor extraction, soil washing, etc., may allow for complete remediation of the chemical contamination at the site. The combined treatment approach is also applicable at sites where soil heterogeneities or other site constraints preclude the use of a single remediation technology to remediate the entire contaminated matrix effectively.

Brown and Sullivan (11) describe the successful use of an integrated system for the remediation of the accidental discharge of gasoline in the Pine Barrens area of southern New Jersey. The soil at the site was sandy and highly permeable. An integrated remediation system was designed for the treatment of the ground water and soil. The system incorporated ground water extraction and treatment using an air stripper, soil vapor extraction, and bioremediation. After 18 months of operation, 34,000 lb of the 47,000 lb of gasoline spilled had been removed or destroyed by the remedial system. This integrated system was effective because it made the best use of the properties of the contaminant and the nature of the site.

Hypothetical Case Study and Design

In order to illustrate the principles that were discussed in the preceding sections, the approach for the design of a hypothetical remediation system will be discussed. The purpose of this exercise is once again to emphasize how important it is to have a multidisciplinary approach for the effective implementation of bioengineered treatment systems.

Scenario Example

Suppose that during a transfer operation of compound X at an industrial facility, a mishap occurs that results in the accidental discharge of approximately 4500 gallons of compound X onto the ground surface. The site is in an industrial area and is in close proximity to a river. A residential area is located approximately 5 miles to the west of the site. The local ground water is not used as a potable water source for the community; however, the river is.

The initial actions after the spill were performed by emergency response crews. They worked to stop the leak and remove as much of the free product as possible via limited soil excavation and the use of a vacuum truck. However, it is apparent that some contamination has leaked into the ground and may eventually affect the river if action is not taken. The owners of the facility have hired an environmental contractor to perform a site assessment and to evaluate potential remedial alternatives.

Site Assessment

The next step in the remediation of the spilled contaminants would be to perform a site assessment and evaluate potential treatment and containment schemes. The purpose of the site assessment is to gather as much information as possible about local site conditions that would allow an understanding of potential migration pathways of the contaminants. Potential remedial alternatives also would be evaluated. Available preliminary information about the site and the contaminant (concerning its physical, chemical, and biological) properties suggests that compound X is a potential candidate for remediation via in situ bioremediation. Review of MSDS sheets for the compound, as well as a literature review, reveals that compound X is biodegradable under laboratory conditions.

The successful implementation and operation of a bioengineered treatment system requires a multidisciplinary approach involving input from personnel with training in hydrogeology, geology, soil science, chemistry, microbiology, and engineering (chemical, civil, and environmental) (18). For this situation, the team has gotten together and outlined the following site-assessment activities:

Hydrology, Geology, Soil Science

- Install monitoring wells to determine the direction of ground water flow.
- Perform additional borings to define the current extent of contamination.
- Perform characterization analyses on representative soil samples collected.
- Perform testing to determine the hydraulic characteristics of the aquifer.

Microbiology

- Collect representative ground water and soil samples to perform a biotreatability study to determine the following:

Presence of an adequate indigenous bacterial population.

Site conditions, i.e., are they toxic or inhibitory?

Extent of contaminant biodegradation.

Nutrients or amendment requirements.

Chemistry

- Determine whether approved analytical measurements are available to measure contaminant concentration in this particular site matrix adequately.

- Oversee sample collection and analysis.

- Evaluate the fate and transport characteristics of compound X in the site matrix.

Engineering

- Evaluate site logistics to design system layout, components, and operations.

The information listed in Table 5-2 was obtained from the site-assessment activities. It is the consensus of the team, based on the results of the site assessment, that in situ bioremediation is the best option for remediation of the site. Also part of the site-assessment activities were meetings with the appropriate environmental regulatory agencies that have jurisdiction in the site locale. Permitting requirements, a monitoring schedule, and cleanup criteria were discussed and agreed on.

As a result of the meeting with the appropriate regulatory personnel, an emergency underground injection permit was approved. A work plan that will describe the design and operation of the proposed remediation system will be submitted to the regulatory agencies for approval prior to installation and operation.

Remediation Design

The team has decided on the remediation approach illustrated in Figure 5-2. The goals of this treatment system are to maintain hydraulic control of the treatment area, to prevent further migration of the contaminant, and to provide optimal conditions for the aerobic microbiological degradation of the contaminant. Because the spill took place in an active industrial area, the owners of the facility have requested that every effort be made to install a remedial system that will have minimal impact on plant activities.

Table 5-2. Results from the Site Assessment

Microbiology: The treatability study has indicated the following:
- Aerobic conditions will be necessary for the biodegradation of the contaminant.
- The indigenous bacterial population was found to be present at high levels ($> 1 \times 10^7$ CFU/g). This indicates that the site matrix does not appear to be inhibitory.
- The pH of the soil was 5.8 and will have to be adjusted to approximately 7.0.
- The available nitrogen and phosphorus mineral concentrations are low. Therefore, an external source will be required to make conditions conducive to biological treatment.
- Bioremediation can achieve the desired cleanup goals of 100 mg/kg in the soil and 10 mg/liter in the groundwater after approximately 9 months of treatment at temperatures >55°F.

Chemistry
- Contaminant is present as the mobile phase and adsorbed phase.
- Contaminant is not volatile.
- Conventional analytical procedures will work to detect the contaminant in the site matrix to the required detection limits.

Geology/Hydrogeology/Soil Science
- Spill has occurred in an area composed of a homogeneous medium to coarse sand.
- Depth of contamination is between the surface and 7 ft below ground surface (BGS).
- Hydraulic conductivity of the material is 1×10^{-2} cm/s.
- Intrinsic permeability of the material is 1×10^1 darcys.
- Grain size of the material is medium to coarse sand.
- Depth to groundwater is 15 ft BGS.
- Ground water is not impacted at this time.
- Mathematical model is used to evaluate system design to maintain hydraulic control.

Engineering
- Institute free product recovery first.
- Recommend the installation of surface and vadose zone injection systems.
- Recommend the use of an above-ground mix tank for nutrient (N & P) additions, pH adjustment, and aeration.
- Recommend the installation of a downgradient trench for ground water recovery.

The overall goal of the in situ remediation is to optimize conditions for the indigenous microorganisms so that they can metabolize the contaminant in the most efficient manner possible. This will entail adjusting the pH and adding mineral nutrients and oxygen. Hydraulic control of the ground water in the contaminated area will be main-

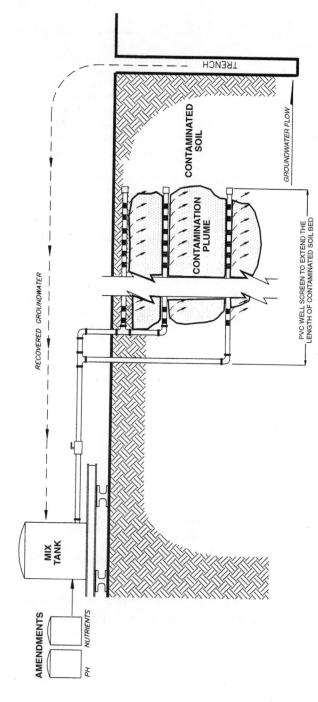

Figure 5-2. Conceptual design of in situ bioremediation system. (*Drawing by Willie E. Jetter.*)

tained to ensure that the contaminant does not migrate off-site. Through this process, ground water will be recovered downgradient of the spill and then reinjected upgradient. The water that is reinjected will be pH adjusted, nutrient supplemented and aerated. This amended water will "flush" through the soil, promoting aerobic microbiological activity.

The proposed bioremediation system is composed of the following components:

1. A physical recovery system initially to remove any free-phase product that may be present.

2. A downgradient trench for ground water recovery.

3. A pump that will supply the recovered ground water to an aboveground tank. This recovered water will be aerated, pH adjusted, and mineral-nutrient amended.

4. The recovered water will then be reinjected into three vadose zone injection trenches.

5. The system will be equipped with appropriate controls that will allow for shutdown in the event of an emergency.

6. Three additional off-site wells will be routinely monitored to verify hydraulic control.

The system is designed to be automated and continuously operated with minimal operator involvement.

Monitoring

Table 5-3 shows the monitoring schedule that will be incorporated for the duration of the treatment system operation. On-site personnel will be trained by the remedial contractor to perform daily oversight activities. These activities include daily inspection of the system for leaks, blockages, pump operation, etc. Nutrient, pH, dissolved oxygen levels, as well as system flow rates, also will be monitored closely. The remedial design contractor will be available as needed for troubleshooting and also will perform biweekly system inspections and amendment (nutrients and pH) additions as required. Adjustments to the operation of the system will be made on an as-needed basis to ensure that optimal conditions are maintained for the microbiological remediation of the contaminants.

Table 5-3. Bioremediation Treatment System Monitoring Requirements

Daily:	Inspect the system components, i.e., piping, pumps, valves, etc. Monitor pH, dissolved oxygen, temperature, and mineral nutrient levels within the treatment system. Monitor flow rates and pumping rates.
Monthly:	Monitor the following parameters within the treatment system and in the off-site monitoring wells: • Contaminant concentration • Aerobic heterotrophic bacterial population density • pH • Dissolved oxygen • Temperature • Available mineral nutrient concentrations
Quarterly:	Perform a series of soil borings and analyze for the following parameters: • Contaminant concentration • Aerobic heterotrophic bacterial population density • pH • Soil moisture • Available mineral nutrient concentrations

Adjustments will be made accordingly to the treatment system based on the results from these analyses.

Closure

Once the designated cleanup goals for soil and ground water for compound X have been obtained and confirmed, the site can be considered for closure. This information will be forwarded to the regulatory agency(s). Once the appropriate approvals have been received, closure activities will commence. Since the treatment system is located in a heavy traffic area at the plant, only the above-ground tanks, pumps, and piping will be dismantled; the underground components will be abandoned in place.

Long-term monitoring of several of the installed monitoring wells will be required to ensure that there are no remaining pockets of contamination left that might potentially migrate and contaminate the ground water.

Potential Problem Areas

Despite numerous investigations documenting the potential for in situ bioremediation for contaminated soils and ground waters, the actual implementation of in situ bioengineered systems is still in its infancy.

One of the major difficulties is the lack of methodologies to verify and monitor the in situ biodegradative activities of the microorganisms (40). However, several researchers have presented innovative approaches to overcoming this problem (4,9,54,61).

Another difficulty with the operation of an in situ biological treatment system is that there is no direct way of assessing how well the system is operating. There is a good possibility that pockets may exist where contamination will still be present. Short of excavating the entire area, there is no way to determine where these pockets are located. However, from a risk standpoint, there may be little potential for this material to migrate and cause harm. Because no site is geologically homogeneous and the distribution of contaminants is not homogeneous either, it is possible that some hot spots may not reach the treatment standard. Through long-term monitoring of the ground water in the area, movement of the pockets of contamination can be observed and appropriate action taken if required.

Conclusion

The bioengineering of contaminated soils and ground water as a remediation option will continue to grow in the future. This technology offers the ability to remediate the contaminant in place, allows the potential completely to break down the contaminant to innocuous end products, can be used in conjunction with other technologies, and is cost-effective. As more and more successful projects are completed, the application of bioengineering will become routine instead of being considered an innovative treatment option. As is often the case in the "real world," it is more important to ask the "right questions" than to know the "right answers" when it comes to addressing remedial options for a site. As was discussed in this chapter, this emphasizes the necessity of a multidisciplinary approach to successful bioengineering implementation.

References

1. American Petroleum Institute. 1986. *Beneficial Stimulation of Bacterial Activity in Ground Waters Containing Petroleum Products.* API Publication No. 44427. American Petroleum Institute, Washington, D.C.

2. American Petroleum Institute. 1986. *Enhancing the Microbial Degradation of Underground Gasoline By Increasing Available Oxygen.* API Publication No. 4428. American Petroleum Institute, Washington, D.C.

3. Aelion, M. C., and P. M. Bradley. 1991. Aerobic biodegradation potential of subsurface microorganisms for a jet fuel–contaminated aquifer. *Applied and Environmental Microbiology* **57(1):**57–63.

4. Aggarwal, P. K., and R. E. Hinchee. 1991. Monitoring in situ biodegradation of hydrocarbons using stable carbon isotopes. *Environmental Science Technology* **26(6):**1178–1180.

5. Atlas, R. M., and R. Bartha. 1987. *Microbial Ecology: Fundamentals and Applications*, 2d ed. The Benjamin/Cummings Publishing Company, Inc., Menlo Park, Calif.

6. Balkwill, D. L., and W. C. Ghiorse. 1985. Characterization of subsurface bacteria associated with two shallow aquifers in Oklahoma. *Applied Environmental Microbiology* **50:**580–588.

7. Bates, R. L., and J. A. Jackson (eds.). 1984. *Dictionary of Geologic Terms*, 3d ed. Prepared under the direction of the American Geologic Institute. Anchor Books, Doubleday, New York, N.Y.

8. Bohn, H. 1992. Consider biofiltration of decontaminating gases. *Chemical Engineering Progress* **88:**4:34–40.

9. Borden, R. C., M. D. Lee, M. Thomas, P. B. Bedient, and C. H. Ward. 1989. In situ measurement and numerical simulation of oxygen limited biotransformation. *Ground water Monitoring Review* **9(1):**83–91.

10. Brown, R. A. 1992. The advantages and disadvantages of bioventing and vapor extraction technology. In *Bioventing and Vapor Extraction: Uses and Applications in Remediation Operations*. A Live Satellite Seminar Jointly Sponsored by the Air and Waste Management Association and the HWAC.

11. Brown, R. A., and K. Sullivan. 1991. Integrating technologies enhances remediation. *Pollution Engineering* **23(5):**62–68.

12. Chaudhry, G. R., and S. Chapalamadugu. Biodegradation of halogenated organic compounds. *Microbiological Reviews* **55(1):**59–79.

13. Christianson, J., B. Irwin, E. Titcomb, and S. Morris. 1989. Protocol development for the biological remediation of a wood-treating site. In Y. C. Wu (ed.), *Proceedings from the 1st International Conference on Physicochemical and Biological Detoxification of Hazardous Wastes*. Technomic Publishing Company, Lancaster, Pa.

14. Cobb, G. D., and E. J. Bouwer. 1991. Effects of electron acceptors on halogenated organic compound biotransformations in a biofilm column. *Environmental Science & Technology* **25(6):**1068–1074.

15. Criddle, C. S., J. T. Dewitt, D. Grbic'-Galic', and P. L. McCarty. 1990. Transformation of carbon tetrachloride by *Pseudomonas* sp. strain KC under denitrification conditions. *Applied and Environmental Microbiology* **56(11):**3240–3246.

16. Cunningham, A. B., W. G. Characklis, F. Abedeen and D. Crawford. 1991.

Influence of biofilm accumulation on porous media hydrodynamics. *Environmental Science & Technology* **25:7**:1305–1311.

17. Dragun, J. 1988. *The Soil Chemistry of Hazardous Materials.* Hazardous Materials Control Research Institute, Silver Springs, Md.

18. Essel, A. A., and M. A. Troy. 1992. Information requirements for the in situ biological treatment of petroleum hydrocarbon contaminated soils, pp. 179–196. In B. A. Reed and W. A. Sack (eds.), *Hazardous and Industrial Wastes: Proceedings of the Twenty-Fourth Mid-Atlantic Industrial Waste Conference.* Technomic Publishing Company, Inc., Lancaster, Pa.

19. Evans, P. J., D. T. Mang, and L. Y. Young. 1991. Degradation of toluene and *m*-xylene and transformation of *o*-xylene by denitrifying enrichment cultures. *Applied and Environmental Microbiology* **57(2)**:450–454.

20. Federle, T. W., D. C. Dobbins, J. R. Thornton-Manning, and D. D. Jones. 1986. Microbial biomass, activity, and community in subsurface soils. *Ground Water* **24(3)**:365–374.

21. Freeze, R. A., and J. A. Cherry. 1979. *Ground Water.* Prentice-Hall, Inc., Englewood Cliffs, N.J.

22. Gerba, C. P., and J. F. McNabb. 1981. Microbial aspects of ground water pollution. *ASM News* **47**:326–329.

23. Ghiorse, W. C., and D. L. Balkwill. 1983. Enumeration and morphological characterization of bacteria indigenous to subsurface environments. *Developments in Industrial Microbiology* **24**:213–224.

24. Godsu, E. M., D. F. Goerlitz, and D. Grbic'-Galic'. 1992. Methanogenic biodegradation of creosote contaminants in natural and simulated ground-water ecosystems. *Ground Water* **30(2)**:232–242.

25. Grady, C. P. L., Jr. 1985. Biodegradation: Its measurement and microbiological basis. *Biotechnology and Bioengineering* **27**:660–674.

26. Grady, C. P. L., Jr. 1989. Biodegradation of toxic organics: Status and potential. *ASCE Journal of Environmental Engineering* **116(5)**:805–828.

27. Heath, R. C. 1989. *Basic Ground-Water Hydrology.* United States Geological Survey Water Supply Paper 2220. Prepared in cooperation with the North Carolina Department of Natural Resources and Community Development. U.S. Geological Survey Federal Center, Box 25425. Denver, CO 80225.

28. Hinchee, R. E., and R. N. Miller. 1992. Bioventing for in situ remediation of petroleum hydrocarbons. In *Bioventing and Vapor Extraction: Uses and Applications in Remediation Operations.* A Live Satellite Seminar Jointly Sponsored by the Air and Waste Management Association and the HWAC.

29. Hirsch, P., and E. Rades-Rohkohl. 1990. Microbial colonization of aquifer sediment exposed in a ground water well in northern Germany. *Applied and Environmental Microbiology* **56(10)**:2963–2966.

30. Hirsch, P., and E. Rades-Rohkohl. 1983. Microbial diversity in a ground water environment in northern Germany. *Dev. Ind. Microbiology* **24:**183–199.

31. Hutchins, S. R., G. W. Sewell, D. A. Kovacs, and G. A. Smith. 1991. Biodegradation of aromatic hydrocarbons by aquifer microorganisms under denitrifying conditions. *Environmental Science Technology* **25(1):**68–76.

32. Johnson, P. C., C. C. Stanley, M. W. Kemblowski, D. L. Byers, and J. D. Colthart. 1990. A practical approach to the design, operation, and monitoring of in situ soil-venting systems. *Ground Water Monitoring Review* **10(1):**159–178.

33. Keely, J. F. 1989. *Performance Evaluations of Pump-and-Treat Remediations.* United States Environmental Protection Agency, Superfund Technology Support Centers for Ground Water, Robert S. Kerr Environmental Research Laboratory, Ada, Okla., EPA/540/4-89/005.

34. Lamar, R. T., and D. M. Dietrich. 1990. In situ depletion of pentachlorophenol from contaminated soil by *Phanerochaete* spp. *Applied and Environmental Microbiology* **56(10):**3093–3100.

35. Lee, M. D., J. M. Thomas, R. C. Borden, P. B. Bedient, J. T. Wilson, and C. H. Ward. 1988. Biorestoration of aquifers contaminated with organic compounds. *CRC Critical Reviews in Environmental Control* **18(1):**29–89.

36. Lee, M. D., and C. H. Ward. 1984. Reclamation of contaminated aquifers: Biological techniques. In *Procedures of the 1984 Hazardous Materials Spills Conference.* Government Institutes, Inc., Rockville, Md.

37. Liu, C., and J. B. Evett. 1992. *Soils and Foundations,* 3d ed. Prentice-Hall, Inc., Englewood, Cliffs, N.J.

38. Long, G. 1992. Innovative technologies for contaminated site remediation: Focus on bioventing and vapor extraction. In *Bioventing and Vapor Extraction: Uses and Applications in Remediation Operations.* A Live Satellite Seminar Jointly Sponsored by the Air and Waste Management Association and the HWAC.

39. Lyman, W. J., W. F. Reehl, and D. H. Rosenblatt. 1990. *Handbook of Chemical Property Estimation Methods. Environmental Behavior of Organic Compounds.* American Chemical Scoiety, Washington, D.C.

40. Madsen, E. L. 1991. Determining in situ biodegradation: Facts and challenges. *Environmental Science & Technology* **25(10):**1663–1673.

41. Mercer, J. W., D. C. Skipp, D. Griffin, and R. R. Ross. 1990. *Basics of Pump-and-Treat Ground-Water Remediation Technology.* United States Environmental Protection Agency, Robert S. Kerr Environmental Research Laboratory, Ada, Okla., EPA-600/8-90/003.

42. Miller, R. N., C. C. Vogel, and R. E. Hinchee. 1991. A field-scale investigation of petroleum hydrocarbon biodegradation in the vadose zone enhanced by soil venting at Tyndall AFB, Florida, pp. 283–302. In R. E. Hinchee and R. F. Olfenbuttel (eds.), *In Situ Bioreclamation Applications and*

Investigations for Hydrocarbon and Contaminated Site Remediation Butterworth-Heinemann, Stoneham, Mass.

43. Mueller, J. G., D. P. Middaugh, S. E. Lantz, and P. J. Chapman. 1991. Biodegradation of creosote and pentachlorophenol in contaminated ground water: Chemical and biological assessment. *Applied and Environmental Microbiology* 57(5):1277–1285.

44. Nyer, E. K., and G. J. Skladany. 1989. Relating the physical and chemical properties of petroleum hydrocarbons to soil and aquifer remediation. *Ground Water Monitoring Review* 9(1):54–60.

45. Parkin, G. F., and C. R. Calabria. 1985. Principles of bioreclamation of contaminated ground waters and leachates. Prepared for the 3rd Annual Symposium on International Industrial and Hazardous Waste, Alexandria, Egypt, June 24–27, 1985.

46. Pedersen, T. A. 1992. Soil vapor extraction fundamentals. In *Bioventing and Vapor Extraction: Uses and Applications in Remediation Operations*. A Live Satellite Seminar Jointly Sponsored by the Air and Waste Management Association and the HWAC.

47. Piwoni, M. D., and J. W. Keeley. 1990. *Basic Concepts of Contaminant Sorption at Hazardous Waste Sites*. United States Environmental Protection Agency Office of Research and Development, Office of Solid Waste and Emergency Response, EPA/540/4-90/053.

48. Raymond, R. L., V. W. Jamison, and J. O. Hudson. 1975. *Final Report on Beneficial Stimulation of Bacterial Activity in Ground Water Containing Petroleum Products*. Committee on Environmental Affairs, American Petroleum Institute, Washington, D.C.

49. Ridgway, H. F., J. Sararik, D. Phipps, P. Carl, and D. Clark. 1990. Identification and catabolic activity of well-derived gasoline degrading bacteria from a contaminated aquifer. *Applied and Environmental Microbiology* 56(11):3565–3575.

50. Roberts, P. V., L. Semprini, G. D. Hopkins, D. Grbic'-Galic', P. L. McCarty, and M. Reinhard. 1989. *In Situ Aquifer Restoration of Chlorinated Aliphatics by Methanotrophic Bacteria*. United States Environmental Protection Agency, Robert S. Kerr Environmental Research Laboratory, Office of Research and Development. Ada, Okla., EPA/600/2-89/033.

51. Sewell, G. W., and S. A. Gibson. 1991. Stimulation of the reductive dechlorination of tetrachloethene in anaerobic aquifer microcosms by the addition of toluene. *Environmental Science & Technology* 25(5):982–984.

52. Short, D. A., J. D. Doyle, R. J. King, R. J. Seidler, G. Sotzky, and R. H. Olsen. 1991. Effects of 2,4-dichlorophenol, a metabolite of a genetically engineered bacterium, and 2,4-dichlorophenoxyacetatone some microorganism-mediated ecological processes in soil. *Applied and Environmental Microbiology* 57(2):412–418.

53. Sims, J. L., R. C. Sims, and J. E. Matthews. 1989. *Bioremediation of Contaminated Surface Soils.* Robert S. Kerr Environmental Research Laboratory, U.S. EPA Office of Research and Development, Ada, Okla., EPA-600/9-89/073.

54. Smith, R. L., B. L. Howes, and S. P. Garabedien. 1991. In situ measurement of methane oxidation in ground water by using natural-gradient tracer tests. *Applied and Environmental Microbiology* **57(7)**:1997–2004.

55. Stegmann, R., S. Lotter, and J. Heeremklage. 1991. Biological treatment of oil-contaminated soils in bioreactors, pp. 188–208. In R. E. Hinchee and R. F. Olfenbuttel (eds.), *On-Site Bioreclamation: Processes for Xenobiotic and Hydrocarbon Treatment.* Butterworth-Heinemann, Stoneham, Mass.

56. Testa, S. M., and D. L. Winegardner. 1991. *Restoration of Petroleum Contaminated Aquifers.* Lewis Publishers, Inc., Chelsea, Mich.

57. Thomas, J. M., and C. H. Ward. 1989. In situ biorestoration of organic contaminants in the subsurface. *Environmental Science & Technology* **23(7)**:760–766.

58. Verschueren, K. 1983. *Handbook of Environmental Data on Organic Chemicals,* 2d ed. Van Nostrand Reinhold Company, New York, N.Y.

59. Vira, A., and S. Fogel. 1991. Bioremediation: The treatment for tough chlorinated hydrocarbons. *Environmental Waste Management Magazine* **9(10)**:34–35.

60. Wilson, J. T., J. F. McNabb, T. H. Wang, M. B. Tomson, and P. Bedient. 1985. Influence of microbial adaptation on the fate of organic pollutants in ground water. *Environmental Toxicology and Chemistry* **4**:721–765.

61. Wilson, J. T., L. E. Leach, J. Michalowski, S. Vandegrift, and R. Callaway. 1989. *In Situ Bioremediation of Spills From Underground Storage Tanks: New Approaches for Site Characterization, Project Design and Evaluation of Performance.* U.S. EPA, Washington, D.C., EPA/600/2-89/042.

62. Wilson, S. B., and R. A. Brown. 1989. In situ bioreclamation: A cost-effective technology to remediate subsurface organic contamination. *Ground Water Monitoring Review* **9(1)**:173–179.

63. United States Environmental Protection Agency. 1990. *Bioremediation in the Field.* EPA/5402-90/004.

64. United States Environmental Protection Agency. 1990. *International Evaluation of In-Situ Biorestoration of Contaminated Soil and Ground Water.* EPA/540/2-90/012.

65. United States Environmental Protection Agency. 1991. *Understanding Bioremediation, A Guidebook for Citizens.* EPA/540/2-91/002.

66. United States Environmental Protection Agency. 1990. *Slurry Biodegradation.* EPA/540/2-90/016.

67. United States Environmental Protection Agency. 1992. *VISITT Vendor*

Information System for Innovative Treatment Technologies. EPA/542/R-92/001.

68. Yare, B. S. 1991. A comparison of soil-phase and slurry-phase bioremediation of PNA-containing soils, pp. 173–187. In R. E. Hinchee and R. F. Olfenbuttel (eds.), *On-Site Bioreclamation. Processes for Xenobiotic and Hydrocarbon Treatment.* Butterworth-Heinemann, Stoneham, Mass.

69. Wang, X., X. Yu, and R. Bartha. 1990. Effect of bioremediation on polycyclic aromatic residues in soil. *Environmental Science & Technology* **24(7):**1086–1089.

6

Bioremediation of Surface and Subsurface Soils

Katherine H. Baker

Environmental Microbiology Associates, Inc., Harrisburg, Pa.

Contamination of soils can occur through the accidental release of materials on the surface or through the direct introduction of contaminants into the subsurface, as in the case of leaking underground storage tanks. From the perspective of remediation, the soil environment can be divided into two zones: shallow surface soils and subsurface (vadose) soils. Shallow surface soils usually include the upper 1 to 3 ft of the soil environment. These soils represent the region of the environment typically included in the agronomic definition of soils. They are easily modified and are generally more amenable to remediation than deeper vadose soils. Operationally, *surface soils* can be defined as those soils which can be excavated or treated by surface amendments not requiring the installation of wells. *Vadose soils* are those soils which lie between the surface soils and the water table or aquifer. Vadose soils are generally unsaturated, although there may be pockets of water saturated soil within the vadose zone, particularly in the area of the root zone and in the capillary fringe at the surface of the water table. In addition, there may be inclusions of low permeability materials such as clay lenses within the vadose zone which can become saturated with water. Unlike surface soils, vadose soils are often not amenable to excavation or surface treatment; rather, modification of such soils usually involves the use of infiltration galleries, injection wells, or other engineered means for introducing materials.

This chapter will briefly review the fate and transport of contaminants within the subsurface environment. Factors which limit microbial activity and hence biodegradation in soils will be covered briefly. Basic techniques for soil treatment both in situ and using engineered systems such as bioreactors will be reviewed. Emphasis will be given to field studies.

Fate and Transport of Contaminants in the Vadose Zone

Soil is not a homogeneous substance. The soil matrix contains inorganic solids, organic (humic) solids, soil gases, and liquids (Paul and Clark, 1989). When a material enters the soil, it can undergo a wide variety of physical, chemical, and biological transformations within the soil matrix on both a micro- and a macroscale. On a microscale, these transformations can result in the partitioning of the contaminant within the soil matrix. The contaminant may be distributed as a gas in the soil atmosphere, dissolved in pore water, or associated with soil particles as well as in the form of free product. Figure 6-1 illustrates the theoretical microscale partitioning of a contaminant within the soil environment. On a macroscale (Fig. 6-2), the material can be transformed via biotic and abiotic processes and transported within the soil environment, or it may leave the soil environment, entering either the atmosphere (volatilization/evaporation) or the ground water (leaching).

Volatilization refers to the partitioning of the contaminant into the gaseous phase. Volatilization can occur within the soil matrix, resulting in the accumulation of organic vapors within the soil atmosphere. The volatilization of low-molecular-weight, highly volatile compounds such as benzene and toluene into the soil atmosphere is the basis of the commonly employed soil-gas survey methods used to delimit contaminant plumes. In addition to the partitioning of contaminants within the soil matrix, volatilization also can result in the transfer of materials out of the soil into the atmosphere. Volatilization

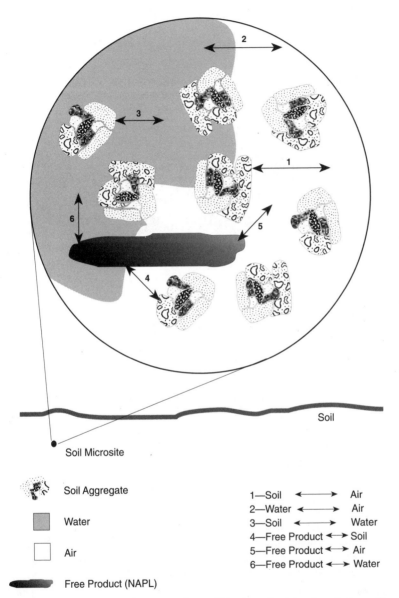

Soil Microsite

Soil Aggregate

Water

Air

Free Product (NAPL)

1—Soil ⟷ Air
2—Water ⟷ Air
3—Soil ⟷ Water
4—Free Product ⟷ Soil
5—Free Product ⟷ Air
6—Free Product ⟷ Water

Figure 6-1. Theoretical microscale partitioning of contaminants in soils. Exchanges among all soil compartments (air, water, soil particles, and free product) are possible. The actual partitioning of the contaminant on a microscale is influenced by the physicochemical characteristics of the contaminants, the soil environment, and biological processes which occur within the soil.

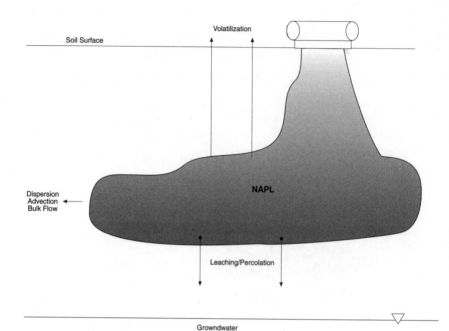

Figure 6-2. Theoretical macroscale partitioning of contaminants in soils. Macroscale processes function to move the contaminants away from the initial spill site. Volatilization and evaporation result in a transfer of the contaminant to the atmosphere. Leaching and percolation move the contaminant downward in the vadose zone and may be responsible for transport to the ground water. Bulk flow, advection, and diffusion result in a horizontal spreading of the contaminant.

of materials from the soil surface can result in air contamination. This can be particularly important in situations in which mixing or tilling of the soil is proposed, since volatilization may result in air emissions which exceed the regulatory limits for compounds such as benzene.

Henry's law constants can provide guidelines in assessing the likelihood of volatilization of materials from the soil environment; however, actual field volatilization rates may differ significantly from these predicted rates. Park et al. (1988) found the rates of volatilization of organic compounds to the atmosphere in soil systems used to treat organic wastes to be 10- to 100-fold less than those predicted on the basis of Henry's law calculations. Fu and Alexander (1992) found that styrene volatilization from a Kendaia loam was significantly less than volatilization from lake water. Thoms and Lion (1992) questioned the validity of using Henry's law constants (usually measured under saturated conditions) to predict the behavior of volatile organic contaminants in unsaturated soils. They note that the partitioning of organic compounds between the soil solids and vapor phase is influenced by such factors as soil moisture content and surface area of the soil solids in addition to physicochemical properties of the contaminant. Spenser and Cliath (1977) noted that volatilization of an organic contaminant from soil systems depended on the physicochemical properties of the contaminant (volatility, solubility, and sorption), the soil environment (soil temperature, water content, bulk density, and organic content), and the concentration of the contaminant present. Thomas (1990) reported that the concentration of the contaminant, soil water content, airflow over the soil surfaces, humidity, temperature, soil organic matter, porosity, density, and clay content can all influence volatilization from soil systems. Finally, Song et al. (1990) have noted that volatilization of organic compounds in soils decreases as biodegradation increases. Thus the tendency of a material to volatilize from the soil is a complex phenomenon, and simple, laboratory-derived measurements such as Henry's law coefficients may not be sufficient to predict real-world situations.

Leaching of contaminants from the vadose zone poses the potential for contamination of aquifers and ground water supplies. Materials tend to leach downward to the ground water under the influence of gravity flow of water. Lateral spread of the contaminants can result from such processes as advection and flow through channels and pores in the soil matrix. The migration of a contaminant away from the source is influenced by such factors as the porosity and permeability of the soil, the solubility of the material in water, the specific gravity of the material, and the bulk velocity of the water. Leaching of contaminants from the vadose zone to the saturated zone is a primary mechanism for the contamination of ground water supplies.

Such phenomena as sorption of contaminants onto soil materials can have profound effects on the final distribution of the material within the environment. Sorption of hydrophobic organic contaminants to soil surfaces, organic matter, and humic materials has been shown to have variable effects on the mobility of the contaminants within and through a soil system (Baker and Herson, 1990; Bollag et al., 1980; Hsu and Bartha, 1974). Sorption is particularly important for highly hydrophobic compounds, including PCBs and chlorinated solvents (Moein et al., 1976; Schwarzenbach and Giger, 1985). Sorption between soil particles and organic contaminants is typically predicted using the organic carbon partition coefficient K_{oc} of the contaminant.

Since K_{oc} values for all contaminants are not available, the octanol-water partition coefficient K_{ow} is often used to estimate K_{oc} (Pussemier et al., 1991). Numerous soil factors, including temperature and relative humidity of the soil pores, have been shown to influence sorption, resulting in deviations from predicted values of K_{ow} (Goss, 1992; Szecsody and Bales, 1991). Kishi et al. (1990) measured the adsorption coefficients for five chemicals (lindane, naphthalene, diphenyl, 1,2,3-trichlorobenzene, and 1,2,3,4-tetrachlorobenzene) in soils with different organic carbon contents. They found significant differences between the absorption (K_{oc}) normalized to soil organic matter between soils differing in their clay content. Meylan et al. (1992) have proposed the use of molecular structure–based methods (molecular connectivity index) for the prediction of soil sorption coefficients for both polar and nonpolar compounds.

Pavlostathis and Jaglal (1991) showed that sorption of trichloroethylene (TCE) to soil could result in a significant retention of the contaminant within the soil matrix. They found that sorption not only was dependent on the organic matter content of the soils but also varied as a function of the contact time between the soil and the contaminant, with increasing contact time resulting in increased resistance to desorption of the TCE. Thus transport of the contaminant from the vadose soils into the ground water would be expected to be lower in soils which had been exposed to the contaminant for a long period of time than in freshly contaminated soils. On the other hand, Magee et al. (1991) have demonstrated that sorption can increase the transport of contaminants away from the source. The adsorption of organic contaminants to dissolved organic matter within the soil matrix was shown to increase transport of phenanthrene by a factor of 1.8 in sand column studies. This process, which they termed "facilitated transport," may explain the observations of contaminant dispersion in excess of that predicted on the basis of sorptive partitioning. Dunnivant et al. (1992) found that mobility of 2,2′,4,4′,6,6′-hexachlorobiphenyl and 2,2′,4,4′,5,5′-hexachlorobiphenyl in laboratory soil columns increased as the dissolved organic matter in the form of dissolved colloidal naturally occurring organic material in the columns increased from 0 to 20.4 mg/liter DOC. Herbert et al. (1993) have demonstrated that the presence of colloidal organic matter within the soil solutions could significantly increase transport of hydrophobic polyaromatic hydrocarbons (PAHs).

Sorption also can influence the bioavailablity and hence the susceptibility to biodegradation of the contaminants. Weissenfels et al. (1992)

have demonstrated that degradation of polyaromatic compounds in soils can be limited by sorption of the PAHs to soil organic materials. Litz et al. (1987) found that ferrous oxides as well as humic materials strongly sorbed linear alkylbenzenesulfonate in soils and that this sorption greatly reduced biodegradation. Miller and Alexander (1991) reported that benzylamine mineralization was slower in the presence of montmorillonite than in the absence of the clay. They ascribed this reduced mineralization to reduced bioavailability of the organic compound due to sorption to the clay particles. In contrast, Fu and Alexander (1992) found that styrene was rapidly sorbed to soil particles in both mineral and organic soils; however, this did not appear to have a significant influence on biodegradation of the styrene. Shimp and Young (1988) examined the influence of sorption on the biodegradation of organic materials in sediments. They found biodegradation of dodecyltrimethylammonium chloride, a quaternary ammonium compound, to be a function of the amount of unadsorbed material present. In the case of phenol, however, biodegradation varied as a function of the total amount (sorbed and free) of the chemical present, indicating that at least a portion of the sorbed material was available for biodegradation.

Due to the complexity and heterogeneity of the soil environment, it is difficult accurately to predict the fate and transport of contaminants within this milieu. Concomitant with the development of risk-assessment methodologies, several fate and transport models for organic materials in vadose soils have been developed recently. Table 6-1 summarizes commonly used vadose zone models currently available. Jury and Ghodrati (1989), Sawhney (1989), and Odencrantz et al. (1992) have recently reviewed the assumptions, mathematical formulations, and limitations of vadose zone transport models.

Biodegradation in Soil Ecosystems

The fundamental unit of structure within a soil is the soil aggregate. Soil aggregates are composed of inorganic soil minerals, organic matter, soil water, soil air, and microorganisms. Typically, less than 50 percent of the volume of a soil aggregate is composed of solid materials (Sims et al., 1986). The remainder of the soil volume is in the form of pore spaces which are occupied either by the soil atmosphere or soil water. Figure 6-3 illustrates a typical soil aggregate. The presence

Table 6-1. Valdose Zone Transport Models

Acronym	Model name	Reference
CMLS	Chemical Movement in Layered Soils	Nofzinger and Hornsby, 1987
GLEAMS	Ground Water Loading Effects of Agricultural Management Systems	Leonard et al., 1986
LEACHMP	LEAching estimation and Chemistry Model—Pesticides	Wagnet and Hutson 1986
MOFAT	Multiphase Organic Flow and Transport	ESTI, 1990
MOUSE	Method of Underground Solute Evaluation	Pacenka and Steenhuis, 1984
PRZM	Pesticide Root Zone Model	Carsel et al., 1984
RITZ	Regulatory and Investigative Treatment Zone	Nofzinger et al., 1987
SESOIL	Seasonal SOIL compartment model	Bonazountas and Wagner, 1984; Hetrick et al., 1989
VIP	Vadose Zone Interactive Processes	Grenney et al., 1987; Stevens et al., 1989

of soil aggregates imposes a heterogeneity on the soil microbial environment not found in aquatic environments or in typical laboratory liquid culture systems. Soil microorganisms are not distributed uniformly throughout the soil matrix; rather, they are found in patchy distributions associated with suitable microhabitats. Stotzky and Burns (1982) have highlighted the importance of microhabitats as follows:

> In soils...a genetically diverse population is competing for a small quantity of unevenly distributed substrate and is subject to fluctuations in pH, redox potential, ionic concentration and level of hydration.

Hamer and Heitzer (1991) presented a discussion of the importance of considerations of subsurface heterogeneity in evaluating the potential for microbial degradation of contaminants in this environment. They stressed the fact that the soil environment may be very different from the microbes' perspective from that which is reflected in terms of mea-

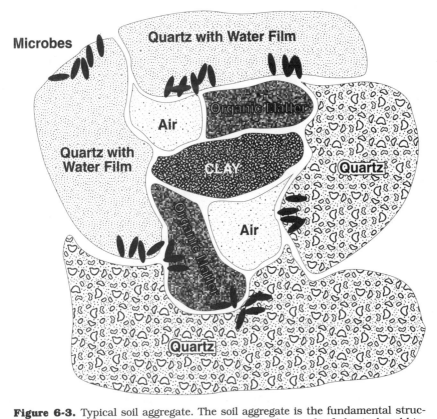

Figure 6-3. Typical soil aggregate. The soil aggregate is the fundamental structural unit of a soil. Formation of soil aggregates is the result of physical and biological activities within the soil. The occurrence of soil aggregates results in the presence of numerous microenvironments within the soil.

surements of physicochemical parameters of the soil made on a gross level. For example, they noted that while denitrification is normally considered to be a process limited to anoxic soils, the presence of anoxic microhabitats within soils can allow this microbial process to proceed in apparently oxic (aerobic) soils.

Factors Limiting Biodegradation

There are numerous factors which can limit the numbers and activity of microorganisms in the soil and other natural environments. These factors most amenable to modification, and therefore, the factors most often of significance in bioremediation are soil moisture, soil pH, temperature, levels of inorganic nutrients, levels of electron acceptors, and types and levels of carbon substrates present.

Soil Moisture. Biodegradation of contaminants is dependent on an adequate supply of water within the soil environment. Water is necessary not only to meet the physiological requirements of microorganisms but also for the transport of nutrients and metabolic by-products to and from the microorganisms and to determine the oxygen status of the soil microenvironment.

Water is present within the soil environment in three forms: gravitational, capillary, and hygroscopic. *Gravitational water* refers to water which can move freely through the soil by gravitational forces (Paul and Clark, 1989). This water primarily occupies the macropores within the soil matrix. Gravitational water may displace air from the soil matrix, leading to the development of anoxic conditions. At contaminated sites, it is the movement of this water which is chiefly responsible for the leaching of materials into lower soil layers and ultimately into aquifers. *Capillary water* is the term applied to the water within the micropores of the soil matrix. The amount of capillary water within a soil will depend on the soil texture (percent of clay, silt, and sand) and structure as well as the input of water into the soil. This water is available to microorganisms. *Hygroscopic water,* or *bound water,* represents water that is bound by hydrogen bonding or dipole interactions to surfaces within the soil matrix. This water is extremely difficult to remove from the soil and generally is not available biologically.

Capillary water determines the field capacity of the soil. Operationally, the *field capacity,* also called the *water-holding capacity,* of a soil is defined as the amount of water remaining within the soil after gravitational water has drained away. A more precise definition of field capacity is the amount of water retained by a soil when a standard pressure ($-\frac{1}{3}$ bars) is applied to the soil. Field capacity is a rough estimate of the amount of water contained within the soil micropores under nonsaturated conditions and is highly dependent on the texture and porosity of the soil. Sandy soils which are well drained with large pores typically have a low field capacity, while medium- and fine-textured soils such as silts/loams and clays have higher field capacities. Medium-textured soils (loams) usually have a high field capacity coupled with a high water availability, reflecting the favorable pore size distribution within the soil matrix.

Although accurate determination of field capacity depends on the measurement of the water content of the soil under a standard pressure regime (typically − ⅓ bar), field capacity of a soil can be estimated easily. A sample of the soil is placed in a porous container such as a ceramic cup. The soil is saturated with water and then allowed to drain under the influence of gravity. The soil is then weighted, dried at 110°C to a constant weight, and reweighed. The difference between these two weights is the amount of water present at field capacity.

In general, biodegradation of contaminants in soil systems is optimal at a soil moisture content between 30 and 80 percent of field capacity (Dibble and Bartha, 1979b; Riser-Roberts, 1992). Stegmann et al. (1991) found optimal biodegradation of diesel fuel in composting systems at 60 percent field capacity. It is essential to emphasize that these moisture levels are in terms of percent field capacity and not in terms of percent water in the soil. Failure to appreciate this fact would lead to the maintenance of excessive water levels in the soil and a concomitant reduction in oxygen availability. Field capacities typically range from 5 to 40 percent of the total soil weight, depending on the texture of the soil. Therefore, a moisture level of 50 percent field capacity might range from 2.5 to 20 percent total moisture by weight.

A number of techniques are available which can be used to modify the moisture content of a soil. In the case of soils which are too dry, water can be introduced using sprinklers or irrigation systems. Care must be used in the design of the sprinkler or irrigation system to prevent leaching of the contaminants from surface soil layers into deeper layers. Sims et al. (1986) and Riser-Roberts (1992) discuss the types of sprinkler and irrigation systems which have been applied in bioremediation projects. They conclude that, in general, sprinkler systems have been most effective in allowing for maintenance of optimal soil moisture without leaching or runoff of the contaminants, although reduced water infiltration due to surface sealing in high-clay-content soils can limit the use of sprinklers in some soils. Portable sprinkler irrigation systems are available. Although these are initially more expensive than traditional permanent or semipermanent installations, the flexibility afforded and the ease of installation (disturbing the soil is not required) of these systems are generally preferable (Riser-Roberts, 1992).

Excess moisture can be a problem in poorly drained soils or in constructed soil treatment cells which are underlain with a geomembrane or similar liner. In this case, excess water from rainfall or moisture modifications can pool at selected spots within the treatment system leading to the development of localized anoxic zones and reduced biodegradation. Excess water can be controlled by improving surface and subsurface drainage or by the installation of wells and well points. Typically, a system of perforated piping is included in the design of a soil treatment cell for the collection of leachate. The collected leachate is treated and then may be either discharged or recirculated through the treatment cell. Finally, the use of mulches or modifications of soil texture can have significant effects on soil moisture (Riser-Roberts, 1992).

Soil pH. Biodegradation of contaminants is typically fastest at neutral or near-neutral pH values. Fu and Alexander (1992) measured volatilization and biodegradation of styrene in lake water and soil samples. Mineralization was rapid in sewage, mineral soil of pH 7.23, and organic soil. Mineralization was slower in ground water and lake water and lowest in aquifer sand, waterlogged soil, and mineral soil of pH 4.87. Dibble and Bartha (1979b) found a soil pH of 7.8 to be optimal for biodegradation of petroleum hydrocarbons. Verstrate et al. (1976) found that adjustment of soil pH from acidic conditions (pH 4.5) to near-neutral conditions (pH 7.4) resulted in a doubling of the rate of biodegradation of gasoline in soil.

Soil pH values for many soils in the world are acidic, although soil pH values ranging from 2.5 (mine spoils) to 11.0 (alkaline deserts) have been reported (Paul and Clark, 1989, Bossert and Bartha, 1984). Therefore, in many bioremediation projects, there is a need to neutralize the soil acidity in order to raise the soil pH closer to a neutral value. The most frequently used technique to increase soil pH is through the application of agricultural limestone, frequently referred to as *liming*. It is not possible to determine the lime requirement of a soil directly from soil pH, since such other factors as soil texture, type and amount of clay, organic matter content, cation exchange capacity (CEC), and exchangeable aluminum influence lime requirements (Follett et al., 1981). In general, the two most important factors in determining the lime requirement of a soil are the soil pH and the buffering capacity of the soil (McLean, 1982). Soil acidity is composed of active and reserve acidity. The *active acidity*, which is only a small fraction of the total acidity and which is in equilibrium with the reserve acidity, is measured by the soil pH value. The *reserve acidity* which arises from adsorbed exchangeable and nonexchangeable Al^{3+} ions, gives rise to the soil's buffering capacity. The finer the soil texture, the greater is the reserve acidity; therefore, the greater is the buffering capacity. Before a soil pH value will increase, the reserve acidity must be neutralized. There are several commercially available test kits that allow for the rapid determination of lime requirements.

Although the term *lime* correctly refers to calcium oxide only, it is commonly applied to any calcium- or calcium- and magnesium-containing compound capable of neutralizing soil acidity such as calcium hydroxide, calcium carbonate, calcium magnesium carbonate, and calcium silicate slag (Riser-Roberts, 1992; Brady, 1972). This ambiguity of terms can lead to problems in the liming of soils, since the compounds used to lime the soil are not equivalent on a weight basis (see Table 6-2). As a rule of thumb, 1 ton of finely ground limestone has the same neutralization capacity as 0.7 ton of commercial calcium hydroxides or slightly over 0.5 ton of calcium oxide (Brady, 1972). Hence care must be taken in a liming program to ensure that the liming requirement calculated is corrected for the type of liming material being used. The effect of a liming agent on soil pH depends on the surface area of the agent in contact with the soil. The more finely the liming agent is ground, the more rapid will be its reaction with the soil and hence its influence on soil pH. In addition, lime, particularly limestone, is not readily soluble and therefore does not migrate within the soil matrix. Thus a liming program may necessitate incorporation into the soil through such processes as plowing or disking to ensure uniform mixing. In some instances, liquid lime can be added using a spray irrigation system (Riser-Roberts, 1992).

The addition of lime to a soil also has indirect effects on soil chemistry and biological activity. Adjustment of the soil pH by liming not only will influence the concentration of hydrogen and hydroxyl ions present but also can have significant effects on the chemical form, solubility, and bioavailability of essential soil nutrients such as nitrogen and phosphorous. In addition, aluminum and manganese are less soluble at higher pH values, and hence potential toxicity from these elements is decreased.

Soil Temperature.   Soil temperature influences biodegradation in both direct and indirect ways. Directly, microbial activity can be related to temperature, with rates of metabolic reactions generally increasing with increasing temperatures. The rule of thumb that has been used classically is that microbial activity doubles with each 10°C increase in temperature up to the optimal temperature for that species. Field studies, particularly in arctic tundra ecosystems, indicate that an even more dramatic response to temperature can occur, with microbial activity quadrupling with each 10°C increase in temperature (Swift et al., 1979). Soil temperature also can have profound effects on both the soil matrix and the physicochemical state of the contaminants. Paul and Clark (1989)

Table 6-2. pH Modification in Soils

Compound	Common name	Neutralizing value, % (Calcium carbonate = 100%)	Notes
Calcium carbonate	Limestone, calcite	100	Usually contains impurities which reduce neutralizing value; must be crushed or pulverized before use
Calcium carbonate + Magnesium carbonate percent (50% each)	Limestone, dolomite	109	Neutralizing value depends on composition (Ca and Mg) and amount of impurities. May range from 87–109; must be pulverized before use
Calcium hydroxide	Hydrate, slaked lime builder's lime	120–135	Caustic; quick acting
Magnesium hydroxide		172	
Calcium oxide	Burned lime; quicklime; oxide; unslaked lime	150–175	Caustic; quick acting; must be rapidly mixed to prevent caking
Magnesium oxide		250	
Marl		50	Usually moist
Blast furnace slag	Basic slag	50–70	Usually contains magnesium; more useful as a source of P than a liming agent

SOURCE: Foth, 1984; Sims et al., 1986; Follett et al., 1981.

note that soil temperature influences soil volume, oxidation-reduction potentials, and water structure within the soil matrix. In addition, soil temperatures can fluctuate widely. Such variations can occur on vertical, diurnal, and seasonal scales. For example, diurnal variations in surface soil temperatures during the midsummer season can exceed 10°C.

While it is well established that temperature is an important factor in the growth and activity of microorganisms, there have been only a few studies that have examined the importance of soil temperature in the degradation of hazardous materials. This probably reflects the fact that the majority of bioremediation projects conducted to date have been in mesophilic regions (temperature between 20 and 70°C). Observations on system performance over winter periods of cold weather indicate that degradation is considerably slowed at lower temperatures. Atlas (1975) found that biodegradation of crude oils by marine isolates was highly dependent on the incubation temperature. Dibble and Bartha (1979b) examined the effect of incubation temperature on microbial degradation of oil sludge in laboratory landfarming systems. They found almost complete inhibition of biodegradation when the samples were incubated at 5°C. Increasing temperatures correlated with increased biodegradation up to 20°C. Little or no increase in biodegradation compared with the 20°C systems was found at higher incubation temperatures (28 and 37°C). Song et al. (1990) compared microbial degradation of gasoline, jet fuel, heating oil, diesel oil, and bunker C oil at 17, 27, and 37°C in laboratory microcosm studies. Hydrocarbon disappearance was maximal at 27°C for all the petroleum hydrocarbons tested. Coover and Sims (1987) examined the effect of temperature on microbial degradation of polyaromatic hydrocarbons (PAHs) in an unacclimated agricultural soil. They found that increasing the soil temperature resulted in a significant increase in the rate and extent of loss of low-molecular-weight PAHs but little or no effect on losses of five- and six-ring molecules. Thornton-Manning et al. (1987) investigated phenol mineralization in samples from different depths in two contrasting soils. They found that temperature affected mineralization differently as a function of soil type. A simple relationship between temperature and biodegradation existed only with one topsoil. On the basis of these observations, these authors concluded that one cannot predict a priori how temperature will affect biodegradation in a given soil environment.

There are several techniques available to modify soil temperature. The most commonly used techniques include mulches, plastic covers, and plant cover. Both mulches and plant cover have been shown to moderate fluctuations in soil temperature (Sims et al. 1986). Vegetation provides an insulating layer to the soil surface and so functions to dampen diurnal fluctuations in soil temperature. Because many bioremediation sites are subjected to frequent tilling, and because the levels of contaminants in the soils may inhibit plant growth, plant cover is not often used in bioremediation.

Soils undergoing bioremediation are frequently covered by some type of a plastic liner to reduce emissions of volatile organics into the atmosphere. These covers can have a significant effect on the temperature of the soil. In general, white or light-colored covers tend to reflect sunlight and thereby prevent excessive heating of the pile during the day. Transparent plastic covers function in a manner similar to a greenhouse and can increase the overall temperature of the soils. Black or dark plastics increase heat absorption during the day and reduce heat loss at night, again resulting in an overall increase in the temperature of the soil pile. Thus the influence of the cover material on temperature of the soils should be considered in choosing a cover for the reduction and control of volatile emissions. Using a black plastic cover for bioremediation of soils in a warm climate (e.g., Florida during the summer) can result in overheating of the soil pile and reduction in the amount of biodegradation (Sims et al. 1986; Riser-Roberts, 1992). In addition, the use of coverings can decrease aeration of the soil and lead to anoxic conditions.

When bioremediation is being conducted in cold climates, it may be possible to increase the temperature of the soil by irrigating the pile with heated water or by injecting steam into the vadose zone. In some instances this approach is worth the added expense of heating the water, since a slight increase in the temperature of the soil can result in dramatic increases in the rate of biodegradation.

Inorganic Nutrients. The primary inorganic nutrients necessary for biodegradation are nitrogen and phosphorous. Nitrogen is necessary for the synthesis of cellular proteins and cell wall components. Phosphorous is necessary for nucleic acids, cell membranes, and ATP. Both nitrogen and phosphorous are frequently limiting in soil environments.

Nitrogen can be present in the soil in numerous different forms. The inorganic forms of soil nitrogen include ammonia, nitrate, nitrite, nitrous oxide, and nitric oxide. The organic forms of nitrogen occur as consolidated amino acids or proteins, free amino acids, amino sugars, and other complex, generally unidentified compounds. In general, the preferred form of nitrogen for growth of soil microorganisms is the reduced form, ammonia (ammonium ions) (Paul and Clark, 1989; Alexander, 1961). Laboratory studies have indicated that ammonium is preferentially used over the other forms of nitrogen which may be present. When other forms of nitrogen are present (e.g., organic nitrogen, nitrate), they are generally converted into ammonia before assimilation by microorganisms. It should be noted, however, that Hadas et al. (1992) have recently presented data indicating that both ammonium and low-molecular-weight organic nitrogen (alanine) can be assimilated concurrently by different populations of soil microorganisms. They noted that the presence of populations of soil microorganisms capable of direct uptake of organic nitrogen can have significant effects on the long-term efficiency of fertilization programs. In addition to biological uptake, nitrogen can be rapidly lost from soils by leaching of ammonia and nitrate and by denitrification in wet soils (FitzPatrick, 1986). In addition, ammonium ions can substitute for potassium within clay lattices, rendering the fixed ammonium unavailable to microorganisms.

Phosphorous is frequently limiting because of its low solubility and hence low bioavailability. Organic soil phosphorous is found mainly in humus, while the inorganic fraction occurs in numerous combinations with Fe, Al, Ca, F, and other elements. The inorganic compounds are usually only very slightly soluble in water. Phosphates also can react with clays to form insoluble clay-phosphate complexes. The common forms of phosphate found in the soil solution at the pH generally found in soils are $H_2PO_4^-$ and HPO_4^{2-}. Minor concentrations of organic phosphorous can be found in the soil solution. In acid soils phosphate is precipitated as iron and aluminum phosphates, while in neutral and alkaline soils it is precipitated as calcium phosphate. In most soils, phosphorous is most soluble in the pH range of 5.5 to 7.0. Mansell et al. (1992) reported that for sandy soils, phosphorous retardation during leaching tends to decrease with increasing concentration in solution. Rubeiz et al. (1992) noted that phosphorous has a limited mobility and a tendency to become fixed via precipitation in calcareous soils. This limits both the availability of phosphorous and the usefulness of phosphorous fertilizers. Aggarwal et al. (1991) have reviewed the geochemical reactions of phosphate salts in deep vadose zone and aquifer areas. They determined that simple soil column transport studies were inaccurate in predicting possible problems with phosphate precipitation and concomitant aquifer plugging. Rather than short-term studies on nutrient transport and mobility, they recommended that studies be run for a minimum of 30 days. They further recommended that phosphorous additions not be made in systems with calcareous soils, since nearly all the added nutrient will be lost by adsorption and precipitation. In sandy soils, they recommended the use of a combination of orthophosphate and trimetaphosphate (TMP). Researchers have reported that humus extracts from soils have increased the solubility of phosphorous. This results from the formation of phosphohumic complexes which are more easily absorbed by plant roots, anion replacement of the phosphate by the humate ions, and the coating of iron oxide particles by humus to form a protective cover and thus reduce the phosphate-fixing capacity of the soil (Cosgrove, 1977).

Numerous authors have demonstrated enhanced biodegradation of xenobiotic compounds when soils are amended with supplemental nitrogen and phosphorous. Biodegradation rates in soils have been stimulated by addition of urea-phosphate, N-P-K fertilizers, and inorganic salts of nitrogen and phosphorous (Jamison et al., 1975; Jobson et al., 1974; Verstrate et al., 1976). Degradation of phenol in soils was reported to be increased by the addition of supplemental nitrogen and phosphorous (Thornton-Manning et al., 1987). Harmsen (1991) reported that the addition of inorganic nitrogen and phosphorous stimulated biodegradation of crude oil and pentachlorophenol in laboratory studies on the bioremediation of soils. Scherrer and Mille (1989) found that the addition of an oleophilic fertilizer to peaty mangrove soils which had been artificially contaminated with petroleum hydrocarbons resulted in an increased rate of biodegradation of the normal alkanes. Song et al. (1990) reported that the addition of nitrogen and phosphorous fertilizers (NH_4NO_3 and K_2HPO_4) enhanced degradation of a variety of petroleum hydrocarbons in microcosm experiments with sand, loam, and clay-loam soils. Block et al. (1990) reported that the biodegradation of petroleum hydrocarbons in soil treatment systems was enhanced by the addition of inorganic fertilizers. They reported that amendment of the soils to a final C:N:P ratio of 25:1:0.5 resulted in optimal biodegradation; however, the form of nutrients used was not specified. Leavitt et al. (1991) found that even in the presence of significant initial soil concentrations of inorganic nutrients (ammonia and phosphate), degradation of coal-coking wastes could be stimulated by additional supplementation with readily available nutrients.

Not all investigations into the effect of nutrient supplementation on the biodegradation of organic contaminants, however, have shown a relationship between nutrient supplementation and increased biodegradation. Thornton-Manning et al. (1987) found that nitrogen and phosphorus addition did not increase phenol mineralization in surface soils but did in subsurface soils. Thomas et al. (1989) found that there was no increased mineralization of naphthalene and 2-methylnaphthalene associated with the addition of supplemental nutrients to samples from a creosote-contaminated site. Hardaway et al. (1991) found variable responses of soil microorganisms from samples within a single study area to nutrient supplementation. They reported that between 13 and 61 percent of the ammonia added to soil slurries was adsorbed to soil particles and hence unavailable to microorganisms. Morgan and Watkinson (1990) found that in some circumstances the addition of fertilizers to soils resulted in a significant inhibition of mineralization of aromatic compounds. This inhibition was not due to changes in the soil pH as a result of fertilizer addition. While the reasons for the observed inhibition are not known, the authors noted that oligotrophic organisms present in the soil can be inhibited by nutrient supplementation.

The relationship between microbial degradation of organic contaminants, therefore, does not appear to be straightforward with respect to inorganic nutrient supplementation. In all likelihood, the conflicting results noted above reflect the complexity and heterogeneity of the soil environment. In addition, they underscore the need for preliminary assessment of microbial response before a treatment regimen is implemented. While biologically available inorganic nutrients may limit microbial growth and activity in some soils, this is by no means universal. Indiscriminate reliance on the addition of inorganic nutrients to stimulate biodegradation may in some cases simply add to the expense of remediation, while in others it may actually impede the remediation process.

Electron Acceptors. In most instances, bioremediation of contaminated soils depends on the activities of aerobic (oxygen-requiring) organisms. Although current research into anaerobic processes may lead to the development of anaerobic or anoxic soil remediation processes, these systems are not practical at this point. Therefore, the presence of an adequate supply of oxygen within the soils is essential for bioremediation. In general, a ratio of approximately 2 to 3 lb of oxygen per pound of petroleum hydrocarbon degraded is necessary to ensure adequate oxygenation for biodegradation (Wilson et al., 1986; Riser-Roberts, 1992; Sims et al., 1986).

Within the soil matrix, oxygen is present in the soil atmosphere within the soil pores. Soil may become oxygen-depleted or anoxic because of limited diffusion of atmospheric oxygen into the soil pores, filling of pore spaces with soil water, and rapid oxygen consumption by soil microorganisms. Soil water content can be used to provide a rapid estimation of soil aeration, since soil water reduces that amount of pore space available for soil gases. Briefly, soil aeration can be estimated by determining the total pore volume of the soil and subtracting the amount of pore space filled by water (percent water). Total pore volume can be estimated by dividing the bulk density of the soil (which can be estimated on the basis of soil texture) by the average density of mineral particulates (approximately 2.6 g/cm^3). In general, a minimum of 10 percent air-filled pore space is necessary to maintain adequate aeration for aerobic microbial activity (Paul and Clark, 1989; Foth, 1984). In addition to the influence of water, the contaminant itself will occupy some of the available pore space in the soil matrix, thus reducing oxygen availability.

The type of aeration technique used for a particular site depends on the depth at which the contaminant is found. Aeration of surface soils can be increased by such physical manipulations of the soil as tilling and the incorporation of mulches and other conditioners. A variety of equipment, including plows, chisels, and harrows, can be used to till the soil surface, thereby increasing aeration. Mulches and other conditioners improve the texture of the soil and thereby increase the aeration. Both of these techniques are applicable only to limited depths in the soil, typically between 6 and 12 in. In addition, physical manipulation of soils such as tilling can result in increased erosion and in volatilization of low-molecular-weight materials such as the aromatic components of petroleum products and thus may result in air pollution. Finally, the use of heavy tilling machinery can result in soil compaction in the areas where the machinery is operated. This will decrease aeration in those areas.

For contamination below the surface layers, oxygen can be introduced using several possible mechanisms. Construction equipment, such as a backhoe, can be used to "till" soils deeper that 2 ft. For deeper soils, oxygen can be added either as a gas through the use of blowers, the application of a vacuum, or direct sparging or by the introduction of water which has been enriched with air, oxygen, or an appropriate oxygen-generating molecule. Water as a carrier and transport compound has been used most often in the remediation of saturated soils and aquifers.

The most commonly used oxygen-generating molecule for bi_remediation purposes is hydrogen peroxide. Since hydrogen peroxide is significantly (7 orders of magnitude) more soluble in water than oxygen, hydrogen peroxide can be used to achieve high concentrations of oxygen in the subsurface. In addition, hydrogen peroxide is relatively inexpensive and does not persist in the environment.

Hydrogen peroxide provides oxygen to the subsurface environment as a result of its decomposition to yield water and oxygen. The decomposition is catalyzed by the microbial enzyme catalase. The reaction for this decomposition is

$$2H_2O_2 \rightarrow 2H_2O + O_2$$

Thus for every mole of hydrogen peroxide introduced into the subsurface environment there is a theoretical yield of 0.5 mol oxygen.

Pardieck et al. (1992), in a review of the use of hydrogen peroxide in subsurface remediation, noted that hydrogen peroxide can undergo a variety of non-oxygen-generating disproportionation reactions in soils and aquifers. These reactions include free-radical oxidations of organic matter (contaminants or soil organic matter), reactions with dissolved metals such as iron and manganese, and reactions with soil halides. Thus the actual yield of oxygen is usually less than the theoretical 0.5 mol/mol hydrogen peroxide.

Laboratory and pilot studies have confirmed the usefulness of hydrogen peroxide in supplying oxygen to both vadose and saturated soils. Flathman et al. (1991) found that the addition of hydrogen peroxide to soil columns contaminated with petroleum hydrocarbons (JP5, diesel fuel, lubricating oil) enhanced biological activity in these soils. Huling et al. (1991) reported on a pilot-scale field study on the use of hydrogen peroxide to bioremediate a seasonal water table contaminated with aviation gasoline. They found that the injection of hydrogen peroxide (50 to 200 ppm) into the ground water increased oxygen concentrations in both the saturated and unsaturated soils at the site.

The use of hydrogen peroxide to supply oxygen to the subsurface is not without its drawbacks. Hydrogen peroxide is widely used as a disinfectant; it is toxic to microorganisms. Inhibition of bacterial cultures has been found to occur at concentrations of hydrogen peroxide in the range of 10 to 1000 ppm. Toxicity is reduced when the number of microbial cells is large and when the microorganisms are growing on a solid substrate. For example, Britton (1985) observed that a mixed biofilm culture of microorganisms was able to tolerate an initial hydrogen peroxide concentration of 500 ppm. Tolerance levels could be increased to 2000 ppm by slowly increasing the hydrogen peroxide dosage over time. The addition of excessive levels of hydrogen peroxide to the subsurface may therefore inhibit microorganisms and decrease the efficiency of a bioremedation project.

The second major drawback to the use of hydrogen peroxide involves its stability in soil systems. Lawes (1990, 1991) studied the stability of hydrogen peroxide in soils. He noted a large variation in the stability of hydrogen peroxide which could not be related to soil properties such as organic matter, iron and manganese concentration, and pH. Because of these findings, he proposed a static batch assay system which can be used to predict the stability of hydrogen peroxide under field conditions.

As already noted, hydrogen peroxide can undergo a variety of oxygen-producing and non-oxygen-yielding reactions under field conditions. The rate and location of these reactions can have a profound effect on the usefulness of hydrogen peroxide in a soil system. Rapid biotic decomposition of hydrogen peroxide in the subsurface may lead to degassing and the formation of oxygen bubbles within the subsurface matrix. This can result in decreased permeability and impede the transport of nutrients to the contaminated site. In addition, reactions between hydrogen peroxide and dissolved iron can lead to the precipitation of iron hydroxides, again resulting in decreased permeability. Such techniques as the addition of phosphate to complex with iron, reduction in the concentration of nutrients added with the peroxide, and the use of pulsed additions of peroxide and/or nutrients have all been helpful in reducing the problems associated with excessive decomposition of hydrogen peroxide and precipitate formation (Spain et al., 1989; Pardieck et al., 1992).

Types and Amounts of Carbon Present/Supplemental Carbon. Carbon is not thought of typically as a limiting nutrient at sites undergoing bioremediation. The organic contaminants present normally supply the carbon requirements of the microbial populations involved in biodegradation. There are, however, several instances in which the organic contaminants may not be sufficient or available for biodegradation by the microorganisms present. First, in the case of compounds which sorb to soil minerals and/or soil organic matter, the compound may not be available biologically (bioavailable) and hence may not be degraded. Second, if the concentration of the compound is below a certain threshold level necessary to supply the carbon and energy requirements of the microorganisms, degradation may not occur. Alternatively, if the concentration of the compound exceeds a certain level, toxicity problems may occur. Finally, the compound may not be amenable to direct degradation; rather, degradation via a cometabolic process may be necessary.

The influence of sorption on biodegradation of organic compounds has already been discussed in this chapter. In general, sorption of organics to soil particles reduces the availability of these materials to microorganisms and hence reduces their biodegradability. Early observations on microbial degradation of such water-insoluble compounds as petroleum hydrocarbons indicated that one of the initial stages in the degradation process was the microbial production of emulsifying agents which functioned to increase the solubility and hence bioavailablity of the compounds (Cooper, 1986). Furthermore, increased biodegradation of sparingly soluble organic compounds in the presence of supplemental surfactants has been demonstrated in several laboratory studies (Thomas et al., 1986; Fogel et al., 1985). Rittmann and Johnson (1989) reported that the addition of surfactants to soil microcosms increased biodegradation of a lubricating oil. Aronstein et al. (1991) found that the addition of surfactants at levels below their critical micelle concentrations enhanced degradation of phenanthrene and biphenyl in soils with high amounts of organic matter. Aronstein and Alexander (1992) reported that nonionic surfactants in low concentrations also could increase both the rate and extent of biodegradation of phenanthrene and biphenyl sorbed to aquifer sands. These observations, coupled with the development of soil washing as a treatment method for contaminated soils, has lead to the suggestion that emulsifying agents such as nonionic detergents or microbially produced emulsifying agents may be useful in bioremediation (Abdul et al., 1992; Berg et al., 1990; Van Dyke et al., 1991). In theory, the addition of such materials to contaminated soils should enhance the aqueous solubility of sorbed organic compounds and thereby increase their bioavailability and susceptibility to biodegradation (Riser-Roberts, 1992). Ying et al. (1990) examined the effect of surfactants on biodegradation. They found that significant degradation (94 percent reduction) of petroleum hydrocarbons in surfactant- and nutrient-supplemented soils occurred within 16 weeks. Ellis et al. (1990) included the addition of surfactants, as well as nutrients and mixed bacterial cultures, in the treatment of oil contaminated soils. Significant reductions in the concentration of oil present were seen within 15- to 34-week treatment times.

Several recent studies, on the other hand, have raised concerns about the use of surfactants in enhancing biodegradation (Falatko and Novak, 1992). Laha and Luthy (1991) examined the influence of non-ionic surfactants on the microbial degradation of phenanthrene in soil-water systems. When surfactants were added at concentrations below the critical micelle concentration, there was no significant influence of the surfactants on degradation. At higher concentrations, however, the presence of surfactants actually inhibited phenanthrene biodegradation. Litchfield et al. (1992) examined the influence of several emulsifiers/surfactants on biodegradation of polyaromatic hydrocarbons in creosote-contaminated soils in microcosm systems. They found that the addition of surfactants resulted in no significant increase in PAH degradation. In the case of one PAH, benzo[a]pyrene, the addition of surfactant to the soils resulted in a statistically significant increase in the apparent concentration of the PAH present. They hypothesized that this was the result of increased solubilization of the PAH by the surfactant and cautioned that such increased solubilization might lead to increased transport of the material with subsequent contamination of ground water. Clearly, the use of surfactants in bioremediation projects must be approached with caution, and bench-scale evaluation of the effect of the surfactant on microbial degradation and the fate and transport of the contaminant need to be considered before the use of surfactants is implemented.

Several researchers have proposed that the addition of nonspecific ancillary carbon compounds to soils may enhance biodegradation of contaminants. Manure, sewage sludge, compost, and straw are often added to soils undergoing bioremediation (Linkenheil, 1988; Stegmann et al., 1991). While reports on sites treated in such a manner are generally favorable, it is not possible to determine whether the enhanced degradation seen with the addition of, for example, manure is the result of the additional carbon available, the presence of additional, possibly adapted, microbial biomass, or alterations in the soil properties, such as soil texture and moisture, resulting from the amendments.

Supplemental carbon additions may be considered under two vastly different sets of circumstances. First, if the concentration of the contaminant is below a threshold level at which microbial growth can occur, there are indications that supplemental carbon may allow for mixed substrate utilization and hence may increase biodegradation. Alternatively, in the presence of high levels of toxic organics, the use of supplemental carbon may increase overall biomass and reduce the per-microbe toxin concentration.

Microbial degradation of contaminants can be limited if the amount of the contaminant falls below a certain threshold value. At levels below this, the carbon requirements of the organism cannot be met with the amounts of organic matter present; the system is carbon-limited. Generally, simultaneous utilization of multiple substrates is favored by microbial growth under carbon-limited conditions (Schmidt et al., 1987). LaPlat-Polasko et al. (1984) found that microbial degradation of methylene chloride was more rapid when acetate was present than when the methylene chloride alone was used as a carbon source. A coryneform bacteria was shown to be able to degrade simultaneously more than 20 carbon sources when grown under carbon-limiting conditions (Lindstrom and Brown, 1989). Schmidt et al. (1987) showed that the presence of glucose as a secondary substrate could increase the rate of degradation of p-nitrophenol (PNP) by a culture of *Pseudomonas*. Despite this demonstrated enhancement of PNP degradation, there was a threshold concentration of PNP below which no degradation was detected, even in the presence of a secondary substrate. They attributed this threshold to the lack of induction of the appropriate degradative enzymes at extremely low concentrations of PNP. In addition, they showed that not all additional substrates resulted in a stimulation of PNP degradation. In fact, the addition of phenol to the culture was found to inhibit degradation of PNP. Thus, care must be taken in selecting a supplemental carbon source for bioremediation because the addition of some supplemental carbon sources can actually result in reduced biodegradation and hence increase the time necessary for remediation to be completed.

Brown et al. (1986) demonstrated that the presence of a supplemental carbon source could increase degradation even at high carbon levels. They investigated the microbial degradation of pentachlorophenol (PCP). Degradation rates were enhanced by the addition of cellobiose to the culture system. In addition, in the presence of cellobiose, the microorganisms were capable of tolerating and degrading higher concentrations of PCP. They attributed this increased tolerance to the increased bacteria numbers present when cellobiose was present, and hence to the lower per cell concentration of PCP in the cultures (Brown et al., 1986; Lindstrom and Brown, 1989).

The final circumstance in which organic matter would be added to the soil for bioremediation is to promote cometabolism through analogue enrichment. Cometabolism is the degradation of compounds which cannot serve as a sole carbon and energy source in the presence of a metabolizable carbon source (Horvath, 1972). The cometabolized substance, typically the contaminant of interest, is fortuitously transformed by microbial enzymes which normally transform a different but structurally similar compound (Rittmann et al., 1992). While cometabolism in its strictest sense is limited to transformations in which the organism derives no carbon or energy from the compound being cometabolized, Spain et al. (1991) have recently noted that there are instances in which the organisms can derive carbon and energy from the cometabolized substrate but are unable to use the substrate as a sole source of carbon and energy.

Analogue enrichment depends on the addition of a metabolizable compound (an inducer) which induces the formation of enzymes responsible for degradation of a secondary compound. Analogue enrichment has been shown to be a significant factor in the biodegradation of a wide variety of organic contaminants, including substituted aromatic compounds, chlorinated hydrocarbons, and polyaromatic hydrocarbons, in soil systems (Keck et al., 1989).

Sims and Overcash (1981) demonstrated that the addition of low amounts of phenanthrene to soils increased the degradation of benzo[a]pyrene. You and Bartha (1982) reported on the degradation of 3,4-dichloroaniline (DCA) by a strain of *Pseudomonas putida*. The microorganism was able to mineralize in soil, but the rate of mineralization was slow. Mineralization was greatly enhanced by the addition of aniline, a structural analogue, to the soil. Similarly, degradation of polychlorinated biphenyls was shown to be enhanced by the addition of biphenyl to soil systems (Focht and Brunner, 1985). Nelson et al. (1987) isolated a *Pseudomonas* strain (G4) that degrades TCE to carbon dioxide and inorganic chloride when the organism is induced for aromatic biodegradation by the presence of phenol or toluene. Degradation of TCE was ascribed to the induction of enzymes associated with the meta-cleavage pathway of aromatic hydrocarbons.

Wilson and Wilson (1985) first reported that aerobic soil microbial communities grown on natural gas also could degrade TCE. They exposed vadose zone soils in a laboratory column to 0.6 percent natural gas in air and observed a reduction in the concentration of TCE present. Fogel et al. (1986), Strand and Shippert (1986), Henson et al. (1988) and Barrio-Lage et al. (1988) extended these observations, documenting the degradation of TCE, c-DCE, t-DCE, 1,1-DCE, chloroform, and vinyl chloride (VC) by soil microorganisms in the presence of natural gas or methane. Tetrachloroethylene (PCE) was not degraded. Based on laboratory studies, Speitel and Closmann (1991) estimated that in soils contaminated with 100 µg of chlorinated solvents per gram soil, microbial degradation could be stimulated by enrichment of the atmosphere with methane. They calculated that soils contaminated with this level of chlorinated solvents could be bioremediated within several months to several years. Janssen et al. (1991) also reported on chloroform degradation in vadose soils. They found increased removal of chloroform in the presence of methane. Broholm et al. (1991) found that the addition of methane to soil could stimulate degradation of TCA and TCE but not of PCE. After 289 days over 90 percent of the TCA and TCE had been removed from the soil. Despite these results, these authors concluded that the implementation of biodegradation of chlorinated hydrocarbons in field situations may be limited by the toxicity of high concentrations of these compounds to microorganisms and by the slow rate of degradation possible.

Degradation of chlorinated hydrocarbons in the presence of methane is the result of cometabolism by a specialized group of microorganisms, the *methylotrophs*. Methylotrophs are microorganisms that grow on one-carbon compounds (methane or methanol) or compounds containing two or more carbon atoms which are not linked to each other via carbon-carbon bonds (dimethylether, dimethylamine) (Stanier et al., 1986). Methylotrophs that degrade methane only are referred to as *methanotrophs* (Hanson, 1980). Methylotrophs are aerobic organisms, although their growth rate is increased under conditions of low oxygen concentration. Metabolism of methane and other growth substrates is mediated by a group of enzymes called *methane monooxygenases*. Methane monooxygenases have a broad substrate range and have been shown to catalyze the oxidation of alkanes to alcohols, alkenes to epoxides, and alkylbenzenes to phenols and alcohols. While methylotrophs are unable to grow on chlorinated hydrocarbons as a carbon and energy source, extracts of methane monooxygenases have been demonstrated to catalyze transformations of chlorinated compounds analogous to those which occur with nonchlorinated hydrocarbons.

Cometabolic reactions are specific, since cometabolism depends on a structural similarity between the normal enzyme substrate and the fortuitously degraded substrate. Therefore, the type of carbon compound added to the soil to stimulate cometabolic processes must be specific for the compounds to be degraded. General organic enrichment with compounds such as manure will not enhance cometabolism. Furthermore, since cometabolism involves enzyme-mediated reactions, the addition of excessive amounts of the primary enzyme substrate can actually reduce cometabolism through competition for the available enzyme. For example, Janssen et al. (1991) noted that high levels of methane inhibited chloroform degradation, presumably through competitive enzyme inhibition.

Types of Soil Treatment Systems

Many names and terms are applied to bioremediation systems for the treatment of soils. For the purposes of this chapter, soil treatment systems will be divided into three major categories: land treatment systems, including landfarms and prepared bed systems; in situ subsurface systems, including biostimulation and bioventing; and bioreactor systems. While the basic microbiology of all three types of treatment systems is the same, they differ in terms of the engineering applied to optimize microbial activity.

Land Treatment. *Land treatment,* or *solid-phase bioremediation* as it is sometimes referred to, is an outgrowth of the landfarming methods that have been used by the petroleum industry for a number of years. In the traditional landfarming technique, petroleum or other organic wastes are added to soils in carefully controlled portions. The soils are tilled to increase aeration, irrigated to optimize moisture levels, and, if necessary, supplemented with inorganic nutrients such as nitrogen and phosphorous. Degradation of the organic material is the result of the activities of the indigenous microbial populations present in the soil. In a review of landfarming technology, the American Petroleum Institute (1980) proposed site characteristics which should be optimized for landfarming. Fundamentally, these site characteristics should be designed to prevent transport or migration of the contaminants to another environmental compartment (atmosphere, ground water, surface water, adjacent land). Such factors as depth to the ground water, soil drainage and runoff collection, distance from residential areas, and climate should all be examined when landfarming is considered. In addition, factors that influence the activity of soil microbial populations and hence the rate and extent of degradation also must be evaluated. Bossert and Bartha (1984) reviewed the importance of loading rate, soil moisture, soil aeration, and inorganic fertilizers in the design and operation of landfarms. While their findings can provide general guidelines concerning soil environmental conditions conducive to biodegradation, site-specific characterization and treatability studies provide the best estimate of what modifications to the site are needed.

There are numerous references and case reports in the literature describing the successful application of landfarming as a treatment process. Estimates indicate that there have been over 100 landfarming units in operation in the United States for the treatment of oily wastes (Smith et al., 1989). Only a few cases will be highlighted here.

Dibble and Bartha (1979a) reported on the successful use of land-farming to remediate a sandy loam soil contaminated by an accidental release of approximately 2 million liters of kerosene contaminating 1.5 ha of land. Surface contamination was removed by physico-chemical means, leaving approximately 200 m^3 of contaminated soil to be remediated. The initial concentration of kerosene in the soil averaged 0.8 percent by weight. Contaminated soils were mixed with pulverized limestone and spread to a depth of 46 cm. The soil was frequently tilled to allow for adequate aeration, and inorganic nutrients in the form of nitrogen, phosphorus, and potassium were added twice during the treatment process. During the 21-month treatment, the concentration of kerosene in the upper 30 cm of the soil decreased to trace levels. The loss of kerosene in the soil was attributed to volatilization and biodegradation. Some kerosene was detected in deeper layers. This was attributed to inadequate aeration of the lower soil layers.

Karmazyn et al. (1989) reviewed the operations and monitoring of an approximately 20-acre landfarming site used for the disposal and treatment primarily of petroleum sludges. They noted that the extremely low permeability of the site soils precluded the possibility of contamination of ground water due to leaching of the wastes. Wastes were treated at this site from 1978 to 1985. The wastes were brought in from off site and deposited directly from trucks onto the treatment area. Loading rates of waste oil at the site were between 5 and 10 percent weight of oil to soil. The wastes were mixed with the soil by plowing or discing. This not only served to mix the wastes with the soil but also ensured adequate aeration for aerobic microorganisms degrading the material. Nutrients were added by yearly application of fertilizer. In addition, rye grass was planted each fall and disced into the soil in the spring to supply an additional source of nutrients. Surface runoff of rainwater was collected and either reapplied to the soils or sent off site for treatment.

Since volatilization of wastes can be significant and in addition may serve as a potential source of air contamination, an air survey was performed at the site during its operation. Ambient air samples collected from the region bordering the site indicated that volatilization was insignificant, with the levels of volatile organics at well below 1 ppb. Analysis of soil samples upon closure of the site indicated that degradation of the oily sludges was achieved for approximately 80 percent of the total applied.

Because of the relatively uncontrolled nature of landfarming, there is a potential that this process will result in a transfer of contaminants from the soil to another environmental medium rather than in degradation of the wastes. Transfers to air can occur through volatilization of the materials either during application or during the treatment process itself. Contaminated solids can be transported off site via wind erosion. The runoff of precipitation can lead to the contamination of surface water supplies. Finally, it is possible that the contaminants may be leached through the treatment zone and enter the ground water. Because of the uncontrolled nature of the landfarming process, federal regulations (RCRA) have been enacted recently which strictly regulate the use of traditional landfarming techniques involving the application of contaminated wastes to clean soil.

Land treatment units (LTUs), also called *solid-phase bioremediation* or *prepared bed treatment,* differ from traditional landfarms in that special treatment cells are constructed to contain the contaminated material during treatment. By using a constructed treatment cell, it is possible to incorporate controls into the cell to prevent migration of the contaminants into the surrounding soil and ground water or contamination of the air via volatilization of the contaminants. Figure 6-4 illustrates a typical LTU such as that described by Baker et al. (1993) for the treatment of a soil contaminated with diesel fuel.

The cell consisted of a high-density polyethylene (HDPE) liner placed onto the pregraded ground surface. A soil aeration and drainage system was placed at the base of the treatment cell, and soil was placed on top of the venting system and bottom liner. Fertilizer nutrients were added directly to the soil during treatment cell construction. The entire cell was then covered and protected by a second HDPE liner. The soil aeration and drainage system was manifolded and connected to a vapor trap and positive-displacement regenerative vacuum blower. Treatment of the off-gas by granular activated carbon (GAC) filtration prior to discharge was provided.

The soil treatment piles were aerated daily to ensure the presence of adequate oxygen for supporting maximum biological activity. This was accomplished by using the treatment cell aeration system to replace oxygen-deficient, high carbon dioxide–containing air within the cell with fresh ambient air. Water was added to the system as needed by removing the top liner and spray irrigating the pile. Remediation of diesel-contaminated soils was completed to closure within 9 to 12 months (Baker et al., 1993).

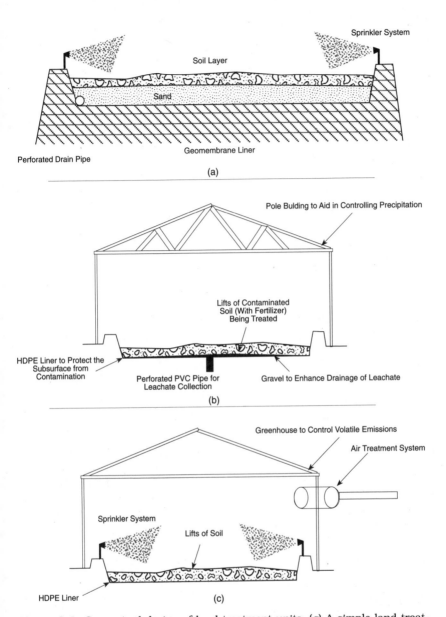

Figure 6-4. Conceptual design of land treatment units. (*a*) A simple land treatment unit consisting of lifts of contaminated soil placed on an impermeable liner. The presence of the liner prevents the leaching of contaminants into the lower soil layers or the ground water. Simple sprinkler systems are used to supply inorganic nutrients and water. Aeration is maintained by periodic tilling of the soil. (*b*) The addition of a covering to a land treatment system can function to reduce the introduction of water into the system from precipitation and the subsequent leaching of the contaminant. (*c*) A fully enclosed land treatment unit allows for control and treatment of fugitive volatile emissions. Enclosure can be achieved by the use of a small greenhouse, as seen here. An alternative method of controlling volatile emissions is to cover a soil treatment pile with a geomembrane liner. Air is introduced into the soil pile through a series of perforated pipes which are manifolded to an air treatment unit such as activated carbon. (*Courtesy of R. E. Wright Associates, Inc., Middletown, Pa.*)

Visscher and Brinkman (1989), in a review of bioremediation projects undertaken in the Netherlands, concluded that gas oil, fuel oil, and cutting oil can be reduced to a residual concentration of 400 to 800 mg oil per kilogram of dry matter in soils within a 4- to 6-month treatment period in land treatment units. They also reported that significant reductions in PAHs (300 to 100 mg/kg) can occur within a 2-month treatment period, although they caution that these treatment rates apply to low-molecular-weight PAHs and that high-molecular-weight PAHs (e.g. benzo[b]fluoranthene and benzo[a]pyrene) are not significantly degraded.

McFarland et al. (1991) reported on the remediation of soils contaminated with PAHs using a prepared bed treatment technique. The soils were mixed with uncontaminated soils (fertile sandy loam). Manure was added to the pile, and the pH was adjusted. Chemical assays indicated that the levels of PAHs in the treatment system were reduced to background levels within approximately 1 year of treatment. In addition, toxicity analysis using Microtox indicated that no toxic leachate was produced.

Compeau et al. (1991) have reported on the full-scale remediation of soil (280,000 yd^3) from a former tank farm facility contaminated with bunker fuel oil. The primary petroleum hydrocarbons found in the soil were in the C_{-10} to C_{-35} chain length, with initial total petroleum hydrocarbon (TPH) concentrations of up to 30,000 ppm. Initial laboratory treatability studies were conducted to determine the optimal nutrient amendments and the maximum starting concentration of TPH which could be treated within a 2- to 3-month time frame. Land treatment units (LTUs) were constructed for the site. Contaminated soils were excavated and added to the LTUs in lifts. Changes in TPH, microbial numbers, and nutrients were monitored during the remediation process. TPH concentrations were reported to decrease from initial values of between 5000 and 6000 ppm to less than 1000 ppm within 12 to 14 weeks of treatment. The rate of decrease was biphasic, with initial rates of decrease slowing down as biodegradation proceeded. Biodegradation rates were slowest between 6 and 9 weeks of the treatment and then increased again toward the end of the treatment process. Decreases in TPH were correlated with changes in the microbial populations present in the soils and with the pattern of inorganic nutrient utilization. The changes in the rate of biodegradation seen as remediation proceeded were felt to reflect changes in the composition of the petroleum hydrocarbon and in the microbial community present.

The use of several commercially available inocula of hydrocarbon-degrading bacteria (L-103 and L-104, Solmar Corporation, and Super-Cee, Microbe Masters, Inc.) for bioaugmentation was evaluated in pilot-scale systems at the site described by Compeau et al. (1991). In no case did the addition of exogenous microbes result in a significant increase in TPH removal despite the higher overall numbers of heterotrophic microorganisms present in these systems. No microbial isolates with the colony morphology of the inoculated strains could be recovered from the soils 4 weeks after the addition of the microorganisms. Based on this observation, the authors postulated that the lack of an effect from the exogenous microorganisms may have reflected an inability of the exogenous strains to colonize the soil.

These same authors (Campeau et al., 1991) also reported on the successful treatment of pentachlorophenol contaminated soils. Because the degradation of chlorinated compounds results in the liberation of chloride ions and the generation of HCl, treatability studies included an assessment of the buffering capacity of the soil. Initially, degradation of the PCP was slow. This was attributed to the unseasonably high precipitation during the first several weeks of remediation. High levels of precipitation could have increased soil moisture content while simultaneously decreasing the amount of oxygen available for aerobic microorganisms in the soils. Degradation of PCP was found to depend on the initial concentration of PCP, with degradation rates increasing with increasing concentrations of PCP up to the maximum level of 700 ppm PCP. Complete remediation to a cleanup level of 150 ppm occurred within 14 weeks of operation.

Ross et al. (1988) reported on the field-scale remediation of 15,000 yd^3 of soils contaminated with petroleum hydrocarbons. Treatability studies indicated that indigenous microorganisms could reduce the concentration of petroleum hydrocarbons in the soil from initial levels of approximately 3000 ppm to below 100 ppm provided sufficient inorganic nutrients were supplied. The soils were treated in a 4-acre treatment facility in which the contaminated soil was spread in 15-in lifts. Inorganic nutrients were added, and the soils were tilled daily to maintain adequate aeration. Within 4 weeks of operation of the field-scale treatment system, the average concentration of petroleum hydrocarbons in the soils had been reduced by greater than 90 percent.

Chemical and petroleum wastes containing aromatic compounds, polyaromatic hydrocarbons, vinyl chloride, and styrene were treated in a land treatment unit in Louisiana (Christiansen et al., 1989). The treatment unit consisted of an excavated treatment area with an 80-ml

HDPE liner system to prevent leaching of the waste constituents into the ground water. Contaminated soils were added to the treatment unit in a series of lifts containing between 4000 and 7000 yd^3 of contaminated soil. Each lift was treated by the addition of a microbial inoculum, inorganic nutrients, and tilling to ensure adequate aeration. After treatment of the first lift to closure levels, subsequent lifts were introduced into the treatment facility and mixed with some of the previously remediated soil to provide microbial inocula. Using this technique, a total of 23,000 yd^3 of contaminated soil was treated to closure levels (final phenol < 10 mg/kg; meets TCLP criteria) within a total of 185 days.

Brubaker and Exner (1988) reported on the use of a bioreactor as a part of the remediation system for surface soils contaminated with formaldehyde. Cleanup of this site was instituted in response to a sudden release of formaldehyde from a train car. Because of the nature of the spill, much of the contaminating formaldehyde could be removed chemically via oxidation with hydrogen peroxide at a pH of between 9 and 10. The initial response resulted in a dramatic decrease in formaldehyde levels in the soils, but chemical methods did not remove all the formaldehyde. Therefore, microbiological polishing also was incorporated into the treatment regime.

Because the remediation was implemented within a short period of time (7 days) after the spill occurred, it was unlikely that an acclimated population of bacteria had been selected in situ. Formaldehyde is toxic to unacclimated bacteria in relatively low doses (150 ppm); however, acclimated strains are able to tolerate and degrade formaldehyde at concentrations over 10 times higher (1750 ppm). Therefore, bioaugmentation of the indigenous microorganisms with an acclimated strain of commercially available microorganisms was considered appropriate for this site. A 1600-gal treatment tank was constructed on site. The tank was equipped with a compressed air generator and diffuser to allow for aeration. A bacterial inoculum and supplemental nutrients were added to the tank and then sprayed over the contaminated soils to enhance microbial degradation. The leachate was collected and returned to the tank, resulting in a closed-loop treatment process. Soil formaldehyde levels decreased from their initial 1750 ppm to less than 1 ppm during a 17-day treatment period.

St. John and Sikes (1991) described a solid-phase land treatment unit used at a Superfund site. Clayey soils at the site were contaminated with styrene, still bottom tars, and chlorinated hydrocarbons. Because of the mixture of contaminants found at the site, a dual treatment system was designed. Air stripping was used to remove the volatile compounds, while a prepared bed treatment facility was used to treat the semivolatile contaminants. The entire treatment facility was enclosed within a plastic greenhouse to control emissions of volatile organic compounds. Leachate was collected and treated in an on-site bioreactor outside the unit. The facility was divided into several treatment lanes so that the relative efficacy of tilling and aeration, biostimulation with added inorganic nutrients, and bioaugmentation could be compared. The authors found no differences between any of the treatment regimes examined. Significant reductions in the concentration of semivolatile contaminants was found to occur, and the authors estimated that removal of the indicator compound phenanthrene could be accomplished within 131 days.

Land treatment units are also being used to treat soils contaminated with wood-treatment wastes at the Libby Montana Superfund site (Piotrowski, 1990; Piotrowski and Carraway,1991). The primary contaminants of concern at this site are polycyclic aromatic hydrocarbons found in creosote and pentachlorophenol. Initial concentrations of contaminants in the soils ranged to over 5000 mg/kg. Pilot-scale studies on approximately 60 yd³ of contaminated soils indicated that constructed land treatment units could be used to treat the soils effectively. During the 3-month pilot study, reductions of 97 percent PCP, 91 percent total PAH, and 74 percent carcinogenic PAHs were documented to occur (Piotrowski, 1990).

Based on these results, a full-scale (1-acre) land treatment unit was constructed. The unit was double lined with compacted clay, geotextiles, and geomembranes and bermed to prevent migration of the contaminants. In addition, the LTU contained a leachate collection system and ground water monitoring wells. Preliminary assessment of system functioning was based on the treatment of two 12-in-thick lifts, each containing approximately 800 yd³ of contaminated soils. The soils were periodically tilled to ensure adequate oxygenation for aerobic degradation of the contaminants. In addition, water and nutrient amendments were added as necessary. Remediation of the soils to U.S. EPA–established goals was achieved within 1 to 3 months. Analysis of the leachate indicated that migration of the contaminants was not significant during the treatment process, and air quality investigations documented that volatilization of the contaminants was minimal. Based on these initial results, remediation of all the contaminated soils at the site is estimated to be completed within 8 years.

Bioremediation of soils (5000 yd³) from an oil gasification plant contaminated with PAHs was examined in a pilot-scale land treatment unit (Warith et al., 1992). Initial concentrations of PAHs in the soil ranged between 300 and 400 ppm. A treatment unit was constructed. Clay soil lifts were compacted to form a clay liner with a hydraulic conductivity of 10^{-7} cm/s. A drainage layer and sump pumps placed above the clay liner allowed for the collection of any leachate. A clay-faced berm was used to surround the entire treatment area. Exogenous acclimated bacteria and inorganic nutrients were added to the soils to be treated on a regular basis. In addition, the soils were tilled on a daily basis to ensure adequate aeration. Over a 3-month treatment period BTX compounds decreased by 73 percent and total PAHs decreased by 86 percent, indicating that bioremediation of the soils could be obtained within a reasonable time period.

Piotrowski et al. (1992) reported on remediation of a diesel fuel–contaminated site in a U.S. Air Force station in arctic Alaska. For contaminated soil fill which could be excavated, lined land treatment units (LTUs) were constructed. Surfactant and micronutrients were added, and the soils were mixed with a bulldozer to oxygenate them at regular intervals. TPH concentrations in the LTU were reduced by 75 percent over two summers. In addition to the LTUs, passive bioremediation was implemented where the presence of permafrost precluded construction of treatment units. Although results with the passive bioremediation system were not as dramatic as those for the LTUs, the authors concluded that the initial results suggested that biodegradation could be an important mechanism for the restoration of diesel fuel–contaminated tundra soils.

In Situ Subsurface Systems: Biostimulation and Bioventing. Treatment of vadose soils several feet below the surface usually involves some type of in situ treatment system. Electron acceptors, inorganic nutrients, and other supplements (i.e., bacterial cultures) are introduced, if necessary, into the subsurface environment to stimulate microbial degradation of the contaminants. Proper controls must be included in the system to ensure that the contaminants do not migrate.

Water was the first medium used to transport materials throughout the vadose zone. In this approach to bioremediation, hydraulic control of the site is first established. This typically involves installation of a series of injection wells or trenches and recovery wells. Alteration of the water table level also may be undertaken. Nutrients and oxygen dissolved in water are injected into the subsurface environment. As the water percolates through the subsurface, nutrients and oxygen are delivered to the microorganisms, stimulating biodegradation of the contaminant. The material which leaches through the vadose zone into the saturated zone is captured and pumped to the surface, where it is treated, if necessary, and recirculated into the system. Figure 6-5 illustrates a typical vadose zone in situ biostimulation system. This type of treatment system is fundamentally the same as treatment systems designed for in situ bioremediation of contaminated aquifers. Often contaminant in the vadose zone and concomitant contamination of ground water are treated as a single unit.

Brubaker and Exner (1988) discussed combined treatment of contaminated soils and ground water using in situ bioremediation. Migration of the contaminant was contained by the installation of recovery wells. Water containing supplemental inorganic nutrients and hydrogen peroxide as a source of oxygen was injected into the subsurface via a series of injection wells. Progress of the remediation was monitored using soil and ground water samples collected on a periodic basis. These samples indicated that the concentration of petroleum hydrocarbons decreased dramatically in both the soils and the ground water.

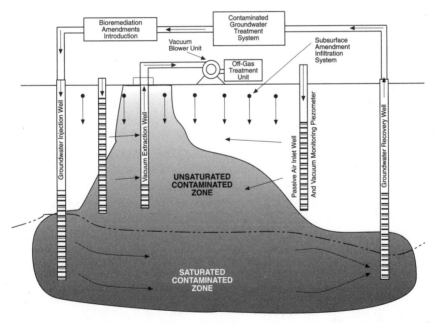

Figure 6-5. Conceptual system design for in situ bioremediation of vadose soils and ground water. (*Courtesy of R. E. Wright Associates, Inc., Middletown, Pa.*)

Brenoel and Brown (1985) used biostimulation for the integrated remediation of petroleum hydrocarbon–contaminated vadose soils and ground water. A series of injection and recovery wells was installed at the site to establish hydraulic control and prevent migration of the contaminants off site. Nutrients and oxygen were introduced into the subsurface. Once adsorbed petroleum hydrocarbons had been degraded, a carbon adsorption system was used as a final polishing treatment for the ground water. Additional examples of in situ bioremediation of vadose soils are discussed in Chap. 7.

An alternative to the use of water as an oxygen carrier is to introduce oxygen into the subsurface in the form of a gas, typically as atmospheric air. This type of bioremediation, often called *bioventing*, is specifically applicable to vadose soils and will not work in saturated zone soils. There are several advantages to introducing oxygen into the subsurface in the form of a gas. The concentration of oxygen present in the atmosphere vastly exceeds the concentration of oxygen which can be achieved in aqueous systems.

Bioventing is conceptually based on the soil treatment technology called *soil vapor extraction* (SVE), also called *soil venting*. In this technique, air is forced through the subsurface, either by pumping air into the subsurface or, more often, through the application of a vacuum to the soil. This movement of air through the soil transfers volatile organic compounds to the surface, where they are collected and treated. Figure 6-6 illustrates a typical bioventing system. The application of SVE as a remediation technique depends on the volatility of the contaminant, the air-filled permeability of the soils, and the appropriate stratigraphy to allow for effective gas flow, as well as cost considerations (Coffin and Glasgow, 1992; Johnson et al., 1990). Hutzler et al. (1988) have reviewed technical aspects of the design and implementation of soil venting systems. Coffin and Glasgow (1992) have analyzed the factors which influence well placement for effective gas flow in soil venting systems. They propose a gas-flow model which can be used to aid in the placement of wells and in the establishing of operating pressures which should be maintained in a soil venting system. Johnson et al. (1990) presented a flowchart that can be used to assess the likely success of a soil venting system and to establish the associated design criteria.

Initial research on bioventing began in the early 1980s with observations by the Texas Research Institute and Dutch researchers that the increased airflows associated with soil venting correlated with increased microbial activity and biodegradation (Hoeppel et al., 1991).

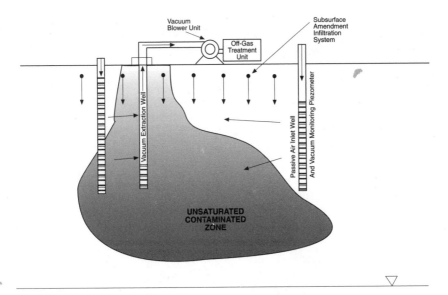

Figure 6-6. Conceptual design of a bioventing system. (*Courtesy of R. E. Wright Associates, Inc., Middletown, Pa.*)

Wilson and Ward (1988) were among the first to propose that the same systems used for the extraction of volatile materials from the subsurface also could be used to transport oxygen into the vadose environment to enhance bioremediation. They noted that the use of air as an oxygen carrier could overcome the solubility limitations associated with water as a carrier medium. Since their observation, several investigators have examined the feasibility of bioventing for the remediation of contaminated vadose soils (Bennedsen et al., 1987; Hinchee et al., 1989; Ely and Heffner, 1988; Hoeppel et al., 1991; Urlings et al., 1991).

High airflow rates in SVE allow for the quickest removal of volatile compounds and hence shorten the time necessary for remediation. In the case of bioventing, on the other hand, the purpose of moving air through the subsurface is to deliver oxygen to the indigenous microorganisms, not to remove volatile contaminants. Thus optimal airflow rates for bioremediation can be considerably lower than those necessary for SVE. DuPont et al. (1991) found that under reduced gas flow rates, 80 percent of the decrease in moderately heavy fuels (diesel, JP-4, fuel oil) could be attributed to biodegradation. Miller et al. (1991) have reported that under flow rates equivalent to 0.5 air void volumes per day, 85 percent of the removal of jet fuel from a sandy soil could be attributed to biodegradation.

Ostendorf and Kambell (1990) have proposed a mathematical model for the evaluation of bioventing. The model includes a one-dimensional balance of storage, sorption, and advection. Biodegradation is modeled using Michaelis-Menton kinetics with degradation constants derived from microcosm studies. These authors subsequently applied this model to analysis of the transport and biodegradation of aviation gas at the Traverse City Superfund site (Ostendorf and Kambell, 1991). Good correlations were found between the model predictions and field observations, although the authors cautioned that further development of the model is necessary. Sleep and Sykes (1991) have described a multiphase mathematical model that can be used to simulate dissolution, volatilization, and biodegradation of contaminants in the subsurface under a variety of airflow regimes. Use of such mathematical models can aid in the design of a bioventing system.

Both soil moisture and temperature are influenced by bioventing systems. Because of the movement of air through the subsurface, bioventing systems increase the evaporation of water from vadose soils. Thus drying of the soils and concomitant inhibition of the microorganisms can develop in these systems. Hinchee and Arthur (1991) examined the effect of moisture-management regimens on bioventing in bench-scale studies on the remediation of JP-4. They found that biological removal of the contaminant was optimized when soil moisture levels were maintained at approximately 75 percent field capacity. DuPont et al. (1991) reported on field pilot studies on the same site studied by Hinchee and Arthur. Water was added to the field plots using a surface spray irrigation system. Maintenance of soil moisture between 30 and 50 percent field capacity resulted in a statistically significant increase in biodegradation compared with control plots in which moisture maintenance was not implemented. Miller et al. (1991), on the other hand, found that moisture addition had no significant effect on the degradation of jet fuel using a bioventing system. They noted, however, that soil moisture content ranged from 6.5 to 9.8 percent (by weight) in both control soils and soils to which additional water was supplied. In addition to the potential drying effects of bioventing, the movement of large amounts of atmospheric air through vadose soils can have a pronounced influence on the temperature of the soils. Miller et al. (1991) noted that under conditions of SVE or bioventing, soil temperature reflects ambient air temperature.

This can be a major limitation on the use of bioventing in cold climates or during winter seasons.

Nutrient supplementation for bioventing systems is usually accomplished by the addition of dissolved nutrients at the same time that water is added to the system. The addition of nutrients can be either via sprinkler systems or through injection wells. In either case, the delivery of nutrients to the microorganisms depends on the migration (percolation and advection) of the nutrient solutions through the subsurface. The design considerations for this type of nutrient delivery are the same as those for traditional vadose zone in site bioremediation (ISB). Nitrogen also can be delivered in the gas phase through the use of anhydrous ammonia as a source of nitrogen (Dineen et al., 1990).

Bioreactors

Bioreactors represent the most highly engineered and hence most highly controlled approach to bioremediation. For the treatment of contaminated soils, there are fundamentally two types of bioreactor units which are commonly used: slurry reactors and compost systems.

Slurry Reactors. In slurry reactor treatment systems, the contaminated soils are excavated, mixed with water to form a slurry, and treated in a constructed reactor. The reactor contents are agitated to promote breakdown of soil aggregates, enhance desorption of contaminants from soil solids, increase contact between the wastes and microorganisms, and enhance oxygenation of the slurry.

In many instances, pretreatment of the soils precedes the actual introduction of the slurry into the reactor. The soil may first be physically sized and graded to a small enough mesh size to allow for maintenance of an acceptable slurry (King et al., 1992). Such fractionation of the soils not only reduces the costs associated with mixing and agitating the final slurry, but since many organic contaminants such as PAHs have a high affinity for clay minerals, fractionation of the soils also may reduce the total volume which needs to be treated and increase the rate of biodegradation of the contaminants (Portier, 1989). Bachmann and Zehnder (1988), for example, found that biodegradation rates were increased significantly when soil particle size was decreased to 30 μm compared with 60 μm. Additional types of pretreatment may be necessary to properly prepare the soil slurry. For example, Black et al. (1991) treated homogenized soils to be used in bench-scale bioreactor evaluations by the addition of sodium hydroxide and sodium chloride to neutralize soil acidity and disperse clay particles which might trap contaminants.

After pretreatment, if necessary, the soils are mixed with water to form a slurry typically containing between 20 and 40 percent solids, although reactors with a solids content as low as 5 percent or as high as 50 percent (dry weight basis) have been reported (Stroo, 1989) and placed into a tank. Oxygen, nutrients, and supplemental bacteria are added as necessary. Temperature and pH can be regulated to maintain optimal ranges for microbial growth and activity. Several package slurry reactor systems are available commercially (Pflug and Burton, 1988). Because of the high degree of engineering involved in the design and construction of slurry bioreactors, the cost of these systems is considerably higher than that of in situ systems.

Proper functioning of a slurry reactor depends on adequate mixing of the slurry. In well-mixed slurries, contact between the microorganisms, contaminant, nutrients, and oxygen is optimized, and the mass-transfer limitations frequently associated with land treatment and ISB are overcome (Black et al., 1991). Mixing can be accomplished through the use of mechanical mixers or as a result of aeration of the slurry. Many of the systems for mixing used in conventional wastewater treatment systems such as activated sludge can be used in slurry bioreactors.

Because many soils are contaminated with mixtures of compounds, some of which are volatile, some type of control for air emissions also must be included in the design of a slurry bioreactor. In many instances, this can be accomplished by simply covering the bioreactor. In other cases, the volatile emissions that are released are captured and treated (Yare, 1991).

Rates of biodegradation in slurry bioreactors are generally faster than rates of degradation of the same compound in situ. This reflects the greater degree of environmental control and more uniform mixing which can be attained in slurry bioreactors. Yare (1991) compared the degradation of PAHs in solid-phase land treatment units and slurry bioreactors. Half-lives of total PAHs and phenanthrene were about four times greater in the land treatment system than in the bioreactors. Mueller et al. (1991) found that biodegradation of creosote components, with the exception of some of the heavy PAHs which were not degraded, was faster in slurry bioreactors than in solid-phase land treatment units. Castaldi and Ford (1992) discussed the design of a slurry bioreactor for the treatment of tarry waste sludges. They found that a reactor with a low waste sludge-to-microorganism ratio was most effective at reducing the concentration of petroleum contaminants. Stroo (1989) compared degradation of refinery wastes in slurry bioreactors with those found in land treatment systems. Slurry-phase treatment was found to degrade oil sludges between 4 and 15 times faster than land treatment. Ross et al. (1988) used a slurry bioreactor to treat more than 750 yd^3 of soils which had been highly contaminated with pesticides (2,4-D and MCPA). Using the slurry treatment system, pesticide levels in the soils were reduced from initial concentrations of 800 mg/kg of soil (equivalent to 400 mg/kg of slurry) to less than 10 mg/kg within a 2-week treatment period.

Compost.   Composting has gained wide acceptance in the treatment of agricultural and municipal wastes. Compost systems range from relatively simple compost piles (windrows) to highly engineered and controlled continuous-feed reactors (Atlas and Bartha, 1987). Regardless of the actual process design, composting is an obligately aerobic and usually thermophilic process. Therefore, design parameters must allow for the maintenance of adequate oxygen levels and an appropriate temperature. This, of course, increases the overall cost of the composting process. In addition, it is frequently necessary to add bulking agents and thermal sources to composting systems. Bulking agents are any materials which increase the void space in the compost and hence increase oxygen flow and transfer within the compost system. Traditionally, such materials as wood chips, straw, and vermiculite have been used as bulking agents. Thermal sources are readily biodegradable materials such as dry molasses or manure which are incorporated into the compost mixture to promote microbial activity and the concomitant increase in temperature necessary for composting.

Recently, composting has been applied to the treatment of contaminated soils, particularly soils contaminated with explosives, heavy oils, and PAHs, although other materials also have been treated using this technique. Composting of soils contaminated with explosives (TNT, RDX, HMX) has been under investigation since the late 1970s (Myler and Sisk, 1991). Initial laboratory studies indicated that composting of explosive-contaminated soils might result in the generation of toxic intermediates (Kaplan and Kaplan, 1982). Subsequent laboratory studies and pilot-scale testing, however, indicated that these compounds were not produced under thermophilic composting conditions but were under mesophilic landfarming and therefore concluded that the initial studies had not achieved the thermophilic conditions necessary for composting (Osmon and Andrews, 1978).

Williams et al. (1992) have recently reported on the use of composting in the treatment of soils contaminated with "pink water" generated during munitions packing and loading. Compost piles were prepared on concrete test pads underlain with drainage channels and pumps for the recirculation of water. The compost mixture was prepared by mixing contaminated soil with horse manure, straw, alfalfa, and horse feed. Inorganic nutrients (fertilizer) were added to attain a C:N ratio of approximately 30:1. The compost pile base and insulating cover were constructed of sawdust, wood chips, and baled straw. The piles were periodically dismantled, remixed, and remoistened when pile temperatures dropped below those considered optimal for composting. Composting resulted in a dramatic decrease in the concentration of explosives in the soils. For example, extractable TNT decreased from 11,840 to 3 mg/kg, although, as the authors noted, not all this reduction was the result of mineralization of the contaminants. In addition, the mutagenicity of TNT-contaminated soils was significantly reduced during 90 days of composting (Tan et al., 1992).

Composting systems also have been proposed for the treatment of soils contaminated with petroleum wastes (Fyock et al., 1991). Van den Munckhof and Veul (1991) reported on the use of a modified DANO composting installation for household garbage for the treatment of oil-contaminated soils. Based on laboratory optimization studies, they determined that mesophilic (30 to 35°C) decomposition of the oil was possible within 2 weeks, provided that moisture and nutrient levels were adequate. Pilot-scale results on soils contaminated with petrol [gasoline], diesel fuel, and a mixture of diesel and lubrication oil were mixed due to the failure of the soils in the reactor to achieve the optimal temperature. The authors concluded that the process has potential; however, unless better process control can be devised and unless a lower final concentration of oil can be achieved, the process would not be feasible within the Dutch regulatory framework.

McMillen et al. (1992) have examined the use of composting in the treatment of oily production pit sludges. These sludges initially contained extremely high (10.8 percent) concentrations of extractable hydrocarbons. Reductions of approximately 92 percent of the extractable hydrocarbons was reported to occur after 4 weeks of composting when the initial sludge was mixed with wood chips and manure to form a compost mixture. Final concentrations of extractable hydrocarbons achieved by this process were sufficiently low that land disposal of the finished compost could be considered as a final option. Taddeo (1989) reported that a compost system could be used to reduce the concentration of total hydrocarbons and PAHs (two, three, and four rings) in soils contaminated with coal tar. Initial concentrations of 20,000 mg/kg coal tar were reduced significantly within an 80-day treatment period in static-pile compost units. Reduction of total hydrocarbons was 94 percent, while PAHs were reduced between 89 and 95 percent, with greater reductions associated with lower-molecular-weight PAHs. Valo and Salkinoja-Salonen (1986) used composting in the field-scale cleanup of chlorophenol-contaminated soils. Wood bark was mixed with the soil as a bulking agent. Water and nutrients were added as necessary to maintain optimal conditions for microbiological activity. Doelman et al. (1988) have reported on the use of composting for the treatment of hexachlorohexane-contaminated soils.

McFarland et al. (1992) examined the use of the white-rot fungi *Phanerochaete chrysosporium* in a composting system for the treatment of benzo[*a*]pyrene in soils. The soils were mixed with corn cobs inoculated with the white-rot fungus. The corn cobs served as a bulking agent, a source of supplemental carbon, and a carrier for the fungi. Initial rates of degradation of the PAHs were faster in systems inoculated with the fungus than in uninoculated controls, although the total amount of benzo[*a*]pyrene ultimately removed from the soils did not differ between the two treatments.

Conclusion

Bioremediation has been shown to be an efficient and cost-effective treatment method for the remediation of soils contaminated with a wide variety of organic contaminants. Remediation systems and technologies are being used routinely to treat soils contaminated with petroleum hydrocarbons and industrial solvents. Bioremediation of vadose zones encompasses a number of treatment options ranging from in situ treatment systems through highly engineered and controlled reactor systems. Despite the variation in the treatment systems possible, all bioremediation depends on optimizing the environment to allow for microbial processes to degrade and detoxify the contaminants.

References

Abdul, A. S., T. L. Gibson, C. C. Ang, J. C. Smith, and R. E. Sobczynski. 1992. In situ surfactant washing of polychlorinated biphenyls and oils from a contaminated site. *Ground Water* **30**:219–231.

Aggarwal, P. K., J. L. Means, and R. E. Hinchee. 1991. Formulation of Nutrient Solutions for In Situ Bioremediation. In R. E. Hinchee and R. F. Olfenbuttel (Eds.), *In Situ Bioreclamation: Applications and Investigations for Hydrocarbon and Contaminated Site Remediation.* Butterworth-Heinemann, Stoneham, Mass., pp. 51–66.

Alexander, M. 1961. *Introduction to Soil Microbiology.* John Wiley and Sons, Inc., New York.

American Petroleum Institute. 1980. *Landfarming: An Effective and Safe Way to Treat/Dispose of Oily Refinery Wastes.* Solid Wastes Management Committee, API, New York.

Aronstein, B. N., and M. Alexander. 1992. Surfactants at low concentrations stimulate biodegradation of sorbed hydrocarbons in samples of aquifer sands and soil slurries. *Environ. Toxicol. Chem.* **11**:1227–1223.

Aronstein, B. N., Y. M. Calvillo, and M. Alexander. 1991. Effect of surfactants at low concentrations on the desorption and biodegradation of sorbed aromatic compounds in soil. *Environ. Sci. Technol.* **25**:1728–1731.

Atlas, R. M. 1975. Effects of temperature and crude oil composition on petroleum biodegradation. *Appl. Microbiol.* **30**:396–403.

Atlas, R. M., and R. Bartha. 1987. *Microbial Ecology: Fundamentals and Applications,* 2d Ed. The Benjamin/Cummings Publishing Company, Menlo Park, Calif.

Bachmann, A., and A. J. B. Zehnder. 1988. Engineering Significance of Fundamental Concepts in Xenobiotic Biodegradation in Soil. In *Contaminated Soil '88.* Kluwer Academic Publishers, Netherlands.

Baker, K. H., and D. S. Herson. 1990. In situ bioremediation of contaminated aquifers and subsurface soils. *Geomicrobiol. J.* **8**:133–146.

Baker, K. H., R. C. Cronce, P. Vercellone-Smith, and D. S. Herson. 1993. Bioremediation of Soils Contaminated with Fuel Oils. In *Biotechnology in Waste Treatment.* Lewis Publishing Company, Chelsa, Minn.

Barrio-Lage, G. A., F. Z. Parsons, and P. A. Lorenzo, 1988. Inhibition and stimulation of trichloroethylene biodegradation in microaerophic microcosms. *Environ. Toxicol. Chem.* **7**:889–895.

Bennedson, M. B., J. P. Scott, and J. D. Hartley. 1987. Use of Vapor Extraction Systems for in Situ Removal of Volatile Organic Compounds from Soil. In *Proceedings of the Conference on Hazardous Wastes and Hazardous Materials.* Washington, pp. 92–95. Haz. Mat. Control. Res. Inst., Washington, D.C.

Berg, G., A. G. Seech, H. Lee, and J. T. Trevors. 1990. Identification and characterization of a soil bacterium with extracellular emulsifying activity. *J. Environ. Sci. Health* **A25**:753–764.

Black, W. V., R. C. Ahlert, D. S. Kosson, and J. E. Brugger. 1991. Slurry-Based Biotreatment of Contaminants Sorbed onto Soil Constituents. In R. E.

Hinchee and R. F. Olfenbuttel (Eds.), *On-Site Bioreclamation: Processes for Xenobiotic and Hydrocarbon Treatment.* Butterworth-Heinemann, Stoneham, Mass., pp. 408–421.

Block, R. N., T. P. Clark, and M. Bishop. 1990. Biological Treatment of Soils Contaminated by Petroleum Products. In P. T. Kostecki and E. J. Calabrese (Eds.), *Petroleum Contaminated Soils,* Vol. 3. Lewis Publishers, Chelsea, Mich., pp. 167–175.

Bollag, J.-M., S.-Y. Liu, and R.D. Minard. 1980. Cross-coupling of phenolic humus constituents and 2,4-dichlorophenol. *Soil Sci. Soc. Am. J.* **44:**52–56.

Bonazountas, M., and J. Wagner. 1984. SESOIL: A Seasonal Soil Compartment Model. A. D. Little, Inc., for the U.S. Environmental Protection Agency, Washington.

Bossert, I., and R. Bartha. 1984. The Fate of Petroleum in Soil Ecosystems. In R. M. Atlas (Ed.), *Petroleum Microbiology.* Macmillan Publishing Company, New York, pp. 434–476.

Brady, N. C. 1972. *The Nature and Properties of Soils.* 8th Ed. Macmillan Publishing Company, New York.

Brenoel, M., and R. A. Brown. 1985. Remediation of a Leaking Underground Storage Tank with Enhanced Bioreclamation. Presented at NWWA Fifth National Symposium and Exposition: Aquifer Restoration and Ground Water Monitoring, May 21–24, Columbus, Ohio.

Britton, L. N. 1985. *Feasibility Studies on the Use of Hydrogen Peroxide to Enhance Microbial Degradation of Gasoline.* American Petroleum Institute, Washington, API Pub. No. 4389.

Broholm, K., T. H. Christensen, and B. K. Jensen. 1991. Laboratory feasibility studies on biological in-situ treatment of a sandy soil contaminated with chlorinated aliphatics. *Environ. Technol.* **12:**279–289.

Brown, E. J., J. J. Pignatello, M. M. Martinson, and R. L. Crawford. 1986. Pentachlorophenol degradation: A pure bacterial culture and an epilithic microbial consortium. *Appl. Environ. Microbiol.* **52:**92–97.

Brubaker, G. R., and J. H. Exner. 1988. Bioremediation of Chemical Spills. In G. S. Omen (Ed.), *Environmental Biotechnology: Reducing Risks from Environmental Chemicals through Biotechnology.* Plenum Press, New York, pp. 163–171.

Carsel, R. F., C. N. Smith, L. A. Mulkey, J. D. Dean, and P. Jowise. 1984. *Users' Manual for the Pesticide Root Zone Model (PRZM): Release 1.* EPA-600/3-84-109. Environmental Protection Agency, Environmental Research Laboratory, Office of Research and Development, Athens, Ga.

Castaldi, F. J. and D. L. Ford. 1992. Slurry bioremediation of petrochemical waste sludges. *Water Sci. Technol.* **25:**207–212.

Christiansen, J. A., T. Koenig, S. Laborde, and D. Fruge. 1989. Topic 2: Land Treatment Case Study. Biological Detoxification of a RCRA Surface Impoundment Sludge Using Land Treatment Methods. In *Proceedings of the 10th National Conference: Superfund '89.* HMCRI, Washington, pp. 362–364.

Claxton, L. D., V. S. Houk, R. Williams, and F. Kremer. 1991. Effect of bioremediation on the mutagenicity of oil spilled in Prince William Sound, Alaska. *Chemosphere.* **23**:643–650.

Coffin, D., and L. Glasgow. 1992. Effective gas flow arrangements in soil venting. *Water Air Soil Pollution* **62**:303–324.

Compeau, G. C., W. D. Mahaffey, and L. Patras. 1991. Full-Scale Bioremediation of Contaminated Soil and Water. In G. S. Sayler, R. Fox, and J. W. Blackburn (Eds.), *Environmental Biotechnology for Waste Treatment.* Plenum Press, New York, pp. 91–110.

Cooper, D. G. 1986. Biosurfactants. *Microbiol. Sci.* **3**:145–149.

Coover, M. P., and R. C. Sims. 1987. The effect of temperature on polycyclic aromatic hydrocarbon persistence in an unacclimated agricultural soil. *Haz. Waste Haz. Mater.* **4**:69–82.

Cosgrove, D. J. 1977. Microbial Transformations in the Phosphorus Cycle. In M. Alexander (Ed.), *Advances in Microbial Ecology,* Vol.1. Plenum Press, New York, pp. 95–134.

Dibble, J. T., and R. Bartha, 1979a. Rehabilitation of oil inundated agricultural land: A case history. *Soil Sci.* **128**:56–60.

Dibble, J. T., and R. Bartha. 1979b. Effect of environmental parameters on biodegradation of oil sludge. *Appl. Environ. Microbiol.* **37**:729–739.

Dineen, D., J. P. Slater, P. Hicks, J. Holland, and L. D. Clendening. 1990. In Situ Biological Remediation of Petroleum Hydrocarbons in Unsaturated Soils. In P. T. Kostecki and E. J. Calabrese (Eds.), *Petroleum Contaminated Soils,* Vol. 3. Lewis Publishers, Chelsea, Mich., pp. 177–187.

Doelman, P., L. Haanstra, and A. Vos. 1988. Microbial sanitation of soil with alpha and beta HCH under aerobic glasshouse conditions. *Chemosphere* **17**:489–492.

Dunnivant, F. M., P. M. Jardine, D. L. Taylor, and J. F. McCarthy. 1992. Cotransport of cadmium and hexachlorobiphenyl by dissolved organic carbon through columns containing aquifer material. *Environ. Sci. Technol.* **26**:360–368.

DuPont, R. R., W. J. Doucette, and R. E. Hinchee. 1991. Assessment of in Situ Bioremediation Potential and Application of Bioventing at a Fuel Contaminated Site. In R. F. Hinchee and R. F. Olfenbuttel (Eds.) *In Situ Bioreclamation: Applications and Investigations for Hydrocarbon and Contaminated Site Remediation.* Butterworth-Heinemann, Stoneham, Mass., pp. 262–282.

Ellis, B., M. T. Balba, and P. Theile. 1990. Bioremediation of oil contaminated land. *Environ. Technol.* **11**:43–455.

Ely, D. L., and D. A. Heffner, 1988. Process for in Situ Biodegradation of Hydrocarbon Contaminated Soil. United States Patent Office, Patent Number 4,765,902. August 23, 1988.

ESTI. 1990. *MOFAT: A Two-Dimensional Finite Element Program for Multiphase Flow and Multicomponent Transport, Program Documentation,* Version 2.0. Environmental Systems and Technologies, Blacksburg, Va.

Falatko, D. M., and J. T. Novak. 1992. Effects of biologically produced surfactants on the mobility and biodegradation of petroleum hydrocarbons. *Water Environ. Res.* **64**:163–169.

FitzPatrick, E. A. 1986. *An Introduction to Soil Science.* John Wiley and Sons, Inc., New York.

Flathman, P. E., J. H. Carson, Jr., S. J. Whitehead, K. A. Khan, D. M. Barnes, and J. S. Evans. 1991. Laboratory Evaluation of the Utilization of Hydrogen Peroxide for Enhanced Biological Treatment of Petroleum Hydrocarbon Contaminants in Soil. In R. E. Hinchee and R.F. Olfenbuttel (Eds.), *In Situ Bioreclamation: Applications and Investigations for Hydrocarbon and Contaminated Site Remediation.* Butterworth-Heinemann, Stoneham, Mass., pp. 125–142.

Focht, D. D., and W. Brunner. 1985. Kinetics of biphenyl and polychlorinated biphenyl metabolism in soil. *Appl. Environ. Microbiol.* **50**:1058–1563.

Fogel, M., A. R. Taddeo, and S. Fogel. 1986. Biodegradation of chlorinated ethanes by a methane-utilizing mixed culture. *Appl. Environ. Microbiol.* **51**:720–724.

Fogel, S., R. Lancione, A. Sewall, and R. S. Boethling. 1985. Application of biodegradability screening tests to insoluble chemicals: Hexadecane. *Chemosphere* **14**:375–382.

Follett, R. H., L. S. Murphy, and R. L. Donahue. 1981. *Fertilizers and Soil Amendments.* Prentice-Hall, Inc., Englewood Cliffs, N.J.

Foth, H. D. 1984. *Fundamentals of Soil Science,* 7th Ed. John Wiley and Sons, Inc., New York.

Fu, M. H., and M. Alexander. 1992. Biodegradation of styrene in samples of natural environments. *Environ. Sci. Technol.* **26**:1540–1544.

Fyock, O. L., S. B. Sordrum, S. Fogel, and M. Findlay. 1991. Pilot Scale Composting of Petroleum Production Sludges. In *Treatment and Disposal of Petroleum Sludges.* University of Tulsa, Center for Environmental Research and Technology, Tulsa, Okla.

Goss, K.-U. 1992. Effects of temperature and relative humidity on the sorption of organic vapors on quartz sand. *Environ. Sci. Technol.* **26**:2287–2294.

Grenney, W. J., C. L. Caupp, R. C. Sims, and T. E. Short. 1987. A mathematical model for the fate of hazardous substances in soil: Model description and experimental results. *Haz. Wastes Haz. Mater.* **4**:223–239.

Hadas, A., M. Sofer, J. A. E. Molina, P. Barak, and C. E. Clapp. 1992. Assimilation of nitrogen by soil microbial populations: NH_4 versus organic N. *Soil Biol. Biochem.* **24**:137–143.

Hamer, G. and A. Heitzer. 1991. Polluted Heterogeneous Environments: Macro-Scale Fluxes, Micro-Scale Mechanisms and Molecular Control. In G. S. Sayler, R. Fox, and J. W. Blackburn (Eds.), *Environmental Biotechnology for Waste Treatment.* Plenum Press, New York, pp. 233–248.

Hanson, R. S. 1980. Ecology and diversity of methylotrophic organisms. *Adv. Appl. Microbiol.* **26**:3–39.

Hardaway, K. L., M. S. Katterjohn, C. A. Lang, and M. E. Leavitt. 1991. Feasibility and Other Considerations for Use of Bioremediation in Subsurface Areas. In G. S. Sayler, R. Fox, and J. W. Blackburn (Eds.),

Environmental Biotechnology for Waste Treatment. Plenum Press, New York, pp. 111–125.

Harmsen, J. 1991. Possibilities and Limitations of Landfarming for Cleaning Contaminated Soils. In T. E. Hinchee and R. F. Olfenbuttel (Eds.), *On-Site Bioreclamation: Processes for Xenobiotic and Hydrocarbon Treatment.* Butterworth-Heinemann, Stoneham, Mass., pp. 255–272.

Herbert, B. E., P. M. Bertsch, and J. M. Novak. 1993. Pyrene sorption by water-soluble organic carbon. *Environ. Sci. Technol.* **27**:398–403.

Henson, J. M., M. V. Yates, J. W. Cochran, and D. L. Shackleford. 1988. Microbial removal of halogenated methanes, ethanes, and ethylenes in an aerobic soil exposed to methane. *FEMS Microb. Ecol.* **53**:193–201.

Hetrick, D. M., C. C. Travis, S. K. Leonard, and R. S. Kinerson. 1989. *Qualitative Validation of Pollutant Transport Components of an Unsaturated Soil Zone Model (SESOIL).* ONRL TM-10672. Oak Ridge National Laboratory, Oak Ridge, Tenn.

Hinchee, R. E., and M. Arthur. 1991. Bench scale studies on the aeration process for bioremediation of petroleum hydrocarbons. *J. Appl. Biochem. Biotechnol.* **28/29**:901–906.

Hinchee, R. E., D. C. Downey, and T. Beard. 1989. Enhancing Biodegradation of Petroleum Hydrocarbons Fuels Through Soil Venting. In *Proceedings of the API/NWWA Conference on Petroleum Hydrocarbons and Organic Chemicals in Ground Water.* American Petroleum Institute, Washington, D.C., Houston, Texas, pp. 235–248.

Hoeppel, R. E., R. E. Hinchee, and M. F. Arthur. 1991. Bioventing soils contaminated with petroleum hydrocarbons. *J. Ind. Microbiol.* **8**:141–146.

Horvath, R. S. 1972. Microbial cometabolism and the degradation of organic compounds in nature. *Bacteriol. Rev.* **36**:146–155.

Hsu, T.-S., and R. Bartha. 1974. Interaction of pesticide-derived chloroaniline residues with soil organic matter. *Soil Sci.* **116**: 444–452.

Huling, S. G., B. E. Bledsoe, and M. V. White. 1991. The Feasibility of Utilizing Hydrogen Peroxide as a Source of Oxygen in Bioremediation. In R. F. Hinchee and R. F. Olfenbuttel (Eds.), *In Situ Bioreclamation: Applications and Investigations for Hydrocarbon and Contaminated Site Remediation.* Butterworth-Heinemann, Stoneham, Mass., pp. 83–102.

Hutzler, N. J., B. E. Murphy, and J. G. Gierke. 1988. *State of Technology Review: Soil Vapor Extraction Systems.* U.S. EPA Hazardous Waste Engineering Research, CR-814319-01-1, Cincinnati, Ohio.

Jamison, V. M., R. L. Raymond, and J. O. Hudson. 1975. Biodegradation of high-octane gasoline in ground water. *Dev. Ind. Microbiol.* **16**:305–312.

Janssen, D. B., A. J. van den Wijingaard, J. J. van der Waarde, and R. Oldenhuis. 1991. Biochemistry and Kinetics of Aerobic Degradation of Chlorinated Aliphatic Hydrocarbons. In R. E. Hinchee and R. F. Olfenbuttel (Eds.), *On-Site Bioreclamation: Processes for Xenobiotic and Hydrocarbon Treatment.* Butterworth-Heinemann, Stoneham, Mass. pp. 92–112.

Jobson, A., M. McLaughlin, F. D. Cook, and D. W. S. Westlake. 1974. Effect on amendments on the microbial utilization of oil applied to soil. *Appl. Microbiol.* **27:**166–171.

Johnson, P. C., C. C. Stanley, M. W. Kemblowski, D. L. Byers, and J. D. Colthart. 1990. A practical approach to the design, operation and monitoring of In Situ soil-venting systems. *Ground Water Monit. Rev.* Spring.

Jury, W. A., and M. Ghodrati. 1989. Overview of Organic Chemical Environmental Fate and Transport Modeling Approaches. In *Reactions and Movement of Organic Chemicals in Soils*. SSSA Special Publication No. 22, Madison, Wis., pp. 271–304.

Kaplan, D. L., and A. M. Kaplan. 1982. Thermophilic biotransformations of 2,4,6-trinitrotoluene under simulated composting conditions. *Appl. Environ. Microbiol.* **44:**747–760.

Karmazyn, J., A. M. Mergenthaler, and L. Sherman. 1989. Operations and Monitoring of Land Treatment of Biodegradable Petroleum Waste. In C. A. Cole and D. A. Long (Eds.), *Hazardous and Industrial Wastes: Proceedings of the Twenty-First Mid-Atlantic Industrial Waste Conference*. Technomic Publishing Company, Lancaster, Pa., pp. 213–228.

Keck, J., R. C. Sims, M. Coover, K. Park, and B. Symons. 1989. Evidence for cooxidation of polynuclear aromatic hydrocarbons in soil. *Water Res.* **23:**1467–1476.

King, B. R., G. M. Long, and J. K. Sheldon. 1992. *Practical Environmental Bioremediation*. Lewis Publishers, Boca Raton, Fla.

Kishi, H., N. Kogure, and Y. Hashimoto. 1990. Contribution of soil constituents in adsorption coefficient of aromatic compounds, halogenated alicyclic and aromatic compounds in soil. *Chemosphere* **21:**867–876.

Laha, S., and R. G. Luthy. 1991. Inhibition of phenanthrene mineralization by nonionic surfactants in soil-water systems. *Environ. Sci. Technol.* **25:**1920–1930.

LaPlat-Polasko, L. T., P. L. McCarty, and A. B. J. Zehnder. 1984. Secondary substrate utilization of methylene chloride by an isolated strain of *Pseudomonas* sp. *Appl. Environ. Microbiol.* **47:**825–830.

Lawes, B. C. 1990. Soil-Induced Decomposition of Hydrogen Peroxide: Preliminary Findings. In P. T. Kostecki and E. J. Calabrese (Eds.), *Petroleum Contaminated Soils*, Vol. 3. Lewis Publishers, Chelsea, Mich., pp. 239–249.

Lawes, B. C. 1991. Soil-Induced Decomposition of Hydrogen Peroxide. In R. F. Hinchee and R. F. Olfenbuttel (Eds.), *In Situ Bioreclamation: Applications and Investigations for Hydrocarbon and Contaminated Site Remediation*. Butterworth-Heinemann, Stoneham, Mass., pp. 143–156.

Leavitt, M. E., D. A Graves, and C. A. Lang. 1991. Evaluation of Bioremediation in a Coal-Coking Waste Lagoon. In G. S. Sayler, R. Fox, and J. W. Blackburn (Eds.), *Environmental Biotechnology for Waste Treatment*, Plenum Press, New York, pp. 71–84.

Leonard, R. A., W. G. Knisel, and D. A. Still, 1986. GLEAMS: Ground water loading effects of agricultural management systems. *Trans. ASAE* **30.**

Lindstrom, J. E., and E. J. Brown. 1989. Supplemental carbon use by microorganisms degrading toxic organic compounds and the concept of specific toxicity. *Haz. Waste Haz. Mater.* **6:**195–200.

Linkenheil, R. 1988. On-Site Biological Treatment of Creosote-Contaminated Soils. In G. S. Omen (Ed.), *Environmental Biotechnology: Reducing Risks from Environmental Chemicals through Biotechnology.* Plenum Press, New York, p. 455.

Litchfield, C. D., D. S. Herson, K. H. Baker, R. Krishnamoorthy, and W. J. Zeli. 1992. Biotreatability Study and Field Pilot Design for a Wood Treating Site. *IGT Symposium on Gas, Oil and Environmental Biotechnology,* Sept. 21–23, 1992.

Litz, N., H. W. Doering, M. Thiele, and H.-P. Blume. 1987. The behavior of linear alkylbenzenesulfonate in different soils: A comparison between field and laboratory studies. *Ecotoxicol. Environ. Safety* **14:**103–116.

Magee, B. R., L. W. Lion, and A. T. Lemley. 1991. Transport of dissolved organic macromolecules and their effect on the transport of phenanthrene in porous media. *Environ. Sci. Technol.* **25:**323–331.

Mansell, R. S., S. A. Bloom, B. Burgoa, P. Nkedi-Kizza, and J. S. Chen. 1992. Experimental and simulated P transport in soil using a multireaction model. *Soil Sci.* **153:**185–194.

McFarland, M. J., R. C. Sims, and J. W. Blackburn. 1991. Use of Treatability Studies in Developing Remediation Strategies for Contaminated Soils. In G. S. Sayler, R. Fox, and J. W. Blackburn (Eds.), *Environmental Biotechnology for Waste Treatment.* Plenum Press, New York, pp. 163–174.

McFarland, M. J., X. J. Qui, J. L. Sims, M. E. Randolph, and R. C. Sims. 1992. Remediation of petroleum impacted soils in fungal compost bioreactors. *Water Sci. Technol.* **25:**197–206.

McLean, E. O. 1982. Soil pH and Lime Requirement. In A. L. Page (Ed.), *Methods of Soil Analysis.* Part 2: *Chemical and Microbiological Properties,* 2d Ed. American Society of Agronomy, Inc., Madison. Wis.

McMillen. S. J., J. M. Kerr, N. R. Gray, and M. Findlay. 1992. Composting of a Production Pit Sludge. *IGT Symposium on Gas, Oil and Environmental Biotechnology,* Sept. 21–23, 1992.

Meylan, W., P. H. Howard, and R. S. Boethling. 1992. Molecular topology/fragment contribution method for predicting soil sorption coefficients. *Environ. Sci. Technol.* **26:**1560–1567.

Miller, M. E., and M. Alexander. 1991. Kinetics of bacterial degradation in a montmorillonite suspension. *Environ. Sci. Technol.* **25:**240–245.

Miller, R. N., C. C. Vogel, and R. E. Hinchee. 1991. A Field-Scale Investigation of Petroleum Hydrocarbon Biodegradation in the Vadose Zone Enhanced by Soil Venting at Tyndall AFB, Florida. In R. E. Hinchee and R. F. Olfenbuttel (Eds.), *In Situ Bioreclamation: Applications and Investigations for Hydrocarbon and Contaminated Site Remediation.* Butterworth-Heinemann, Stoneham, Mass. pp. 283–302.

Moein, G. J., A. J. Smith, and P. L. Stewart. 1976. Follow-Up Study of the Distribution and Fate of Polychlorinated Biphenyls and Benzenes in Soil and Ground Water Samples after an Accidental Spill of Transformer Fluid.

In *Proceedings of the 3rd. National Conference Control of Hazardous Material Spills*. Information Transfer, Inc., Rockville, Md.

Morgan, P., and R. J. Watkinson. 1990. Assessment of the potential for in situ biotreatment of hydrocarbon-contaminated soils. *Water Sci. Technol.* **22**:63–68.

Mueller, J. G., S. E. Lantz, B. O. Blattmann, and P. J. Chapman. 1991. Bench-scale evaluation of alternative biologicalal treatment processes for the remediation of pentachlorophenol- and creosote-contaminated materials: Slurry-phase bioremediation. *Environ. Sci. Technol.* **25**:1055–1061.

Myler, C. A., and W. Sisk. 1991. Bioremediation of Explosives Contaminated Soils (Scientific Questions/Engineering Realities). In G. S. Sayler, R. Fox, and J. W. Blackburn (Eds.), *Environmental Biotechnology for Waste Treatment*. Plenum Press, New York, pp. 137–146.

Nelson, M. J. K., S. O. Montgomery, W. R. Mahaffey, and P. H. Pritchard. 1987. Biodegradation of trichloroethylene and the involvement of an aromatic biodegradative pathway. *Appl. Environ. Microbiol.* **53**:949–954.

Nofzinger, D. L., and A. G. Hornsby. 1987. *Chemical Movement in Layered Soils: User's Manual*. University of Florida Soil Science Department, Gainesville, Fla.

Nofzinger, D. L., J. R. Williams, and T. E. Short. 1987. *Interactive Simulation of the Fate of Hazardous Chemicals During Land Treatment of Oily Wastes: RITZ User's Guide*. Robert S. Kerr Environmental Research Laboratory, U.S. EPA, Ada, Okla.

Odencrantz, J. E., J. M. Farr, and C. E. Robinson. 1992. Transport model parameter sensitivity for soil cleanup level determinations using SESOIL and AT123D in the context of the California Leaking Underground Fuel Tank Field Manual. 1992. *J. Soil Contam.* **1**:159–182.

Osman, J. L., and C. C. Andrews. 1978. The Biodegradation of TNT in Enhanced Soil and Compost Systems. In *U.S. Army Armament Research and Development Command Report* ARLD-TR-77032. Dover, N.J.

Ostendorf, D. W., and D. H. Kambell. 1990. Bioremediated soil venting of light hydrocarbons. *Haz. Waste and Haz. Mater.* **7**:319–334.

Ostendorf, D. W., and D. H. Kambell. 1991. Biodegradation of hydrocarbon vapors in the unsaturated zone. *Water Resources Res.* **27**:453–462.

Pacenka, S., and T. Steenhuis, 1984. *User's Guide for the MOUSE Computer Program*. Cornell University Agriculture Engineering Department, Ithaca, N.Y.

Pardieck, D. L., E. J. Bouwer, and A. T. Stone. 1992. Hydrogen peroxide use to increase oxidant capacity for in situ bioremediation of contaminated soils and aquifers: A review. *J. Contam. Hydrol.* **9**:221–242.

Park, K. S., D. L. Sorenson, J. L. Sims, and V. D. Adams. 1988. Volatilization of wastewater trace organics in slow rate land treatment systems. *Haz. Waste Haz. Mater.* **5**:219–229.

Paul, E. A., and F. E. Clark. 1989. *Soil Microbiology and Biochemistry*. Academic Press, Inc., San Diego.

Pavlostathis, S. G., and K. Jaglal. 1991. Desorptive behavior of trichloroethylene in contaminated soil. *Environ. Sci. Technol.* **25**:274–279.

Pflug, D. A., and M. B. Burton. 1988. Remediation of Multimedia Contamination from the Wood-Preserving Industry. In G. S. Omen (Ed.), *Environmental Biotechnology: Reducing Risks from Environmental Chemicals through Biotechnology*. Plenum Press, New York, pp. 193–201.

Piotrowski, M. R. 1990. U.S. EPA-Approved, Full-Scale Biological Treatment for Remediation of a Superfund Site in Montana. In E. J. Calabrese and P. T. Kostecki (Eds.), *Hydrocarbon Contaminated Soils*, Vol. 1. Lewis Publishers, Chelsea, Mich. pp. 433–457.

Piotrowski, M. R., and J. W Carraway. 1991. Full-Scale Bioremediation of Soil and Ground Water at a Superfund Site: A Progress Report. Presented to HazMat South '91, Georgia World Congress Center, Atlanta, Ga.

Piotrowski, M. R., R. G. Aaserude, and F. J. Schmidt. 1992. Bioremediation of Diesel Contaminated Soil and Tundra in an Arctic Environment. In P. T. Kostecki and E. J. Calabrese. (Eds.), *Contaminated Soils: Diesel Fuel Contamination*. Lewis Publishers, Chelsea, Mich. pp. 115–141.

Portier, R. J. 1989. Examination of Site Data and Discussion of Microbial Physiology with Regard to Site Remediation. In *Proceedings of the 10th National Conference: Superfund '89*. Washington, D.C. Haz. Mat. Control. Res. Inst.

Pussemier, L., G. Szabó, and R. A. Bulman. 1991. Prediction of the soil adsorption coefficient K_{oc} for aromatic pollutants. *Chemosphere*. **21**:1199–1212.

Riser-Roberts, E. 1992. *Bioremediation of Petroleum Contaminated Sites*. C. K. Smoley/CRC Press, Inc., Boca Raton, Fla.

Rittmann, B. E., and N. M. Johnson. 1989. Rapid biological clean-up of soils contaminated with lubricating oil. *Water Sci. Technol.* **21**:209–219.

Rittmann, B. E., A. J. Valocchi, E. Seagren, C. Ray, B. Wrenn, and J. R. Gallagher. 1992. *A Critical Review of in Situ Bioremediation*. Gas Research Institute and U.S. Department of Energy, Chicago, Ill.

Ross, D., T. P. Maziara, and A. W. Bourquin. 1988. Bioremediation of Hazardous Waste Sites in the USA: Case Histories. In *Proceedings of the 9th National Conference: Superfund '88*. HMCRI, Washington, D.C., pp. 395–397.

Rubeiz, I. G., J. L. Streohlein, and N. F. Oebker. 1992. Availability and fixation of phosphorous from urea phosphate incubated in a calcareous soil column. *Commun. Soil Sci. Plant Anal.* **23**:331–335.

Sawhney, B. L. 1989. Movement of Organic Chemicals through Landfill and Hazardous Waste Disposal Sites. In *Reactions and Movement of Organic Chemicals in Soils*. SSSA Special Publication No. 22, Madison, Wis., pp. 447–474.

Scherrer, P., and G. Mille. 1989. Biodegradation of crude oil in an experimentally polluted peaty mangrove soil. *Mariné Pollution Bull.* **20**:430–432.

Schmidt, S. K., K. M. Scow, and M. Alexander. 1987. Kinetics of *p*-nitrophenol mineralization by a *Pseudomonas* sp.: Effects of second substrates. *Appl. Environ. Microbiol.* **53**:2617–2623.

Schwarzenbach, R. P., and W. Giger. 1985. Behavior and Fate of Halogenated Hydrocarbons in Ground Water. In C. H. Ward, W. Giger, and P. L.

McCarty (Eds.), *Ground Water Quality*. John Wiley and Sons, Inc., New York, pp. 114–127.

Shimp, R. J., and R. L. Young. 1988. Availability of organic chemicals for biodegradation in settled bottom sediments. *Ecotoxicol. Environ. Safety* **15**:31–45.

Sims, R. C., and M. R. Overcash. 1981. Land Treatment of Coal Conversion Wastewaters. In *Environmental Aspects of Coal Conversion Technology VI: A Symposium on Coal-Based Synfuels*. EPA Report No. EPA-600-9-82-017. Environmental Protection Agency, Washington, D.C., pp. 218–230.

Sims, R., D. Sorensen, J. Sims, J. McLean, R. Mahmood, R. Dupont, J. Jurinak, and K. Wagner. 1986. *Contaminated Surface Soils In-Place Treatment Techniques*. Noyes Publications, Park Ridge, N.J.

Sleep, B. E., and J. F. Sykes. 1991. Biodegradation of Volatile Organic Compounds in Porous Media with Natural and Forced Gas-Phase Advection. In R. F. Hinchee and R. F. Olfenbuttel (Eds.), *In Situ Bioreclamation: Applications and Investigations for Hydrocarbon and Contaminated Site Remediation*. Butterworth-Heinemann, Stoneham, Mass., pp. 245–261.

Smith, J. R., J. Lynch, and G. R. Brubaker. 1989. Overview of Selected Bioremediation Technologies. Hazardous Waste and Hazardous Materials Conference, April 12–14.

Song, H.-G., X. Wang, and R. Bartha. 1990. Bioremediation potential of terrestrial fuel spills. *Appl. Environ. Microbiol.* **56**:652–656.

Spain, J. C., C. A. Pettigrew, and B. E. Haigler. 1991. Biodegradation of Mixed Solvents by a Strain of *Pseudomonas*. In G. S. Sayler, R. Fox, and J. W. Blackburn (Eds.), *Environmental Biotechnology for Waste Treatment*. Plenum Press, New York, pp. 175–184.

Spain, J. C., J. D. Milligan, D. C. Downey, and J. K. Slaughter. 1989. Excessive bacterial decomposition of H_2O_2 during enhanced biodegradation. *Ground Water* **27**:163–167.

Speitel, G. E., Jr., and F. B. Closmann. 1991. Chlorinated solvent biodegradation by methanotrophs in unsaturated soils. *J. Environ. Eng.* **117**:541–558.

Spenser, W. F., and M. M. Cliath. 1977. The Solid-Air Interface: Transfer of Organic Pollutants Between the Solid-Air Interface. In I. H. Suffet (Ed.), *Fate of Pollutants in Air and Water Environments: Part 1. Advances in Environmental Science and Technology*. John Wiley and Sons, Inc., New York.

Stanier, R. Y., J. L. Ingraham, M. L. Wheelis, and P. R. Painter. 1986. *The Microbial World*, 5th Ed., Prentice-Hall, Englewood Cliffs, N.J.

Stegmann, R., S. Lotter, and J. Heerenklage. 1991. Biological Treatment of Oil-Contaminated Soils in Bioreactors. In R. E. Hinchee and R. F. Olfenbuttel (Eds.), *On-Site Bioreclamation: Processes for Xenobiotic and Hydrocarbon Treatment*. Butterworth-Heinemann, Stoneham, Mass., pp. 188–208.

Stevens, D. K., W. J. Grenney, Z. Yan, and R. C. Thompson. 1989. Dispersion in the vadose zone due to nonequilibrium adsorption kinetics. *Haz. Waste Haz. Mater.* **6**:407–419

St. John, W. D., and D. J. Sikes. 1991. Complex Industrial Waste Sites. In G. S.

Sayler, R. Fox, and J. W. Blackburn (Eds.), *Environmental Biotechnology for Waste Treatment*. Plenum Press, New York, pp. 237–252.

Stotzky, G., and R. G. Burns. 1982, The Soil Environment: Clay-Humus-Microbe Interactions. In R. G. Burns and J. H. Slater (Eds.), *Experimental Microbial Ecology*. Blackwell Scientific Publications. Oxford, England, pp. 105–133.

Strand, S. E., and L. Shippert. 1986. Oxidation of chloroform in an aerobic soil exposed to natural gas. *Appl. Environ. Microbiol.* **52:**203–205.

Stroo, H. F. 1989. Bioremediation of Hydrocarbon-Contaminated Soil Solids Using Liquid/Solids Contact Reactors. In *Proceedings of the 10th National Conference: Superfund '89*. HMRCI, Washington, D.C., pp. 331–337.

Swift, M. J., O. W. Heal, and J. M. Anderson. 1979. *Decomposition in Terrestrial Ecosystems*. University of California Press, Berkeley.

Szecsody, J. E., and R. C. Bales. 1991. Temperature effects on chlorinated-benzene sorption to hydrophobic surfaces. *Chemosphere* **22:**1141–1151.

Taddeo, A. 1989. Field Demonstration of a Forced Aeration Composting Treatment for Coal Tar. In *Proceedings of the Second National Conference on Biotreatment: The Use of Microorganisms in the Treatment of Hazardous Material and Hazardous Waste*. HMRCI, Washington, D.C., pp. 123–128.

Tan, E. L., C. H. Ho, W. H. Griest, and R. L. Tyndall. 1992. Mutagenicity of trinitrotoluene and its metabolites formed during composting. *J. Toxicol. Environ. Health* **36:**165–175.

Thomas, J. M., M. D. Lee, M. J. Scott, and C. H. Ward. 1989. Microbial ecology of the subsurface at an abandoned creosote waste site. *J. Ind. Microbiol.* **4:**109–120.

Thomas, J. M., J. R. Yordy, J. A. Amador, and M. Alexander. 1986. Rates of dissolution and biodegradation of water-insoluble organic compounds. *Appl. Environ. Microbiol.* **52:**290–296.

Thomas, R. G. 1990. Volatilization from Soil. In W. J. Lyman, W. F. Reehl, and D. H. Rosenblatt (Eds.), *Handbook of Chemical Property Estimation Methods: Environmental Behavior of Organic Compounds*. American Chemical Society, Washington, D.C., pp. 16-1–16-50.

Thoms, S. R., and L. W. Lion. 1992. Vapor-phase partitioning of volatile organic compounds: A regression approach. *Environ. Toxicol. Chem.* **11:**1377–1388.

Thornton-Manning, J. R., D. D. Jones, and T. W. Federle. 1987. Effects of experimental manipulation of environmental factors on phenol mineralization in soil. *Environ. Toxicol. Chem.* **6:**615–621.

Urlings, L. G. C. M., F. Spuy, S. Coffa, and H. B. R. J. van Vree. 1991. Soil Vapor Extraction of Hydrocarbons: In Situ and On-Site Biological Treatment. In R. E. Hinchee and R. F. Olfenbuttel (Eds.), *In Situ Bioreclamation: Applications and Investigations for Hydrocarbon and Contaminated Site Remediation*. Butterworth-Heinemann. Stoneham, Mass., pp. 321–336.

Valo, R., and M. Salkinoja-Salonen. 1986. Bioreclamation of chlorophenol-contaminated soil by composting. *Appl. Microbiol. Biotechnol.* **25:**68–75.

Van den Munckhof, Ger P. M., and M. F. X. Veul. 1991. Production-Scale Trials on the Decontamination of Oil-Polluted Soil in a Rotating Bioreactor at Field Capacity. In R. E. Hinchee and R. F. Olfenbuttel (Eds.), *On-Site Bioreclamation: Processes for Xenobiotic and Hydrocarbon Treatment.* Butterworth-Heinemann, Stoneham, Mass., pp. 443–451.

Van Dyke, M. I., H. Lee, and J. T. Trevors. 1991. Applications of microbial surfactants. *Biotech. Adv.* 9:241–252.

Verstrate, W., R. Vanloocke, R. DeBorger, and A. Verlinde. 1976. Modeling of the Breakdown and the Mobilization of Hydrocarbons in Unsaturated Soil Layers. In J. M. Sharpley and A. M. Kaplan (Eds.), *Proceedings of the 3rd International Biodegradation Symposium.* Applied Science Publishers, Ltd., London, pp. 99–112.

Visscher, K., and J. Brinkman. 1989. Biologicalal degradation of xenobiotics in waste management. *Haz. Waste Haz. Mater.* 6:201–212.

Wagnet, R. J., and J. L. Hutson. 1986. Predicting the fate of nonvolatile pesticides in the unsaturated zone. *J. Environ. Qual.* 15:315–322.

Warith, M. A., R. Ferehner, and L. Fernandes. 1992. Bioremediation of organic contaminated soil. *Haz. Waste Haz. Mater.* 9:137–147.

Weissenfels, W. D., H.-J. Klewer, and J. Langhoff. 1992. Adsorption of polycyclic aromatic hydrocarbons (PAHs) by soil particles: Influence on biodegradability and biotoxicity. *Appl. Microbiol. Biotechnol.* 36:689–696.

Williams, R. T., P. S. Ziegenfuss, and W. E. Sisk. 1992. Composting of explosives and propellant contaminated soils under thermophilic and mesophilic conditions. *J. Ind. Microbiol.* 9:137–144.

Wilson, J. T., and C. H. Ward. 1988. Opportunities for bioremediation of aquifers contaminated with petroleum hydrocarbons. *J. Ind. Microbiol.* 27:109–116.

Wilson, J. T., and B. H. Wilson. 1985. Biotransformation of trichloroethylene in soil. *Appl. Environ. Microbiol.* 49:242–243.

Wilson, J. T., L. E. Leach, M. Henson, and J. N. Jones. 1986. In situ biorestoration as a ground water remediation technique. *Ground Water Mon. Rev.* 1:56–64.

Yare, B. S. 1991. A Comparison of Soil-Phase and Slurry-Phase Bioremediation of PNA-Containing Soils. In R. E. Hinchee and R. F. Olfenbuttel (Eds.), *On-Site Bioreclamation: Processes for Xenobiotic and Hydrocarbon Treatment.* Butterworth-Heinemann, Stoneham, Mass., pp. 173–187.

Ying, A., J. Duffy, G. Shepherd, and D. Wright. 1990. Bioremediation of Heavy Petroleum Oil in Soil at a Railroad Maintenance Yard. In P. T. Kostecki and E. J. Calabrese (Eds.), *Petroleum Contaminated Soils,* Vol. 3. Lewis Publishers, Chelsea, Mich. pp. 227–238.

You, I.-S., and R. Bartha. 1982. Stimulation of 3,4-dichloroaniline mineralization by aniline. *Appl. Environ. Microbiol.* 44:678–682.

7

Subsurface Aerobic Bioremediation

Carol D. Litchfield

Biology Department
George Mason University
Fairfax, Va.

In general, the subsurface environment can be divided into three zones: the vadose zone, the capillary fringe, and the saturated zone. The *vadose zone* is that region which has as its only source of water moisture percolating from the surface; i.e., it is unsaturated and unaffected by fluctuations in the water table. The *capillary zone,* on the other hand, is influenced both by the moisture percolating from the surface and by water table fluctuations. As such, if light nonaqueous-phase liquids (LNAPLs) are present (such as gasoline), then the contaminants can be smeared over the capillary zone as the water table rises and falls. This region is also sometimes called the *smearing zone* because of this distribution of contaminants into the soil airspaces. Finally, that area which is always saturated with respect to moisture is the *saturated zone,* and this moisture is the ground water. There are slightly different modifications of bioremediation depending on which region is to be treated, although one can usually combine the capillary zone and the saturated zone in designing the bioremedial response.

Before bioremediation is initiated, it is important to determine whether light nonaqueous-phase liquids (LNAPLs) or dense nonaqueous-phase liquids (DNAPLs) are present. Removal of the LNAPLs as free product will be quicker and cheaper than attempting to biotreat the concentrated contaminants. Similarly, if extensive pools of DNAPLs are present, it may be possible to remove them while pumping to control ground water flow, although DNAPLs tend to "pancake" at the top of the impermeable layer and are difficult to remove completely before initiation of biotreatment.

Bioremediation of ground water can be accomplished by treating the ground water in place (in situ bioremediation) or by treating extracted ground water (often called pump and treat). The advantage of the latter process is that the microbiologist and the engineer can design a bioreactor unit which will operate under optimal conditions for the microorganisms, in a controlled environment, and they can even seed the bioreactor with especially selected microbial populations to enhance the rate, extent, and type of biodegradation.

However, such pump and treat technologies do not address the soil portion of the ground water matrix. This is the advantage of using in situ bioremediation (ISB), wherein the native microbial population is enhanced to degrade the contaminants in place in the soils and ground water. Obviously, under such circumstances, one does not have direct control of the microbial population, the recharge rate to the aquifer, or other environmental parameters. One can control, to some extent, the pH, inorganic nutrient levels, and the bulk redox state.

In some instances, it may be useful to combine in situ biotreatment with an above-ground treatment system. Conditions which make this the selected design include more soluble contaminants, high concentrations of the constituents in the ground water, an extremely stringent cleanup goal, restrictions by the regulatory agencies against recycling contaminated ground water back into the aquifer, and the comparative ease of manipulating the microbial population, redox conditions, temperature, and pH of the system to shorten the time for the remediation. All these factors, of course, must be weighed against the increased costs from the construction of an above-ground bioreactor and the operation and maintenance costs associated with such a unit.

This chapter describes laboratory studies and in situ remediation efforts in ground water aquifers, vadose zones, harbors, and lagoons and includes examples of the combined treatment-train approach described above.

Preliminary Information Needed Before Selecting the Bioremediation System

Prior to the design, construction, and installation of an in situ system, several basic facts must first be obtained. The gathering of this information should be a cooperative effort among the hydrogeologist assigned to the project, the field/design engineer, the chemist, and the microbial ecologist. Each of these individuals has an important role to play in the successful ISB system. The hydrogeologist should obtain information on the direction of ground water flow, the hydraulic conductivity of the aquifer, the physicochemical composition of the aquifer, the presence of LNAPLs or DNAPLs (depending on the contaminants), and the yield of the aquifer (i.e., how much water can be pumped from and back into the aquifer). This latter information is useful to the design engineer, who must size pumps, piping, and tanks. The yield of the aquifer and the hydraulic conductivity are also needed by the microbiologist to estimate the length of time the remediation will take. Obviously, if a nutrient is limiting (oxygen, nitrogen, phosphorous, etc.), then the rate at which that nutrient can be supplied to the microbes will determine the duration of the remediation effort. Based on the data obtained during pump tests, drilling of bore holes or wells, the chemical analyses, and the physicochemical data, the hydrogeologist then can model the aquifer using standard models such as FLOWPATH[14] or BIOPLUME II[29] for optimal placement of the injection system and the recovery system for the ISB system.

The field/design engineer prepares the plans for the unit. Even if the ISB system is a skid-mounted, off-the-shelf unit, connection to the recovery wells and to utilities, the concrete pad or housing for the skid, and the skid configuration that is best suited to the site must all be designed and approved by the client and regulatory agencies.

The chemist's role is to determine the concentrations of the contaminants in the various matrices. This includes both ground water and soils in the saturated and unsaturated zones. If the geologist who is at the site logging the borings does not send soil samples from within the saturated zone, the borings will have to be repeated. It is essential that the capillary fringe and between at least 5 to 10 ft into the saturated zone be sent for analysis, since many contaminants have limited aqueous solubilities and may be trapped in the interstitial pores as the aquifer rises and falls. Therefore, the bulk of the contaminants tends to be within the soil matrix. The chemist must provide information that is acceptable to the regulatory agencies, and this usually means performing the analyses according to U.S. EPA methods[33,34] for the appropriate matrix.

The chemist also will analyze for the loss of the constituents resulting from the microbiological testing, and it is at this point that it may be necessary to modify the standard methods or obtain duplicate runs on the gas chromatograph to avoid "loosing" information on the less concentrated contaminants. An example of where this should be applied is with polynuclear aromatic hydrocarbons (PAHs). The level B carcinogenic PAHs are generally present in concentrations at least an order of magnitude less than the two-, three-, and four-ringed compounds. If the gas chromatograph is set only for the most concentrated compounds, no useful data will result on the fate of the constituents of primary regulatory concern. Thus the chemist and microbiologist must communicate and decide on a plan of action to avoid wasting time, money, and effort.

The microbial ecologist must evaluate the extent of the microbial population, determine its ability to degrade the contaminants, evaluate metabolic intermediates if necessary, determine how to optimize this degradation based on the limitations provided by the chemist and hydrogeologist, and monitor the effects of the ISB system. Once the optimal nutrient package has been developed, one must then evaluate the impact of those nutrients on the soil matrix, determining the effect on flow rate and cation exchange capacity.

Too often it is assumed that microorganisms are indeed present and able to degrade the contaminants. Most of the time this is true, but on three or four occasions, the author has observed situations in which no microorganisms could be detected or enhanced. The apparent toxicity of the water or soils in each instance also meant that bioaugmentation (adding cultures) would not have succeeded either. The small additional expense incurred in proper microbial testing is insignificant compared with the overall costs of the remediation. However, because this is an "up-front" cost, clients are often unwilling to evaluate the potential success of bioremediation. Despite this, every effort should be made to convince the client that he or she will save money in the long run by performing at least minimal testing prior to the initiation of a field program.

Evaluation of the microbial population and its ability to degrade the contaminants is typically performed using both culture techniques and microcosms. The preferred culture techniques involve surface-spread plating on a low-nutrient medium, surface-spread plating on a nonorganic medium which is then supplemented with the contaminants of concern, and/or most probable number techniques such as the Sheen Screen used by Brown and Braddock for petroleum hydrocarbons.[4] The low-nutrient medium is used to estimate the numbers of total heterotrophic bacteria present in the soils and ground water, while the contaminant-specific medium is designed to estimate the numbers of contaminant-degrading microorganisms present. Because of the high temperature required for pour plates, this technique is not recommended, since aquifer organisms are typically exposed to temperatures below 20°C.

While there is as yet no standard protocol for performing treatability tests, general patterns are emerging. For a landfarming operation, most microbial ecologists are using pan tests to which nutrients can be added along with water and/or raking to simulate rototilling. For proposed bioreactor applications, small fermentors or stirred minireactors are used, while for in situ treatments, either a biometer flask or a stationary soil slurry system is most frequently used. If leaching of the contaminants from the soils during an ISB project is of concern, then a soil column is most commonly used. However, soil columns require longer time periods (up to a year) and more attention than a biometer flask, so soil columns have seen limited application. A review of treatability protocols was recently published by Morgan et al.,[24] who compared soil slurry tests with pan tests for the biodegradation of PAHs. They concluded that for soils containing over 10 percent fines, the soil slurry test was just as accurate for showing PAH biodegradation. All these tests are based on loss of the chemicals of interest.

One additional testing protocol is sometimes reported. This involves measurements of the amount of oxygen taken up by the microorganisms and/or the amount of carbon dioxide respired. The apparatus can involve the manual titration of carbon dioxide, as is done in a biometer flask, or one of the several computerized or automatic instruments on the market. The assumption in these systems is that the oxygen taken up is used for the degradation of the contaminants only or primarily, and therefore, the carbon dioxide evolved is only from this biodegradation.

In reality, there are many other organic constituents in soil and ground water, so physiologically based methods must still be validated with chemical analyses at time zero and at the end, if not at two to three points during the study. Such correlations were performed by Tabak et al.,[32] who showed that oxygen uptake and carbon dioxide evolution correlated with the loss of various fractions of Alaskan weathered crude oil, especially following supplementation with nutrients. In fact, respirometry may be most useful in demonstrating the effects on microbial activities resulting from the additions of various compounds (surfactants, nutrients, polymers) to in situ operations or bioreactors.

The exact system configuration used in the microbial degradation test should provide optimal conditions for microbial degradation, be cost-effective, and provide information on the extent of degradation possible under these ideal conditions. Columns, biometer flasks,[2] special microcosms to study volatile organic degradation,[37] jars, pans, serum bottles, etc. have all been used. There is no one right method, since they all have their advantages and limitations. However, an important consideration in selecting one configuration over another is the amount of material available for chemical analysis. Making certain that sufficient material is sent to the chemist again requires close communication among team members. If at all possible, all microcosms should be constructed in duplicate and preferably in triplicate. At least three time periods, including zero time, should be sampled. Again, if possible, there should be four sampling periods if rate information is to be produced. However, this increases the analytical costs by 25 percent, and the additional information, which is of value only for that microcosm configuration, may not be worth the added expense.

As in all microbial ecological studies, the incubation temperature should approach that of the aquifer. Site ground water should be used for dilutions and slurry preparation, and static incubations will mimic the natural situation better than incubation on a shaker. If the material is shaken or a pan study is selected, then air emissions must be monitored to distinguish between bioremediation and volatilization of the contaminants. Incubation times vary from 1 to 2 weeks for rapidly degraded materials such as gasoline in biometer flasks to 1 year in soil columns. The time constraints of the client also will be a factor in selecting the microcosm configuration.

An interesting approach to large-scale laboratory studies has recently been proposed by Lund and Gudehus.[20] These authors developed a technique for freezing in situ with liquid nitrogen a large sample (approximately 0.7 m in diameter) of unsaturated soil. Once frozen, the core is removed with a crane and transported to the laboratory for further studies. The intact frozen sample is placed in a "testing stand" which maintains the integrity of the sample after thawing and allows for injection of nutrients and subsampling as required. The authors used this test to demonstrate that a combination of injection and vacuum wells to supply oxygen could be used in conjunction with percolation of water from the surface to stimulate the indigenous microorganisms in the vadose zone to decompose the contaminating less volatile petroleum hydrocarbons. This increased contaminant destruction by approximately 90 percent. Unfortunately, no field data have been published yet, and there is no information on the correlation between these extremely large-scale laboratory studies and actual field results. The authors have, however, described in detail the field test design.[21]

Case Studies of In Situ Bioremediation

Two basic approaches have been developed for in situ bioremediation depending on whether the contamination is in a shallow unsaturated zone and the ground water or there is a deep unsaturated zone with or without ground water contamination. In the former case "traditional" in situ bioremediation can be performed, while in the latter case one should examine the potential for in situ bioventing. Case studies of each of these approaches will be presented in this section along with some of the advantages and disadvantages of each technology.

In Situ Bioremediation (ISB)

The more traditional approaches of enhancing the native microbial population through nutrient and electron acceptor additions are based on the work of Raymond and coworkers in the 1970s for the Suntech.[17,27,28] Raymond's success with petroleum hydrocarbons led others into utilizing the concept of ISB on a wide scale. An example of how this process is being applied can be found in the case study reported by Ellis et al.[11] They described a site at a coal tar distillery near the center of Stockholm, Sweden. The contaminants were the usual mixture from a creosote operation: phenolics, polyaromatic hydrocarbons (PAHs), and petroleum hydrocarbons. The cleanup targets were less than 200 mg creosote per kilogram of soil and drinking water standards for the ground water of less than 10 mg phenols, PAHs, and cyanide per liter. In addition, development of the site was to proceed during the remediation.

ISB was initiated in June of 1990 with the extraction of ground water from the 3- to 4-m-thick aquifer (which is underlain by clay). The ground water was supplemented with nutrients, oxygen, and inoculum and returned to the aquifer through a series of five infiltration galleries constructed on the surface using gravel to improve water distribution. It is not clear whether surfactants were being added to the in situ system or not, although they were tested during the laboratory feasibility studies. The system was designed to permit reversal of the flow, i.e., injection into the lower horizontal pipes and withdrawal from the surface. The idea here was to prevent plugging of the aquifer by reversing the flow. The total flow was 22 to 44 gal/min, with excess fluid pumped into an up-flow, aerated activated carbon column.

Within the first 4 months of operation, the average level of creosote had decreased by 60 percent, with individual compounds ranging from 100 percent removal for naphthalene to only 3 percent reduction for acenaphthalene. Because of the short treatment season (7 months), the system is still operating and is monitored both by Biotreatment, Ltd., and the Swedish Environmental Protection Agency and the Stockholm Public Health Department.[11] These levels of reduction approximate what many have observed both in laboratory microcosms and in the field.

One of the early applications of Raymond's technology was described by Yaniga and Smith in 1984.[39] This site had an unknown amount of gasoline leak from an underground storage tank. The tank pit was underlain by 6 to 7 ft of heavy, silty loam which was underlain by fractured red-brown shale and silt stone. The ground water table was within the bedrock at 20 to 25 ft below land surface. The plume of dissolved organics extended approximately 200 to 250 ft in the north-south direction and 300 to 350 ft in an east-west direction. Concentrations ranged from the detection limit (< 10 ppb) to over 15 ppm.

An air stripper was installed along with a series of wells to retard further plume migration. After removal of the leaking storage tank, the tank pit was converted into an infiltration gallery. A nutrient solution was added to the recovered ground water after it had passed through the air stripper. Oxygen, as the electron acceptor, was introduced through air sparging. The authors then followed Raymond's lead and added hydrogen peroxide as the oxygen source. This resulted in more than a doubling of the oxygen concentrations in the ground water, and the hydrocarbon concentrations in the ground water fell to 2 to 2.5 ppm. That is, during the first 11 months of operation, 50 to 85 percent reductions in total dissolved hydrocarbons occurred.[39]

Similar results have been reported by Kohlmeier[19] for a field pilot test of ISB at a chemical factory in Heidelberg, Germany, where the contaminants were a mixture of aromatic and aliphatic compounds, including halogenated chemicals. The contamination has penetrated 40 m into the soils, while the water table is at 6 m. The plume extends over approximately 2 km. During the previous 5 years over 132 million gallons of water had been pumped and treated. In the pilot study reported in this paper, an air stripper was used along with an infiltration gallery to reinject nutrient-supplemented water. There was an approximately 66 percent reduction in the contaminants within the pilot-study area during the first 140 days of operation. Contaminant concentrations varied widely but were essentially unchanged in a monitoring well outside the test pilot area. Remediation is continuing at this site.[19]

Dey et al.[7] also have used integrated treatment trains to treat contaminated sites. In 1988 they were involved in the remediation of a petroleum storage plant in southern New Jersey where approximately 8400 gal of unleaded gasoline had been spilled.[7] The site is underlain by the Cohansy sand, so even though plant personnel recovered as much free product and surficial contamination as possible, the material had leaked into the sole-source aquifer. The spill zone was approximately 33 × 28 m and had penetrated to the aquifer.

A three-prong attack was developed which involved ground water extraction and treatment via an air stripper, soil vapor extraction in the vadose zone and treatment of the off-gas in a thermal treatment system, and in situ bioremediation of the saturated soils and ground water. Laboratory studies were performed to optimize the nutrient mixture to be added. The nutrient/peroxide-supplemented ground water was injected through five wells and a series of shallow infiltration lines placed in the vadose zone. This latter method allowed for biodegradation within the vadose zone along with soil vapor extraction. After 400 days of operation, approximately 38 percent of the almost 15,400 kg of recovered gasoline could be attributed to ISB.[7]

Another site involving both vadose zone and ground water contamination was the Naval Air Station at Lakehurst, New Jersey.[13] A break in a lined storage lagoon containing approximately 25 percent ethylene glycol occurred in 1982. The vadose zone is comprised of porous sandy soils, which resulted in a plume approximately 60 × 15 m trending to the east and containing approximately 1440 ppm ethylene glycol. The bioremediation of the ethylene glycol contamination was performed using a two-step system. Contaminated ground water was recovered from several recovery wells and passed through an aerobic above-ground activated sludge system. This treated water was allowed to settle and then was reintroduced to the vadose zone, supplying nutrients and inoculum and thereby creating a closed-loop system. The flow rate was approximately 20 gal/min, and inorganic nitrogen and phosphate (as diammonium phosphate) along with oxygen were added to the system. Only the native microbial population was used.[13]

Within 26 days of the initiation of treatment, two of the monitoring wells within the plume had 92 and 88 percent reductions in ethylene glycol, from 690 and 420 ppm to below the detection limit of 50 ppm, respectively. At two wells closer to the storage lagoon, concentrations fell from 3400 to 1100 ppm to below the detection limit, representing 85 and 92 percent removals, respectively, within 26 days. After 435 days, ethylene glycol levels were below the detection limit in all recovery wells.[13]

A similar system also was used by Flathman and Githens[12] to biotreat vadose zone soils and ground water contaminated with acetone, tetrahydrofuran, and isopropanol. Initially, an air stripper was used to remove the free product which resulted from leakage of several buried isopropanol tanks into a basin containing approximately 2800 m^3 of sand and gravel. In addition, the client had recently had a spill of 500 gal of tetrahydrofuran at another similar-sized basin. Based on chemical analyses, these authors demonstrated the conversion of isopropyl alcohol to acetone in both laboratory microcosms and during the field implementation of ISB. In the field, the acetone was quickly biodegraded so that by day 44, acetone and isopropyl alcohol were below the detection limits. The tetrahydrofuran was below detection limits within 35 days after initiation of the ISB-activated sludge treatment system.[12]

Another example that included the addition of microorganisms to the subsurface involved the bioremediation of gasoline-contaminated ground water in West Springfield, Massachusetts.[25] This is a very complex aquifer showing three aquifers and a perched water table within the vadose zone. During the site assessment, a vapor extraction system was installed to stop migration of the plume, and a free product recovery system was developed. Wells were installed between 1.6 and 3 m into the saturated zone and included 41 injection wells placed within the zone of influence of the recovery well system.

These authors used a different approach from Flathman and coworkers. A bioreactor was used to grow bacteria at the site, and the bacteria were injected under pressure along with nutrients and hydrogen peroxide, also under pressure, into the subsurface. This method was tried in the field operations which began in May of 1989 and continued until the end of November 1989. During this phase of the remediation, commercially available mixtures of bacteria were used. During the next treatment period, however, a mixture prepared from the bacteria indigenous to the site was used. Various nutrient mixtures were added to the subsurface to increase the likelihood of biodegradation. The major nitrogen source for the organisms was nitrate.[25]

By the end of November 1989, both benzene and ethylbenzene concentrations had decreased 68 and 38 percent, respectively, while toluene, xylenes, and total petroleum hydrocarbon concentrations appeared to increase. The bioremediation improved during the 1990 treatment season so that benzene had decreased to 93 percent of the initial levels, while ethylbenzene, toluene, and xylenes had decreased by 53, 81, and 53 percent, respectively. The authors reported that the addition of hydrogen peroxide directly at the injection wells, instead of the bioreactor, greatly increased the amount of oxygen down hole and thus increased the microbial population and biodegradation.[25]

Another treatment-train approach has been developed at Karlsruhe, Germany, for a drinking water aquifer polluted by many oil spills around a railroad marshalling yard near the Durlacher Wald waterworks.[30] In addition, cyanide also was identified in the ground water from an abandoned chemical plant 3 km upstream. Unfortunately, no specific contaminant concentrations either before or after treatment were cited by the authors. An in situ system was designed in which water from the recovery well was pretreated with ozone (to provide an oxygen source and to "soften" the organic contaminants) at 0.5 to 1.5 ppm prior to injection into seven infiltration wells. The treatment system operated for 7 years, and in 1989 the authors reported that the four drinking water supply wells were pumping water for the community without any further treatment of the extracted water.[30]

Not all investigations of ISB have involved the addition of nutrients to the subsurface. In a case study reported by Agertved et al.,[1] the natural degradation of two added herbicides, MCPP [(+/) − -2-(4-chloro-o-methylphenoxy)propionic acid], and Atrazine (2-chloro-4-ethylamino-6-isopropylamino-S-triazine), was examined at a Canadian armed forces base in Borden, Ontario, Canada.[1]

The area for the aerobic investigation consisted of uniform fine- to medium-grained glaciofluvial sand from the surface to approximately 9 m below land surface. The water table began approximately 0.5 to 1.0 m below land surface with a velocity of approximately 8 cm/day. The dissolved oxygen concentrations ranged from 2.1 to 8.0 mg/liter, while total heterotrophic bacterial counts in the ground water ranged from 1.0 to 1.9 × 10^4 CFU/liter.[1]

Solutions of the herbicides were injected into a central well, and passage of the contaminants through the aquifer was monitored at a series of multilevel piezometers. Chloride tracer tests also were performed for estimating the migration rates and differentiating retardation, dispersion, and transformation. Data indicated that over 42 to 56 days, the concentration of MCPP had declined, and based on the tracer data, this decline was interpreted as being due to biotransformation. There was no similar transformation of Atrazine. The authors attributed the lag period to adaptation of the organisms to the MCPP and found no adaptation to the Atrazine over a 96-day period.[1]

Another example of naturally occurring biodegradation was reported by Klečka et al.,[18] who examined an area previously used for the disposal of charcoal manufacturing wastes and therefore contaminated with phenolics, polycyclic compounds, and cresol. Oxygen levels were depleted in the affected area but were substantial throughout the rest of the site, implying microbial degradation of the contaminants until oxygen became the limiting factor. This hypothesis was confirmed through a combination of laboratory microcosm studies and computer modeling which indicated greater than 90 percent degradation within 75 to 120 m of the source.[18]

In Situ Bioventing

The field work of Lund and Gudehus,[21] mentioned previously, and that of Dey et al.[7] are modifications of the concept of bioventing, which recently has been proposed by Hinchee and coworkers.[10,16] For aerobic degradative processes, the availability of oxygen is among the more critical factors in determining the rate and extent of bioremediation. Hinchee et al.[10,16] suggested that the introduction of air into the vadose zone would permit the activation of the native bacteria and allow them to degrade the contaminants.

This is essentially a refinement of soil venting, but with bioventing the rate of movement of air through the aquifer is adjusted to the rate of microbial respiration so that emissions are kept to a minimum while microbial degradation is optimized.

Bioventing was combined with ISB and oil-water separation and air stripping at a site in Missouri with an unsaturated overburden and a fractured bedrock.[3] Other studies have successfully demonstrated bioventing at Tyndall Air Force Base, Florida,[23] at a coke oven plant in Germany,[21] at a Coast Guard Station in Travis City, Michigan,[15] at a number 2 fuel spill at a Burlington National Railroad fueling pump house,[9] and at the Hill Air Force Base.[10] Furthermore, similar treatments can be applied to excavated soils by adding nutrients and pulling a vacuum on the above-ground soil piles with the same rapid treatment effect. Some specific case studies are described below.

Dineen et al.[8] have described the laboratory feasibility testing, field testing, and design for three bioventing projects in southern California. They found that treatment of the vadose zone with ammonia and air resulted in a one to two orders of magnitude increase in the microbial counts and presumably in the amount of degraded hydrocarbons. No specific field data were reported.[8]

In a paper describing the modeling of in situ bioventing in both ground water and vadose zones, McCullough[22] summarized work by Marley and Middleton et al. which indicated remediation times of from 30 days for BTEXs, to 60 to 240 days for trichloroethylene, to 540 days for 1,1,1-trichloroethane. He developed an equation for the mass transfer of the contaminants from the water or soil phases to the air phase and emphasized that for in situ bioventing to be effective, the contaminants need to be relatively insoluble and in an aquifer with high conductivity.[22]

At the Kelly Air Force Base in San Antonio, Texas, JP-4 fuel contaminants were found in a 25-ft vadose zone which consisted of unconsolidated alluvial clays, silts, and silty clays. Hydraulic conductivities ranged from 7×10^{-5} to 6.4×10^{-7}, and the average TPH and BTEX concentrations were 700 and 45 ppm, respectively.[5] Coho and Larken[5] noted that with time the amount of oxygen in the recovered gases was increasing and the carbon dioxide was decreasing. They attributed this to one of or a combination of several factors: too rapid pumping causing excessive dehydration of the subsurface, hydrocarbon concentrations reaching a limiting level, or decreased hydrocarbon availability due to sorption onto the surface of the clay particles. They calculated the rate of biodegradation as 1 mg hydrocarbon per kilogram of soil based on an oxygen uptake rate of 0.08 to 0.008 percent oxygen per minute. The relatively slow rate of oxygen uptake may be a reflection of the tighter soils at this site. Most previous reports have dealt with bioventing in more sandy soils.[5]

Bioventing at a paint factory where toluene was contaminating the vadose zone close to drinking water supply wells was reported by Urlings et al.[35] in The Netherlands. The subsurface consists of fine sand to gravel with a water table at approximately 7 m below land surface. Toluene concentrations ranged up to 2200 mg/kg of soil. Although the system was operated primarily as a soil vapor extraction unit, approximately 2 mg C/kg per day was calculated to have been biodegraded. This was based on microbial numbers, oxygen uptake, and carbon dioxide production.[35]

In Situ Treatment of Harbor Sediments

De Brabandere and David[6] recently reported on the in situ remediation of Zeebrugge, Belgium, harbor sediments contaminated with PAHs and tributyltin. PAH concentrations ranged from 2 to 1391 ppm, with tributyltin levels ranging from 8 to 895 ppb. Not surprisingly, the sedimentary redox was less than -280 mV, and the contaminants were tightly bound to the sedimentary particles. After successful laboratory testing showing that bioaugmentation resulted in biodegradation of the PAHs, tributyltin, and the total petroleum hydrocarbons, it was decided to demonstrate this success in the field. An approximately 1000-m^2 area of the harbor sediment area was screened off using a turbidity curtain. The sediments were mixed with a cutter-dredger, and combinations of no additional treatment or added microbial culture with or without nutrients were tested. The sediments were mixed to about 1 m depth. The mixing was monitored via computer program so that precise additions and locations were known at all times. The results during the reported 8-week study showed that concentrations of bacteria and oxygen were much higher initially in the two treated areas as compared with the untreated zone. However, by the end of the 8-week test period, both had approximately the same redox potential and only two orders of magnitude difference in total bacterial numbers compared with the control plot (approximately 10^6 versus 10^4, respectively).[6]

Furthermore, concentrations of the PAHs and tributyltin increased under both treatment regimes, while TPH levels decreased by 14 and 67 percent in the area simply mixed and the area inoculated and supplemented with nutrients, respectively. An ex situ treatment cell showed substantially improved biodegradation of all the target compounds during the same time interval.[6]

In Situ Lagoon Treatments

The French Limited site in Crosby, Texas, is a 7.3-acre lagoon that was used as a waste disposal facility between 1966 and 1971. The lagoon contained industrial waste oils, acidic galvanizing wastes, phenols, pickling acids, polychlorinated biphenyls, and heavy metals.[31, 38] Preliminary testing demonstrated the existence of a native microbial population with the necessary degradative properties. Initial field pilot testing was conducted on an isolated portion of the lagoon comprising about 5 percent of the total area/volume to be treated. Aeration pipes were installed on the bottom of the lagoon on 6-ft centers and connected to the oxygen source. This effectively killed the anaerobic biomass which had populated the bottom sludges. Because of the high rate of aeration, air emissions were monitored on an hourly basis.[31]

Within the first 4 months of operation, the removal of organic priority pollutants was reported to be over 99 percent effective in the sludge. Chloride ion concentrations increased as the volatile organic pollutant concentrations decreased, indicating transformation and/or mineralization of the chlorinated constituents. Benzene, toluene, and ethylbenzene also disappeared, as did the PAHs, during this period. The most slowly degraded materials were the PCBs. Subsoil remediation was initiated after 134 days of sludge aeration. The EC_{50} as determined by the Microtox Bioassay also was reduced by over 50 percent, confirming a reduction in toxicity caused by decreased concentrations of the contaminants.[31] This site is currently undergoing full-scale in situ bioremediation.

Another case study of in situ lagoon bioremediation was reported by Vail and Ramsden[36] for approximately 9000 yd^3 of oil and grease sludge which had accumulated in a storm basin at a refinery. Again, following careful laboratory treatability studies which demonstrated the potential for biodegradation of the sludges, a full-scale system was designed. The sludge contained over 2000 mg PAHs per kilogram and approximately 200 mg BTEXs per kilogram. Initially, after isolation of a portion of the basin, the existing sludges were aerated and supplemented with nitrogen and phosphorus (at a ratio of 5:5). This resulted in significant decreases during the first 2 months of operation in the PAHs, oil and grease, and BTEX, although the latter could have been air stripped from the system. After the initial treatment period, sludges from the remainder of the basin were added to the treatment unit, and the aeration and nutrient addition continued for another 7 weeks. Final concentrations for oil and grease were 5.1

percent from a beginning value of approximately 17 percent, and PAH concentrations went from approximately 2000 mg/kg to less than 1 mg/kg at the end of phase 2, indicating that microbial biodegradation had occurred.

One further example of in situ lagoon treatment has been reported by Piotrowski[26] for the impoundment basin of the Alyeska Ballast Water Treatment Facility in Valdez, Alaska. The major problems to be resolved there were the variations in temperatures, salinities, and flow rates of the ballast water depending on the season. The treatment facility was effective in reducing the overall oil and grease, but the dissolved BTEXs remained fairly constant in the discharge water. Consequently, an aeration system was designed which optimized oxygen transfer to the impounded water while reducing air-stripping effects. Approximately 85 percent of the observed decreases in BTEXs could be accounted for via biodegradation, and during the summer months the values for the dissolved constituents were typically less than 0.5 ppm. PAH concentrations also decreased to around the detection limit due to this in situ lagoon aeration system.[26]

Conclusion

In situ bioremediation has been shown to be an effective treatment technology for hazardous waste sites. It has been applied most effectively to petroleum hydrocarbons either in the vadose zone or the saturated zone and has recently been used to treat creosote-contaminated locations successfully. To date, however, no creosote or PAH sites have been closed due to completion of the bioremediation. This may reflect the recent application of ISB to these sites and the longer time frames it takes for the microbial degradation of these more complex molecules. Similarly, the application of ISB to lagoons is a relatively recent event, and the mass balances necessary to show that indeed biodegradation and not just air stripping occurred have not been performed. Based on laboratory treatability work, however, aeration of lagoons and sediments with or without nutrient addition should result in successful degradation of the organic contaminants. We are already seeing the inclusion of bioremediation as one of the technologies used at complex sites, and this should increase as more knowledge is gained about microbial degradation in the "real world" and confidence is developed in the efficacy of this approach.

References

1. Agertved. A., K. Rügge, and J. F. Baker. 1992. Transformation of the herbicides MCPP and Atrazine under natural aquifer conditions. *Ground Water* **30(4)**:500–506.

2. Bartha, R., and D. Pramer. 1965. Features of a flask and method for measuring the persistence and biological effects of pesticides in soil. *Soil Sci.* **100**:68–70.

3. Bell, R. A., and A. H. Hoffman. 1991. Gasoline Spill in Fractured Bedrock Addressed with in Situ Bioremediation. In R. E. Hinchee and R. E. Olfenbuttel (Eds.), *In Situ Bioreclamation: Applications and Investigations for Hydrocarbon and Contaminated Site Remediation.* Butterworth-Heinemann, Stoneham, Mass., pp. 437–443.

4. Brown, E. J., and J. F. Braddock. 1990. Sheen Screen, a miniaturized most-probable-number method for enumeration of oil-degrading microorganisms. *Appl. Environ. Microbiol.* **56(12)**:3895–3896.

5. Coho, J. W., and R. G. Larkin. 1992. Preliminary Results of a Vapor Extraction Remediation in Low Permeability Soils, Session 92-7.08. In *Proceedings of the 85th Annual Meeting and Exhibition of the Air & Waste Management Association, June 21–26, 1992,* AWMA, Pittsburgh, Pa., 92-7.08–16.

6. De Brabandere, J., and C. David. 1991. In situ treatment of contaminated harbour sludge. *Med. Fac. Landbouww. Rijksuniv. Gent* **56(4a)**:1483–1497.

7. Dey, J. C., R. A. Brown, and W. E. McFarland. 1991. Integrated site remediation combining ground water treatment, soil vapor recovery, and bioremediation. *Haz. Mater. Control* **4**:31–39.

8. Dineen, D., J. P. Slater, P. Hicks, and J. Holland. 1990. In Situ Biological Remediation of Petroleum Hydrocarbons in Unsaturated Soils. In P. T. Kostecki and E. J. Calabrese (Eds.), *Hydrocarbon Contaminated Soils and Ground Water: Analysis, Fate, Environmental and Public Health Effects, Remediation,* Vol. 1. Lewis Publishers, Chelsea, Mich., pp. 177–187.

9. Downey, D. C., P. R. Guest, and C. A. Culley. 1991. Physical and Biological Treatment of Deep Diesel-Contaminated Soils. In *Proceedings of the Conference on Petroleum Hydrocarbons and Organic Chemicals in Ground Water: Prevention, Detection, and Restoration, Nov. 20–22, 1991.* National Water Well Association, Dublin, Ohio, pp. 361–376.

10. Dupont, R. R., W. J. Doucette, and R. E. Hinchee. 1991. Assessment of In Situ Bioremediation Potential and the Application of Bioventing at a Fuel-Contaminated Site. In R. E. Hinchee and R. E. Olfenbuttel (Eds.), *In Situ Bioreclamation: Applications and Investigations for Hydrocarbon and Contaminated Site Remediation.* Butterworth-Heinemann, Stoneham, Mass., pp. 262–282.

11. Ellis, B., P. Harold, and H. Kronberg. 1991. Bioremediation of a creosote contaminated site. *Environ. Technol.* **12**:447–459.

12. Flathman, P. E., and G. D. Githens. 1985. In-Situ Biological Treatment of Isopropanol, Acetone, and Tetrahydrofuran in the Soil/Ground Water Environment. In E. K. Nyer (Ed.), *Ground Water Treatment Technology*. Van Nostrand Reinhold Company, New York, pp. 173–185.

13. Flathman, P. E., D. E. Jaeger, and L. S. Bottomley. 1989. Remediation of contaminated ground water using biological techniques. *Ground Water Monit. Rev.* **9**:105–119.

14. Franz, T., and N. Guiguer. *FLOWPATH*, Ver. 4: *Steady-State Two-Dimensional Horizontal Aquifer Simulation Model*. Waterloo Hydrogeologic Software, 200 Candlewood Crescent, Waterloo, Ontario, Canada.

15. Griffin, C. J., J. M. Armstrong, and R. H. Douglass. 1991. Engineering Design Aspects of an in Situ Soil Vapor Remediation System (Sparging). In R. E. Hinchee and R. E. Olfenbuttel (Eds.), *In Situ Bioreclamation: Applications and Investigations for Hydrocarbon and Contaminated Site Remediation*. Butterworth-Heinemann, Stoneham, Mass., pp. 517–528.

16. Hinchee, R. E., D. C. Downey, and T. Beard. 1989. Enhancing Biodegradation of Petroleum Hydrocarbon Fuels Through Soil Venting. In *Proceedings of the Conference on Petroleum Hydrocarbons and Organic Chemicals in Ground Water: Prevention, Detection and Restoration, Nov. 15–17, 1989*. National Water Well Association, Dublin, Ohio, pp. 235–248.

17. Jamison, V. W., R. L. Raymond, and J. O. Hudson, Jr. 1975. Biodegradation of high-octane gasoline in ground water. *Dev. Ind. Microbiol.* **16**:305–312.

18. Klečka, G. M., J. W. Davis, D. R. Gray, and S. S. Madsen. 1990. Natural bioremediation of organic contaminants in ground water: Cliffs-Dow Superfund site. *Ground Water* **28**:534–543.

19. Kohlmeier, E. 1990. In-situ-Sanierung auf einem Areal der chemischen Industrie. WLB Wasser, Luft, und Boden.

20. Lund, N.-Ch., and G. Gudehus. 1990. Large Scale Sample Tests for a Biological in Situ Remediation of Soils Polluted by Hydrocarbons. In F. Arendt, M. Hinsenveld, and W. J. van den Brink (Eds.), *Contaminated Soil '90*. Kluwer Academic Publishers, Dordrecht, The Netherlands, pp. 463–472.

21. Lund, N.-Ch., J. Swinianski, G. Gudehus, and D. Maier. 1991. Laboratory and Field Tests for a Biological in Situ Remediation of a Coke Oven Plant. In R. E. Hinchee and R. E. Olfenbuttel (Eds.), *In Situ Bioreclamation: Applications and Investigations for Hydrocarbon and Contaminated Site Remediation*. Butterworth-Heinemann, Stoneham, Mass., pp. 391–412.

22. McCullough, M. L. 1992. Modelling Innovative Vapor Extraction Techniques Applied for In Situ Remediation of Volatile Organic Compounds in Soil and Ground Water, Session 92-32.03. In *Proceedings of*

the 85th Annual Meeting and Exhibition of the Air & Waste Management Association, June 21–26, 1992. AWMA, Pittsburgh, Pa., pp. 1–10.

23. Miller, R. N., C. C. Vogel, and R. E. Hinchee. 1991. A Field-Scale Investigation of Petroleum Hydrocarbon Biodegradation in the Vadose Zone Enhanced by Soil Venting at Tyndall AFB, Florida. In R. E. Hinchee and R. E. Olfenbuttel (Eds.), In Situ Bioreclamation: Applications and Investigations for Hydrocarbon and Contaminated Site Remediation. Butterworth-Heinemann, Stoneham, Mass., pp. 283–302.

24. Morgan, D. J., A. Battaglie, J. R. Smith, A. C. Middleton, and D. V. Nakles. 1991. Evaluation of a Biodegradation Screening Protocol for Contaminated Soil from Manufactured Gas Plant Sites. In C. Akin and J. Smith (Eds.), Gas, Oil, Coal and Environmental Biotechnology III. Institute of Gas Technology, Chicago, Ill., pp. 55–74.

25. Myers, J. M., K. A. Minkel, and Brian S. Schepart. 1991. In-Situ Bioaugmentation of Gasoline Contaminated Soil. In Proceedings of the Conference on Petroleum Hydrocarbons and Organic Chemicals in Ground Water: Prevention, Detection, and Restoration, November 20–22, 1991. National Ground Water Association, Dublin, Ohio, pp. 443–456.

26. Piotrowski, M. 1991. Bioremediation of Hydrocarbon Contaminated Surface Water, Ground Water, and Soils: The Microbial Ecology Approach. In P. T. Kostecki and E. J. Calabrese (Eds.), Hydrocarbon Contaminated Soils and Ground Water: Analysis, Fate, Environmental and Public Health Effects, Remediation, Vol. 1. Lewis Publishers, Chelsea, Mich., pp. 203–238.

27. Raymond, R. L., V. W. Jamison, and J. O. Hudson, Jr. 1976. Beneficial stimulation of bacterial activity in ground waters containing petroleum products: I. Physical, chemical wastewater treatment. AIChE Symp. Series **73**:390–404.

28. Raymond, R. L., V. W. Jamison, J. O. Hudson, R. E. Mitchell, and V. E. Farmer. 1978. Field Application of Subsurface Biodegradation of Gasoline in Sand Formation. Final report submitted to American Petroleum Institute, Washington, D.C.

29. Rifai, H. S., and P. B. Bedient. 1987. BIOPLUME II: Two-Dimensional Modeling for Hydrocarbon Biodegradation and in Situ Restoration. In Proceedings of the Conference on Petroleum Hydrocarbons and Organic Chemicals in Ground Water: Prevention, Detection, and Restoration, Nov. 17–19, 1987. National Water Well Association, Dublin, Ohio, pp. 431–450.

30. Sontheimer, H., G. Nagel, and P. Werner. 1989. Restoration of Aquifers Polluted with Hydrocarbons. In Toxic Organic Chemicals in Porous Media, Ecology Studies Series 73, pp. 320–334. Springer-Verlag, Inc., Berlin.

31. Sloan, R. 1987. Bioremediation demonstrated at hazardous waste site. Oil Gas J. Sept. 14, 1987(Technology):61–66.

32. Tabak, H. H., J. R. Haines, A. D. Venosa, J. A. Glaser, S. Desai, and W. Nisamaneepong. 1991. Enhanced Degradation of the Alaskan Weathered Crude Oil Alkane and Aromatic Hydrocarbons by Indigenous Microbiota Through Application of Nutrients. In C. Akin and J. Smith (Eds.), *Gas, Oil, Coal, and Environmental Biotechnology III*. Institute of Gas Technology, Chicago, Ill., pp. 3–38.

33. United States Environmental Protection Agency. 1979. *Methods of Analysis of Water and Wastes*. EPA/600/4-79/020. U.S. EPA, Washington, D.C.

34. United States Environmental Protection Agency. 1986. *Test Methods for Evaluating Solid Waste*, 3d Ed. SW846. U.S. EPA, Washington, D.C.

35. Urlings, L. G. C. M., F. Spuy, S. Coffa, and H. B. R. J. van Vree. 1991. Soil Vapour Extraction of Hydrocarbons: In Situ and On-Site Biological Treatment. In R. E. Hinchee and R. E. Olfenbuttel (Eds.), *In Situ Bioreclamation: Applications and Investigations for Hydrocarbon and Contaminated Site Remediation*. Butterworth-Heinemann, Stoneham, Mass., pp. 321–336.

36. Vail, R. L., and D. K. Ramsden. 1992. A Case Study: Biodegradation of Oily Sludges from a Refinery Process Water/Stormwater Basin, Session 92-30.01. In *Proceedings of the 85th Annual Meeting and Exhibition of the Air & Waste Management Association, June 21–26, 1992*, AWMA, Pittsburgh, Pa., pp. 1–16.

37. Wilson, J. T., G. B. Smith, J. W. Cochran, J. F. Barker, and P. V. Roberts. 1987. Field evaluation of a simple microcosm simulating the behavior of volatile organic compounds in subsurface materials. *Wat. Resources Res.* **23**:1547–1553.

38. Woodward, R. E., and D. K. Ramsden. 1989. Bioremediation of Hazardous Wastes at the French Limited Superfund Site. In *Proceedings of the 1989 Environmental Conference*. pp. 397–401.

39. Yaniga, P. M., and W. Smith. 1984. Aquifer Restoration via Accelerated in Situ Biodegradation of Organic Contaminants. In *Conference Proceedings of the Petroleum Hydrocarbon and Organic Chemical in Ground Water*, Houston, Texas, November 5–7, 1984. Water Well Publishing Company, Worthington, Ohio, pp. 451–472.

8

Bioremediation in Freshwater and Marine Systems

D. A. Wubah
Department of Biology, Towson State University, Towson, Md.

D. D. Hale
Technology Applications, Inc., Athens, Ga.,

J. E. Rogers
Environmental Research Laboratory, U.S. Environmental Protection Agency, Athens, Ga.

Bioremediation in Aquatic Environments

A major challenge in today's world is to maximize the benefits that accrue from the industrial, agricultural, and domestic use of xenobiotic chemicals while minimizing any adverse effects on natural ecosystems. Some of the means to achieve this goal are (1) waste minimization, (2) proper management of waste disposal, and (3) cleanup of sites already contaminated. Contaminated sites can include surface

Note: The material presented in this chapter has not been reviewed by the EPA. It should not be interpreted as representing EPA policy or endorsement.

and subsurface soil, ground water, and freshwater and marine sediments with their associated water. In this chapter the use of biological techniques for remediating contaminated aquatic environments will be discussed. Often the cleanup or remediation of such sites must employ techniques from a number of disciplines, including biology, chemistry, geology, hydrology, and engineering.

For decades, biological treatment has been utilized in composting and in treatment of municipal and industrial wastewater. Within the past decade, interest has grown in the use of microorganisms for treatment of hazardous wastes which contaminate a variety of environments. Biological treatment, either alone or combined with other technologies, compares favorably with physical and chemical treatments on the basis of cost and completeness of chemical transformation. The cost of bioremediation is usually less than the cost of other methods of waste disposal. For example, the cleanup of a contaminated lagoon near Houston, Texas, was estimated at $40 to $50 million by bioremediation compared with $120 million by incineration (Sloan, 1987).

The use of indigenous microorganisms to clean waste sites also has made bioremediation a more cost-effective technology. This is especially true for cases in which only minor adjustments to the environment, such as addition of nutrients, are required. Biological treatment also offers the possibility of complete transformation of organic waste. Physical treatment of wastes, including extraction, adsorption, solidification, encapsulation, and filtration, only separates but does not destroy wastes. Thermal processes, such as incineration, either completely destroy wastes or reduce their volume but may produce toxic ash and volatile compounds. Similarly, chemical treatment of wastes may result in hazardous residual components. Thus, in contrast to biological treatment, each of these alternate technologies may require further treatment of wastes or extended storage of treated wastes.

Bioremediation in aquatic environments often requires treatment of both surface waters and sediments. Sediments can serve as long-term sources of pollutants. Contaminated water or sediments may be biologically remediated in situ or in bioreactors. Although in situ treatment may be less expensive than bioreactor treatment, it often requires longer time periods (months to years) to achieve cleanup standards. Treatment in bioreactors may be complete within a shorter time (on the order of weeks) because optimal conditions (i.e., pH, temperature, nutrient concentrations, etc.) for biodegradation can be more easily maintained. However, physical removal of the contaminated material from the site to the bioreactor is required, and operating expenses and maintenance costs are an additional consideration.

Factors Affecting Bioremediation

Biodegradation of hazardous wastes occurs naturally in aquatic systems but often at an unacceptably slow rate to be considered as a cleanup option. Bioremediation is founded on the premise that the natural rate at which biodegradation occurs can be enhanced. The success or failure of bioremediation rests on our ability to optimize biological, physical, and engineering parameters at a specific site to achieve the targeted cleanup standards.

Effective bioremediation requires extensive site characterization, especially for those factors which have been demonstrated to affect the rates of microbial processes. An assessment of the following factors is therefore necessary in order to modify conditions with various engineering options: temperature, pH, redox potential, nutrient ratios, potential microbial activity, availability of electron acceptors, type of contaminants, toxicity of contaminants to microorganisms, and contaminant bioavailability. A description of these factors and how they may be optimized for the purposes of bioremediation follows.

Temperature.   Temperature affects the rates of microbially mediated reactions. Most bacteria are mesophilic and are most active between 5 and 40°C. Optimal activity can vary but generally occurs at 35°C. A few mesophilic bacteria survive at up to 60°C, but only thermophilic microorganisms grow and reproduce above this temperature. Thus, in some environments having extreme temperatures, in situ biological activity may be hindered. Bioremediation in these cases would be limited to bioreactors where the temperature can be controlled.

pH.   The optimal pH for microbial activity is usually between 5.5 and 8.5. However, biodegradation at some sites with extreme pH values may occur as a result of the activity of unique microorganisms which grow in very acidic or alkaline environments. In some in situ treatment systems, the pH is maintained at approximately 7 by adding lime, whereas in some bioreactor systems, the pH is maintained between 7 and 7.5 by adding commercial-grade carbon dioxide.

Redox Potential.   Most aerobic and facultative anaerobic microorganisms require a redox potential of 50 mV or more, but the optimal redox potential for most obligately anaerobic microorganisms is much less than 50 mV. In aerobic treatment scenarios, mechanical aeration or the addition of oxidants (i.e., H_2O_2) is used to maintain the redox potential at an appropriate level for microbial activity. In anaerobic treatment scenarios, organic compounds may be added to effect oxygen removal, which can result in a reduction in redox potential.

Nutrient Ratios. Carbon, nitrogen, and phosphorus are required for cell growth and biosynthesis of most microbial constituents. When wastes are degraded by cometabolism, the addition of a secondary carbon source is often necessary for cellular growth and maintenance as well as maximum microbial activity (Alexander, 1981). When wastes are used as carbon and energy sources, nitrogen and phosphorus may be required (often in the ratio of 5 or 10:1). In many aquatic environments, nitrogen and phosphorus are the limiting nutrients. Sulfur and trace nutrients such as Fe, K, Mg, Ca, Zn, Mo, Cu, and Mn also may be needed by indigenous microorganisms. The ratio and types of nutrients needed to achieve optimal microbial activity are often determined by the microorganisms present at a specific site. Nutrients may be added to bioreactors or contaminated sites in the form of fertilizers.

Electron Acceptors. Microorganisms require a sink for electrons generated during biodegradation of waste compounds. Under aerobic conditions, oxygen is used as the terminal electron acceptor; often, more than 0.2 mg/liter of dissolved oxygen is needed. To prevent oxygen transfer limitations for microbial activity, aerobic treatment systems are designed to allow a sufficient supply of oxygen in both reactor and in situ treatments. Floating or submerged aerators or compressors and spargers are often used to provide oxygen to bioreactors, but H_2O_2 is usually added as a source of oxygen to natural systems. Nitrate, sulfate, and carbonate, which may serve as terminal electron acceptors in anoxic systems, may be added to stimulate anaerobic microbial activity. The design of some anaerobic systems allows treatment at various redox potentials or anaerobic-aerobic cycling.

Type of Contaminants. The structure and complexity of a compound determine its potential for aerobic or anaerobic degradation. Generally, hydrocarbons are degraded under aerobic conditions, with simple hydrocarbons being biodegraded faster than complex hydrocarbons (Grubbs, 1986). Two- and three-ring polycyclic aromatic hydrocarbons (PAHs) are also degraded more rapidly than PAHs with four or more rings (Bradford and Krishnamoorthy, 1991). Some halogenated organic compounds can be degraded under aerobic conditions, but the rate of degradation decreases with an increase in the degree of halogenation. Highly halogenated compounds are often transformed under anaerobic conditions by a process of reductive dehalogenation which replaces the halogen atom(s) with hydrogen atom(s).

Bioavailability of Contaminants. Microbes usually have to come into contact with waste compounds in order to carry out efficient degradation. In most aquatic systems, the availability of a contaminant is influenced by both its solubility and it partitioning between the aqueous and organic phases. For a compound with ionizable hydrogen, pH may affect its partitioning by favoring the ionic or nonionic species, respectively. Contaminants that are not soluble in water may reside in greases or oils in the aquatic system, and microorganisms that are capable of degrading greases and oils may be able to degrade these contaminants also.

Toxicity. Biodegradation can be inhibited by the inherent toxicity of waste material or its transformation products. Initially, waste material may be diluted to a less toxic level by mixing with noncontaminated material. Toxic products of degradation also may be removed as they are produced. Alternatively, bacteria that are resistant to toxicity and that are able to convert the waste into a less toxic form may be introduced to the contaminated material.

Potential Microbial Activity. Although some bacteria readily degrade waste materials following exposure, bacteria often require a period of acclimation or adaptation before effective degradation can occur (McGinnis, 1988; Spain and Van Veld, 1983; Spain et al., 1980; Wiggins et al., 1987). If a mixture of contaminants is present in an aquatic system, degradation may require the presence of several acclimated organisms. Acclimation may result from the time required for (1) growth of the initially small degrading population(s) to a size large enough to effect detectable degradation, (2) induction, derepression, or activity of the requisite microbial enzyme(s), (3) genetic change in the degrading population(s), and (4) the use of alternate substrates by the degrading microorganisms. Each of these phenomena may be influenced by a number of factors which may need to be controlled so as to allow acclimation to occur. Growth of the degrading population(s), for example, may be inhibited by the presence of predators or competitors or toxic compounds, and a reduction of these inhibitory factors may be necessary to allow acclimation to occur by growth.

When natural bacterial populations fail to acclimate to the presence of the contaminant(s) or when acclimation is too slow, other microorganisms may be added. Such microorganisms may be isolated from the contaminated site, enriched in the laboratory, and reintroduced to the site. Alternatively, organisms isolated from other sites or genetically engineered microorganisms may be introduced to a contaminated site. The effectiveness of introduced microorganisms reflects their ability to compete successfully with indigenous microorganisms. A disadvantage in the use of genetically engineered microorganisms is the need to contain the organisms, and the cost of containment often can constitute a major portion of the biotreatment budget. Genetically engineered organisms are therefore generally used in a more cost-effective manner in bioreactor treatments, where containment is easily achieved.

In addition to the preceding factors, choosing appropriate remediation methods for a site is determined to a large extent by the physical and chemical characteristics of the site, remediation goals to be achieved, materials handling and equipment requirements, and cost (Dasappa and Loehr, 1991). Based on these considerations, bench-scale or laboratory studies are often conducted to ascertain the feasibility of biologically treating a specific site. Results from bench-scale studies are further tested in scale-up or pilot studies. Such treatability studies allow optimization of operating conditions for remediation in the field.

The progression of bench-scale, pilot-scale, and field-scale studies is important for several reasons. First, bench-scale studies are necessary to determine the potential for biodegradation in samples from a specific environment and to identify limiting factors. Second, pilot-scale studies validate or invalidate laboratory-derived data in a procedure similar to microcosm validation by correlative analysis to the environment (Portier et al., 1983; Portier, 1985). In addition, data from pilot-scale studies provide necessary information for emissions or residual management as well as process control and operation of the field work. These studies indicate the extent to which laboratory-derived biodegradation data correlate with the natural environment and thus the applicability of bioremediation techniques. Pilot-scale studies also may identify factors affecting bioremediation which were not noted in bench-scale studies. Finally, field-scale endeavors integrate and implement the information from bench- and pilot-scale studies to optimize biodegradation on a larger scale.

Studies have been conducted on several freshwater systems contaminated with various pollutants in the progression described above. Examples of bench-scale, pilot-scale, and field-scale case studies are presented here.

Bench-Scale and Pilot-Scale Studies

Spain et al. (1984) demonstrated the potential for p-nitrophenol (PNP) degradation in freshwater ponds using laboratory and field test systems. Although degradation rates determined in the two test systems differed, the potential for in situ PNP biodegradation was indicated by laboratory test systems. These included shake flasks containing water with and without sediment, ecocores containing water and intact pond sediment, and two sizes of microcosms. The field test system consisted of a horseshoe-shaped pond that was divided in half by a wooden dam. Water was recirculated with portable pumps. Biodegradation rates of PNP were determined for each type of test system following sequential additions of PNP. The biodegradation rate (-0.11 µg/liter/h) for PNP (200 to 300 µg/liter) following the first addition of the compound to the field system was lower than in most laboratory systems (-0.02 to -0.48 µg/liter/h). However, the biodegradation rate (-10.41 µg/liter/h) following the second addition of the compound to the field system was higher than in most laboratory systems (-2.56 to -17.82 µg/liter/h). A closer agreement between laboratory- and field-derived degradation rates may have been possible if the same mixing rates and sediment surface area/volume ratios occurred in the different test systems.

Shimp (1989) and Shimp and Schwab (1991) also assessed the potential biodegradation of a quaternary ammonium surfactant in riverine systems using laboratory and field studies. The surfactant, dodecyltrimethylammonium chloride (C12-TMAC), was added continuously to laboratory microcosms containing sediments from the Little Miami River, Ohio. Substrate concentrations in microcosms were as high as 10 mg/liter, whereas those in channels of a flow-through environmental chamber in the river were 0 (control), 0.25, and 25 mg/liter initially. The surfactant dose in each of the channels was subsequently decreased one order of magnitude (i.e., 0.025 and 2.5 mg/liter). Data obtained from the laboratory studies could have allowed an approximate determination of in situ C12-TMAC biodegradation because although absolute degradation rates in the laboratory and the field differed, relative changes in degradation were similar. Biodegradation potential, for example, in laboratory microcosms and in the field increased 60-fold and from 30- to 60-fold, respectively, following exposure to C12-TMAC. Likewise, first-order biodegradation rate constants increased 10- and 20-fold, respectively, in laboratory and field systems.

The work of Lewis et al. (1986) demonstrates the importance of pilot-scale studies in identifying factors that may affect biodegradation in an unexpected manner in field-scale studies. These researchers initially studied the microbial transformation of the herbicide 2,4-dichlorophenoxyacetic acid butoxyethyl ester (2,4-DBE) under laboratory conditions. In particular, the effects of pH, water velocity, and herbicide concentration on 2,4-DBE transformation rates were evaluated. Subsequently, transformation rates were determined in a small freshwater stream to which the herbicide (less than 100 µg/liter) was periodically added over a 9-month period. Although water temperatures and velocities in the stream decreased during the cold season, 2,4-DBE transformation rates increased up to threefold over those determined in the warm season. Higher biomass concentrations occurred during the cold season due to the accumulation and subsequent colonization of leaf litter. This apparently compensated for the lower herbicide transformation rates expected during periods of lower temperature. However, without a pilot-scale study, these unexpected factors affecting biodegradation would not have been identified.

Several laboratory investigations have been carried out to study aerobic and anaerobic biodegradation of polychlorinated biphenyls (PCBs) in contaminated sediments. Under aerobic conditions, biodegradation is limited to congeners with one to five chlorine atoms and often occurs by cometabolism, where another substrate such as biphenyl is required as a source of energy and to induce the degradative enzymes (Bedard et al., 1987; Furukawa et al., 1979). Anaerobic biodegradation of PCBs by reductive dechlorination frequently occurs by removal of chlorine from PCB congeners with four or more chlorine atoms. Under anaerobic conditions, carbon substrates are required as sources of electrons.

Bench-scale work concerning the effects of different carbon sources on biodegradation of PCBs in contaminated sediments has been reported by Nies and Vogel (1990). The effect of reduced organic substrates on the microbial degradation of PCBs was studied by adding acetate, glucose, methanol, and acetone to PCB-contaminated Hudson River sediment in various culture vessels. These substrates served as carbon and energy sources for the naturally occurring microbial communities in the sediment. The pattern of dechlorination was similar in all the batches amended with organic substrates, but the rates and extents of dechlorination were different. The relative rates and extents of dechlorination were in the order methanol > glucose > acetone > acetate. Dechlorination occurred at the meta and para positions of the highly chlorinated congeners. No significant dechlorination was observed in the sterile controls and batches that did not receive organic substrates.

The rates of dechlorination in this study were much more rapid than those in the natural environment; e.g., 80 and 70 percent of the pentachlorobiphenyls and tetrachlorobiphenyls, respectively, were degraded in the acetone- and glucose-amended batches after 22 weeks. In this bench-scale work, the PCBs that were degraded were added to the sediments and may not be controlled by the same physicochemical sorption mechanisms that occur in a contaminated sediment. Therefore, a pilot-scale study is essential to establish that organic amendments can influence PCB biodegradation at a contaminated site.

Quensen et al. (1990) investigated biologically mediated degradation of four Aroclor mixtures by anaerobic microorganisms enriched from contaminated sediments. Aroclor 1242, 1248, 1254, and 1260, with an average of three, four, five, and six chlorines, respectively, per biphenyl molecule, were used in this study. The Aroclors were added to culture vessels containing microorganisms enriched from sediments collected at an Aroclor 1242–contaminated site (Hudson River) and an Aroclor 1260–contaminated site (Silver Lake). The Silver Lake inoculum dechlorinated Aroclor 1260 at a faster rate than the Hudson River inoculum and showed a preferential removal of chlorine atoms at the meta position. Likewise, the Hudson River inoculum dechlorinated Aroclor 1242 more rapidly than the Silver Lake inoculum. The rate and extent of dechlorination was observed to be inversely proportional to the degree of chlorination of the Aroclor for each inoculum. Results from this study suggest that different PCB-dechlorinating microorganisms occur at different sites.

Field-Scale Studies

In Situ and Bioreactor Treatment of Sediments from the Sheboygan and Hudson Rivers

Bioremediation in freshwater environments is currently the focus of a 5-year study and demonstration program administered by the U.S. Environmental Protection Agency's Great Lakes National Program Office as a part of the Assessment and Remediation of Contaminated Sediments (ARCS) Program. The program, established by Section 118 of the 1987 Amendments to the U.S. Clean Water Act, currently centers on five sites in the Great Lakes. Bioremediation has been selected as one of the treatments to be used in pilot-scale demonstrations at one of the sites, Sheboygan River/Harbor, Wisconsin, which is contaminated with PCBs and heavy metals.

An on-site confined treatment facility (CTF) is being used to treat sediments containing 640 to 4500 ppm PCBs dredged from the upper part of the Sheboygan River. The CTF was built with steel sheet pilings (14,000 ft^2) lined with welded sheets of high-density polyethylene and has a total capacity of approximately 2500 yd^3 of sediment. It is divided into two equal sections to allow two treatments to occur simultaneously. Each section is partitioned into a large study cell and a small control cell, with the latter cell being about one-third the size of the study cell. Each cell has a double liner with a leachate removal/leak-detection system between the liners. Perforated pipes in each cell may be utilized for aeration, addition of nutrients, drainage, etc. A carbon adsorption system is used to ensure that aqueous discharges meet state and federal effluent standards. The four different cells are being used to determine the extent of natural PCB biodegradation under various conditions (i.e., pH, temperature, etc.). Treatment cells will receive additions (i.e., carbon sources, nutrients, etc.) and undergo periods of aerobiosis and anaerobiosis. The control cells will not receive additions or be modified in any way.

In Situ Remediation of a Contaminated Lagoon in Houston

Bioremediation has been invested as an efficient and cost-effective method to clean up a lagoon contaminated with a mixture of organic and inorganic wastes (Sloan, 1987). The site was a waste lagoon formed by a sand mining operation in Texas. It later became a licensed waste disposal facility that operated for several years. The lagoon covered 7.3 acres and contained 5 to 6 million gallons of organic waste material. The major pollutants were industrial waste oils, acidic galvanizing wastes, phenols, pickling acids, PCBs, and heavy metals. An initial survey of the site indicated that there were approximately 70,000 yd^3 of sludges and 50,000 to 70,000 yd^3 of contaminated soil. Contamination was observed in the lagoon water and down to 30 ft in soil in the immediate area, but shallow wells about one-half mile downstream from the lagoon were not contaminated.

Several bacterial strains isolated from the sludge and lagoon water were able to degrade the major organic pollutants under laboratory conditions. Data obtained from studies carried out to determine the factors that maximize bacterial degradation of the pollutants indicated that the highest level of bacterial activity was achieved by adding phosphates and nitrogen in the ratio 3:1, respectively.

A pilot-scale study was conducted at the site for 2 months using 20,000-gal capacity tanks. The tanks were fitted with aerators, mixers, and a sampling system. Although the data collected from the pilot-scale study suggested that bacterial activity was similar to that observed in the laboratory, several questions related to the field application were not answered. These questions included (1) What was the reaction rate of the entire system, and how would it be affected by other microbes in the lagoon? (2) How would the lagoon be aerated? and (3) How will any emitted gases from biodegradation be controlled? Such questions could only be answered by performing in situ demonstration in the lagoon.

A section representing 5 percent of the lagoon was segregated by a wall of interlocking steel sheets for in situ demonstration. This isolated section contained a high level of pollutants. Sampling was conducted by a 24-point grid on the sludges and the sediments. A cable-mounted sampling system which could simultaneously sample slurry and sediment at four different points was used. Samples also were taken at 20-ft intervals along the edges of the experimental site with a slurry bucket. Aeration pipes and air injectors were installed near the bottom of the lagoon on 6-ft centers. Air compressors supplied the needed oxygen through the injectors, which were connected to the main header running all the way around the shore of the experimental section. All emissions from the lagoon surface were passed through a 100-ft^2 floating hood and stack which was placed over the experimental section. The emitted gases were monitored quantitatively at the stack at regular intervals. Gas emissions also were measured at the upwind and downwind edges of the lagoon to detect any volatile materials escaping into the surrounding area. Shallow ground water wells near the contaminated site also were sampled and monitored weekly to determine changes in the contaminant levels.

Initially, there was an increase in the population of yeasts in the experimental section (pH = 6.0). After 2 weeks, the yeasts disappeared, and an increase in bacterial activity was observed. The high bacterial activity and a corresponding increase in chloride in concentration at the experimental section suggested that biodegradation of the organic waste materials was taking place. The accumulated chloride ions caused a pH drop to approximately 4.0 in the experimental section; since the optimal pH for bacterial activity was 7, 100 lb of lime was added daily to maintain the pH at this level. The anaerobic microbiota at the bottom of the lagoon were killed, and subsequently, a 10- to 12.5-cm-thick biomass accumulated on the surface. This biomass persisted for the first 5 days of the test.

Bioremediation of the organic materials in the experimental section occurred within 6 months (Sloan, 1987). Based on results from this demonstration, the lagoon was divided into sections for bioremediation. This method of cleanup was suitable for this site because of the nature of the waste and the presence of acclimated indigenous microorganisms.

Bioreactor Treatment of Extracted Contaminants from a Lagoon

Biological treatment is being used to remediate organic and inorganic extracts and effluents from contaminated sludge and leachate in a lagoon. The lagoon was created by disposing sludges from a chemical manufacturing plant into an unlined surface impoundment (Kosson and Ahlert, 1988). The resulting 4.1-acre lagoon contained 23,000 m³ of primary and secondary sludges. The primary sludge was made up of lime-neutralized inorganic matter, spent catalysts, and solid residues from various chemical manufacturing processes. The secondary sludge contained biomass from aerobic treatment of aqueous effluent from the same manufacturing activities.

Preliminary studies showed that leachate from the lagoon could potentially affect local ground water resources. After excavating sludges from the lagoon, a bench-scale study was conducted to determine the most efficient method to clean the lagoon. This study included extraction of sludge samples followed by biological treatment with mixed aerobic bacteria. A sequential aerobic-anaerobic soil-based microbial treatment of leachate from the lagoon also was carried out. Data collected from the bench-scale study led to the design of a pilot plant. The plant used in situ extraction of sludge deposits with aqueous sodium hydroxide solution. Recovered extracts were then transferred into a tank containing mixed aerobic mircoorganisms. The pilot-scale plant was operated for 140 days, and design criteria were then developed for full-scale cleanup of the lagoon.

The full-scale system consisted of several sequential steps. The first step involved extracting with 0.05 N sodium hydroxide solution by injecting it into the sludge through a series of well points. The resulting extract, containing organic contaminants from the sludge, was recovered and accumulated in an extract tank. After carbonation with commercial-grade carbon dioxide to adjust the pH to between 7 and 7.5, the extract was diluted as necessary for total organic carbon concentration. The nutrients, calcium nitrate and potassium biphosphate, were added. Extracts were applied to the surface of treatment beds and allowed to percolate through the soil system. Each treatment bed consisted of a 12 × 37 × 1.5 m deep trench lined with two layers of PVC liners. Drainage was provided by embedding drainage pipes, which were connected to a central collection pump, between the liners and above the upper liner. Each treatment bed, with a capacity of 5700 liters/day, was filled with local top soil, sand, and granular activated carbon. The treatment beds were seeded with activated sludge at the beginning of each operating season. Treatment bed gases consisting mainly of methane and nitrogen were collected and flared intermittently. Based on initial design estimates, bioremediation of the lagoon will require 5 years of operation.

In Situ Treatment in a
Marine System

Bioremediation in marine environments is exemplified by the recent cleanup of Alaskan beaches contaminated with crude oil released by the *Exxon Valdez* in March of 1989. This effort was the largest successful bioremediation project to date. Within weeks of the oil spill, the U.S. EPA recognized that the degradation of oil on the beaches would be limited by available nutrients. Soon after, a field study was initiated to examine and test the effects of adding nitrogen- and phosphorous-based nutrients to enhance the degradation of oil on contaminated beaches.

Fertilizers were selected for particular application strategies. Three application strategies were identified: slow-release formulations that could be distributed over the site, oleophilic fertilizers that dissolved in the oil, and fertilizer dissolved in water and repeatedly applied to the contaminated site. An encapsulated fertilizer formulation, Customblen, produced by Sierra Chemicals (Milpitas, Calif.), was selected as the slow-release fertilizer. The oleophilic fertilizer, Inipol EAP 22, a product of Elf Aquataine Company (Atrix, France), was selected as the oil-soluble fertilizer. The water-soluble fertilizer was prepared with ammonium nitrate, triple phosphate, and seawater.

Approximately 2 to 3 weeks following application of the oleophilic fertilizer, a visible reduction in the oil content could be observed on the cobblestone surfaces. Differences between treated and untreated portions of the beach were dramatic. Subsurface oil persisted for several more weeks but subsequently also disappeared. There was little visual change in untreated control areas.

In a second series of experiments, a combination of oleophilic and slow-release fertilizer was tested along with a water-soluble fertilizer treatment. Both treatments lead to similar visual results to those observed with the first study with oleophilic fertilizer alone. Analysis of total oil indicated, however, that the soluble-fertilizer treatment was the more successful. The enhanced biodegradation rate was two- to threefold.

References

Alexander, M. 1981. Biodegradation of chemicals of environmental concern. *Science* **211**:132–138.

Bedard, D. L., L. M. Haberl, R. J. May, and M. J. Brennan. 1987. Evidence for novel mechanisms of polychlorinated biphenyl metabolism in *Alcaligenes eutrophus* H850. *Appl. Environ. Microbiol.* **53**:1103–1112.

Bradford, M. L., and R. Krishnamoorthy. 1991. Consider bioremediation for waste site cleanup. *Chem. Eng. Prog.* **87**:80–85.

Dasappa, S. M., and R. C. Loehr. 1991. Toxicity reduction in contaminated soil bioremediation processes. *Water Res.* **25**:1121–1130.

Furukawa, K., N. Tomizuka, and A. Kamibayashi. 1979. Effect of chlorine substitution on the bacterial metabolism of various polychlorinated biphenyls. *Appl. Environ. Microbiol.* **38**:301–310.

Grubbs, R. B. 1986. Enhanced Biodegradation of Aliphatic and Aromatic Hydrocarbons Through Bioaugmentation. Report from the 4th Annual Hazardous Materials Management Conference, Atlantic City, N.J.

Kosson, D. S., and R. C. Ahlert. 1988. Design criteria for in-situ and on-site renovation of an industrial sludge lagoon. *Haz. Waste Haz. Mater.* **5**:31–52.

Lewis, D. L., L. F. Freeman, III, and M. E. Watwood. 1986. Seasonal effects on microbial transformation rate of an herbicide in a freshwater stream: Application of laboratory data to a field site. *Environ. Toxicol. Chem.* **5**:791–796.

McGinnis, D. G. 1988. *Characterization and Laboratory Soil Treatability Studies for Creosote and Pentachlorophenol Sludges and Contaminated Soil.* U.S. EPA Report No. EPA/600/2–88/055.

Nies, L., and T. M. Vogel. 1990. Effects of organic substrates on dechlorination of Arochlor 1242 in anaerobic sediments. *Appl. Environ. Microbiol.* **56**:2612–2617.

Portier, R. J. 1985. Comparison of environmental effect and biotransformation of toxicants on laboratory microcosm and field microbial communities. In T. P. Boyle (ed.), *Validation and Predictability of Laboratory Methods for Assessing the Fate and Effects of Contaminants in Aquatic Ecosystems.* ASTM STP 865. American Society for Testing and Materials, Philadelphia, pp. 14–30.

Portier, R. J., H. M. Chen, and S. P. Meyers. 1983. Environmental effect and fate of selected phenols in aquatic ecosystems using microcosm approaches. *Dev. Ind. Microbiol.* **24**:409–424.

Quensen, J. F., S. F. Boyd, and T. M. Tiedje. 1990. Dechlorination of four commercial polychlorinated biphenyl mixture (Aroclors) by anaerobic microorganisms from sediments. *Appl. Environ. Microbiol.* **56**:2360–2369.

Shimp, R. J. 1989. Adaptation to a quaternary ammonium surfactant in aquatic sediment microcosms. *Environ. Toxicol. Chem.* **8**:201–208.

Shimp, R. J., and B. S. Schwab. 1991. Use of a flow-through in situ environmental chamber to study microbial adaptation processes in riverine sediments and periphyton. *Environ. Toxicol. Chem.* **10**:159–167.

Sloan, R. 1987. Bioremediation demonstrated at hazardous waste site. *Oil Gas J.* **85**:61–66.

Spain. J. C., P. H. Pritchard, and A. W. Bourquin. 1980. Effects of adaptation on biodegradation rates in sediment/water cores from estuarine and freshwater environments. *Appl. Environ. Microbiol.* **40**:726–734.

Spain, J. C., and P. A. Van Veld. 1983. Adaptation of natural microbial communities to degradation of xenobiotic compounds: effects of concentration, exposure time, inoculum, and chemical structure. *Appl. Environ. Microbiol.* **45**:428–435.

Spain, J. C., P. A. Van Veld, C. A. Monti, P. H. Pritchard, and C. R. Cripe. 1984. Comparison of *p*-nitrophenol biodegradation in field and laboratory test systems. *Appl. Environ. Microbiol.* **48**:944–950.

Wiggins, B. A., S. H. Jones, and M. Alexander. 1987. Explanations for the acclimation period preceding the mineralization of organic chemicals in aquatic environments. *Appl. Environ. Microbiol.* **53**:791–796.

9

Anoxic/Anaerobic Bioremediation

Steven N. Liss

*School of Chemical Engineering, Ryerson
Polytechnic University, Toronto, Ontario,
Canada*

Katherine H. Baker

*Environmental Microbiology Associates,
Inc., Harrisburg, Pa.*

Biological methods for the reclamation of contaminated soils and aquifers and for the treatment of oil spills have been based largely on aerobic processes. The prevalence of aerobic technologies is related to the historical observations that the initial steps in the biodegradation of hydrocarbons by microorganisms, including bacteria and fungi, involves oxidation of the substrates by oxygenases (Rehm and Reiff, 1981; Boulton and Ratledge, 1984; Dagley, 1984; Moat and Foster, 1988) and by the recognition of oxygen as a limiting factor in many natural environments (Alexander, 1980; Lee and Ward, 1985, Wilson and Ward, 1987; Lee et al., 1988; Baker and Herson, 1990; Leahy and Colwell, 1990). Notwithstanding the growing evidence that anoxic/anaerobic biodegradation of aromatic and nonaromatic hydrocarbons by microorganisms also occurs (Aeckersberg et al., 1991; Kuhn et al., 1985; Vogel and Grbic-Galic, 1986; Zeyer et al., 1986), the rates of degradation have been considered to be low and of minor ecologic significance (Ratledge, 1978; Atlas, 1991; Horowitz et al., 1982; Schinck, 1985; Aeckersberg et al., 1991).

Studies of the microbial ecology (e.g., community structure, diversity, numbers, and activity) of soil and subsurface habitats have largely focused on aerobic microorganisms (Ehrlich et al., 1983; Ghiorse and Balkwill, 1983; White et al., 1983; Wilson and McNabb, 1983; Harvey et al., 1984; Federle et al., 1986; Smith et al., 1986; Stetzenbach et al., 1986). Many pure cultures of aerobic bacterial strains capable of degrading pollutants have been isolated from these habitats and extensively characterized. Furthermore, both uncontaminated and contaminated subsurface habitats have been found to contain an active aerobic microflora. Where anoxic/anaerobic bacteria and their activities have been identified, these have represented a small proportion of the total microflora (Balkwill and Ghiorse, 1985; Ward, 1985). With respect to the significance of anoxic/anaerobic activities in situ, it is difficult to extrapolate from laboratory studies to field conditions (Kuhn et al., 1985). For example, some groups of denitrifying bacteria are facultative in their relationship to oxygen. Therefore, the presence of these organisms does not conclusively demonstrate if a given environment is anoxic or aerobic (Murray et al., 1990).

There have been only a few examples of pure cultures of anoxic/anaerobic bacteria which degrade recalcitrant compounds. Most of these degrade alkyl halides (Williams and Evans, 1975; Taylor et al., 1979; Shelton and Tiedje, 1984b; Schennen et al., 1985; Blake and Hegeman, 1987; Egli et al., 1987, 1988; Galli and McCarty, 1989; Criddle et al., 1990a, 1990b; Kamal and Wyndham, 1990) Rather, there is a great deal of specialization and interspecies interaction in anoxic/anaerobic habitats. Very few microorganisms degrade a single compound completely. Degradation of organic compounds in these environments is often the result of a community of interacting microbial populations, generally termed a *consortium* (Kaiser and Hanselmann, 1982; Zhang and Wiegel, 1990, Colberg, 1990). The roles of the groups of bacteria that make up anoxic/anaerobic consortia important in biodegradation are not fully understood (Horowitz et al., 1983; Sharak Genthner et al., 1989a, 1989b; Suflita et al., 1982; Tiedje et al., 1987; Urbanek et al., 1989; Bisaillon et al., 1991; Zhang and Wiegel, 1990). For example, syntrophic relationships are common in anoxic/anaerobic consortia. In syntrophy, one (or more) of the members of the consortium produces one or several metabolites necessary to the other members of the consortium. Syntrophy is the basis of interspecies hydrogen transfer, where hydrogen produced by the fermentation of organic compounds by one member of the consortium is used by anaerobic hydrogen-utilizing microorganisms such as methanogens, acetogens, or sulfate-reducers (Ehrlich, 1985). In many instances, the hydrogen-producing organism(s) depend on the activities of the hydrogen-utilizing microorganism(s) because they cannot continue to metabolize the organic substrate in the presence of hydrogen. The metabolic pathways and hence the end products produced also may vary depending on the presence or absence of hydrogen. Syntrophic relationships in anaerobic environments are not limited to interspecies hydrogen transfer but also may involve organic metabolic products such as acetate (Ehrlich, 1985).

The conditions (nutritional and process requirements) for optimal activities of such consortia are complex and difficult to elucidate. As a result, anoxic/anaerobic bioremediation has received little attention in the past. Recently, however, anoxic/anaerobic-based bioremediation has been shown to have considerable potential for remediation of soils and subsurface environments. Anoxic/anaerobic bioremediation may be particularly desirable in the latter case because many of these sites are already anoxic/anaerobic and the effectiveness and cost of oxygen delivery may be formidable.

The Anoxic/Anaerobic Environment

Anoxic/anaerobic bioremediation generally involves processes that do not use oxygen as the terminal electron acceptor (i.e., bioremediation without oxygen). The terms *anoxic* and *anaerobic* can be confusing. These terms are often used interchangeably and sometimes inappropriately in descriptions of processes because they relate to entirely different sets of environmental conditions. Both the level of oxygen present and the redox potential of the environment need to be considered in distinguishing between these conditions. The term *anoxic* generally means without oxygen and with nitrite or nitrate present but can involve conditions where the levels of dissolved oxygen become limiting in an aerobic process. The term *anaerobic*, on the other hand, from a microbiological perspective, implies an oxygen-free and reducing environment such as that typically associated with fermentations or achieved in culture media through the use of chemical reducing agents. Methanogenic and sulfate-reducing habitats are examples of highly reducing anaerobic environments. Based on the use of the terms *anoxic* and *anaerobic* in the literature, describing either field or laboratory bioremediation experiments, the term *anoxic* describes conditions that do not involve oxygen achieved by the addition of alternative electron acceptors such as nitrate or stimulation of redox pairs where the E'_0 is more than 0 V (Table 9-1), whereas *anaerobic* conditions involve redox pairs where E'_0 is less than 0 V. The anoxic redox pairs generally tend to involve both facultative anaerobic microorganisms and obligate anaerobes, whereas the anaerobic redox pairs generally involve obligate anaerobes, although the line of demarcation between the two conditions is not always clear.

Potential of Anoxic/Anaerobic Bioremediation

Recent interest in anoxic/anaerobic bioremediation results from several discoveries, including recognition that compounds which are highly substituted with halides (more oxidized) are more likely to undergo reductive dehalogenation under anoxic/anaerobic conditions, forming less halogenated intermediates. Compounds such as carbon tetrachloride (CT), tetrachloroethene (PCE) (Vogel et al., 1987; Semprini et al., 1992), and highly chlorinated PCBs (i.e., Arochlor 1260) (Brown et al., 1987) persist under aerobic conditions. Nonhalogenated compounds

Table 9-1. Redox Pairs Important in Anoxic/Anaerobic Environments

Redox pair	P_e	Eh (volts)	Microbiological process	Soil type
CO_2/CH_2	− 7.3	− 0.42	Methanogenesis	Highly reduced
CO_2/acetate	− 4.7	− 0.28	Methanogenesis/ fermentation	Highly reduced
SO_4^{2-}/H_2S	− 3.7	− 0.22	Sulfate reduction	Highly reduced
NO_3^-/NO_2^-	+ 7.11	+ 0.42	Nitrate reduction	Moderately reduced
NO_3^-/N_2	+ 12.4	+ 0.74	Denitrification	Anoxic/oxidized
Fe^{3+}/Fe^{2+}	+ 12.9	+ 0.76	Iron reduction	Anoxic/oxidized
O_2/H_2O	+ 13.9	+ 0.82	Aerobic respiration	Oxidized

such as nitroaromatic and aminoaromatic compounds, which include herbicides and hazardous energetic organonitro compounds, also may persist under aerobic conditions and only decompose under anoxic/anaerobic conditions (Kaglan, 1990). Finally, reductive dehalogenation in anoxic/anaerobic environments has been found to be an important mechanism in the fate of organochlorine pesticides (Kuhn and Suflita, 1989b).

The disadvantages associated with aerobic technologies and the potential advantages of anoxic/anaerobic methods are well known (USEPA, 1979; Lee et al., 1988; Major et al., 1988). The delivery of oxygen may involve considerable cost. Hydrogen peroxide, an alternative source of molecular oxygen, decomposes rapidly, resulting in poor oxygen distribution. Furthermore, the formation of hydroxyl free radicals from the reduction of peroxide may cause oxidation of cellular components and lead to death of the biodegrading bacteria. Both oxygen and peroxide can lead to oxidation reactions that can precipitate oxides, which may decrease aquifer permeability.

Depending on the oxygen availability in soils, sediments, and aquifers, anoxic/anaerobic conditions may prevail, and the appropriate conditions for anoxic/anaerobic bioremediation may already exist. Many sites will have already become anoxic/anaerobic because microorganisms will have depleted the dissolved oxygen due to the infiltration of water from landfills (Ehrlich et al., 1983; Reinhard et al. 1984) or contaminated rivers (Piet and Smeenk, 1985; Schwarzenbach et al., 1983) and the subsequent aerobic microbial mineralization of the organic constituents of water. Degradation of high concentrations of organic compounds may lead to a rapid depletion of oxygen and to an eventual decrease in the redox potential in aquifers (Thauer et al., 1977; Ehrlich et al., 1983; Wilson and McNabb, 1983; Bouwer and McCarty, 1984; Zeyer et al., 1986; Lyngkilde et al., 1991). Significant levels of nitrate and sulfate also may be present in ground water which will support denitrifying and sulfate-reducing activities and may promote anoxic/anaerobic biodegradation.

The use of nitrate as an alternative electron acceptor may offer advantages over oxygen in situations where bioremediation is to be promoted under denitrifying conditions, particularly for aromatic hydrocarbons (Bouwer and McCarty, 1983; Kuhn et al., 1985; Zeyer et al., 1986; Major et al., 1988; Hutchins et al., 1991a). Unlike oxygen, nitrate has a high solubility in water and thus will not be retarded and will mix readily into the contaminated ground water. Consequently, single applications of nitrate may be appropriate.

Anoxic/anaerobic processes require less inorganic nutrient supplementation because less energy and, therefore, less biomass are produced. For these reasons, anoxic/anaerobic bioremediation has the potential to overcome some of the limitations associated with traditional biorestoration technologies. Anoxic/anaerobic processes could potentially offer the following advantages:

1. Prevention of biofilm buildup
2. Prevention of fouling of nearby surface waters with excess nutrients
3. Lower maintenance and labor requirements than aerobic systems
4. May be more appropriate where there is low permeability to permit proper circulation of nutrients

In soil and subsurface environments, diffusional constraints promote the physical establishment of a greater diversity of microorganisms, resulting in sequential rate processes (e.g., nutrient concentration gradients, oxygen levels, redox) (Focht, 1992) that may be more favorable for overall biodegradation and remediation. Mineralization of contaminants often depends on the activity of a succession of microbial communities, each requiring different environmental conditions. For example, a large distribution in specificity in reductive dehalogenation pathways is seen in anoxic/anaerobic consortia obtained in individual investigations, and many highly chlorinated organic compounds are dehalogenated sequentially, indicating that these activities are carried out by distinct groups of bacteria (Mohn and Tiedje, 1992). The suggestion that populations that carry out reductive dehalogenation of chlorinated aromatic compounds are rare in nature is in part supported by the fact that only one pure culture of a strict anaerobe that is capable of reductive dehalogenation has been studied in detail (Shelton and Tiedje, 1984b; Mohn and Tiedje, 1992). However, energy may be gained from dehalogenating activity (Dolfing, 1990), and this is supported by Colberg (1990) and Dolfing and Harrison (1992), who have determined the Gibbs free energy of formation for a variety of halogenated aromatic compounds. Although free energies of formation do not allow for a prediction of the kinetics of biodegradation, the potential to conserve energy for growth by directing electrons (H_2) to a wide variety of halogenated aromatic compounds as electron acceptors under anoxic/anaerobic conditions suggests that this process may be common. Aerobic microorganisms and their enzymes also may participate in these processes under appropriate redox conditions (Vogel et al., 1987; Kastner, 1991). Steiert and Crawford (1986) have reported that a *Flavobacterium* species which mineralizes PCP is capable of reductive dechlorination in addition to hydrolytic dehalogenation. Van den Tweel et al. (1987) also reported reductive dehalogenation under aerobic conditions involved in degradation of 2,4-dichlorobenzoate by *Alcaligenes denitrificans*. The cytoplasm of these cells may have a low oxidation-reduction potential which could support reductive reactions.

Anoxic/Anaerobic Processes Important to Bioremediation

Anaerobic Respiration

Anaerobic metabolism of organic compounds by chemotrophic microorganisms involves one of two modes of metabolism: anaerobic respiration and fermentation. *Anaerobic respiration* involves the coupling of the oxidation of organic compounds with the transfer of electrons to organic or inorganic electron acceptors other than molecular oxygen. Anaerobic respiration, involving the dissimulative utilization of inorganic compounds including nitrate, sulfate, carbonate, ferric iron, and manganese oxide in energy metabolism, is environmentally significant (e.g., agricultural soils and ground water, marine sediments) and can be carried out by a variety of bacteria in anoxic and anaerobic habitats. The energy yields from anaerobic respiration are lower than those from aerobic metabolism.

In addition to their availability, the terminal electron acceptor used, and hence the microbial community which develops, will depend to a large extent on the redox conditions of the environment. The half-reaction reduction potentials of the electron acceptors can be ordered from most oxidized to most reduced (see Table 9-1). Each of these conditions may support entirely different pathways (Criddle et al., 1990b) or affect the rate of reductive transformations (Bouwer and Wright, 1988). Criddle et al. (1990b) found that the rate of reaction and the pathway involved in the metabolism of carbon tetrachloride by *Escherichia coli* K12 was dependent on the electron acceptor condition of the medium. Carbon tetrachloride was not metabolized in the presence of oxygen and nitrate. When the concentration of oxygen was reduced to approximately 1%, there was some formation of carbon dioxide from carbon tetrachloride. Under conditions of fumarate respiration, however, carbon tetrachloride was converted to carbon dioxide, chloroform, and a nonvolatile fraction.

A number of organic electron donors have been shown to support anoxic/anaerobic processes, including simple organic acids and alcohols (Gibson and Sewell, 1992; Freedman and Gosset, 1989), hydrogen (DiStefano et al., 1992), water (Nies and Vogel, 1991), toluene (Sewell and Gibson, 1991), and dichloromethane (DCM) (DiStefano et al., 1992). The oxygen requirement for anoxic/anaerobic oxidation reactions is provided by water, as in the case of aromatic hydrocarbon metabolism (Vogel and Grbic-Galic, 1986).

The denitrifiers, sulfate-reducers, and methanogens may not necessarily be involved directly in the transformation of organic pollutants under anoxic or anaerobic conditions, but they can support conditions where the flow of electrons may be diverted to these compounds during their reductive transformation by other microorganisms.

Fermentation

Fermentation involves ATP formation by substrate-level phosphorylation as a result of oxidizing one organic compound and reducing another (Moat and Foster, 1988). The products of fermentation, which include simple organic acids and alcohols, may provide a source of reducing power and protons for reductive metabolism in anoxic/anaerobic bioremediation (Gibson and Sewell, 1992; Criddle et al., 1990a,b).

Degradation of Xenobiotic Compounds

Aromatic Compounds. Tarvin and Buswell (1934) provided the first evidence that aromatic compounds could be degraded under anoxic/anaerobic conditions. Since then, degradation of benzoates and phenols has been described under photosynthetic (anoxygenic), denitrifying, sulfate-reducing, and methanogenic conditions (Evans and Fuchs, 1988). Initially, the aromatic ring is reduced and becomes fully saturated. This destabilizes the ring, which subsequently undergoes fission, yielding a saturated fatty acid or dicarboxylic acid (Fig. 9-1). Degradation of phenols can occur via several initial transformation path-

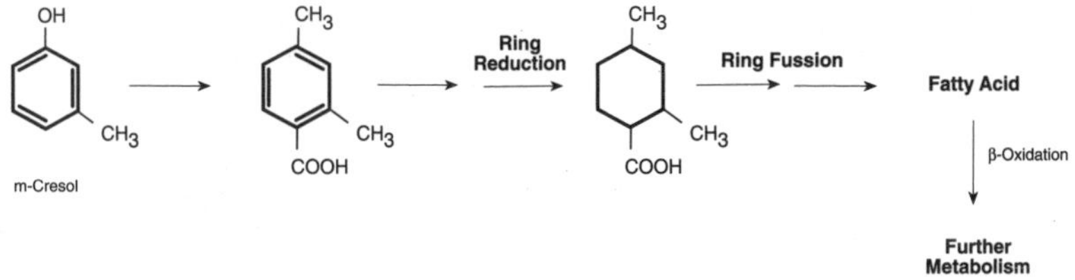

Figure 9-1. Anoxic/anaerobic degradation of benzoates and phenols. The initial reactions in the pathway result in reduction of the aromatic nucleus, which becomes fully saturated. Fission of the aromatic ring yields a saturated fatty acid or dicarboxylic acid which feeds into intermediary metabolic pathways.

ways. Introduction of a carboxyl group onto the aromatic ring appears to facilitate further anaerobic degradation (Kobayashi et al., 1989; Tschech and Fuchs, 1987). The carboxylation occurs at the para position relative to the phenolic hydroxyl group, yielding *p*-hydroxybenzoate (Gallert et al., 1991; Zhang et al., 1990). Alternatively, the initial attack on phenol can be through reduction of the phenol to cyclohexanone, followed by alicyclic ring cleavage (Schink and Tschech, 1988). Aromatic compounds possessing substitutes such as arylmethyl groups (e.g., cresols and toluene) undergo an initial oxidation reaction to carboxyl groups prior to ring cleavage (Roberts et al., 1990; Kuhn et al., 1988). The oxygen for these reactions is derived from water, not molecular oxygen.

In methanogenic consortia, carboxylation of phenol occurs even when methanogenesis is inhibited, indicating that a nonsyntrophic population is involved in the carboxylation of the aromatic compounds (Bechard et al., 1990; Kobayashi et al., 1989). Benzene and *o*-xylene are not considered to be readily degraded under anoxic or anaerobic conditions; however, there are conflicting reports in the literature regarding their susceptibility to anoxic and anaerobic biodegradation. Generally, aromatic compounds with carboxyl and/or hydroxyl substitutions are more susceptible to anoxic/anaerobic degradation.

Highly substituted methyl phenols [e.g., 2,4-dimethylphenol (2,4-DMP) and 3,4-dimethylphenol (3,4-DMP)] and polycyclic aromatic hydrocarbons (PAHs) have been considered to be resistant to degradation under anoxic and anaerobic conditions. Flyvbjerg et al. (1991) have reported that 2,4-DMP and 3,4-DMP in addition to toluene, phenol, and cresols (*o*-, *m*-, and *p*-) were degraded by bacteria in creosote-contaminated ground water under anoxic (denitrifying) conditions. Benzene, xylene, naphthalene, 2,3-DMP, 2,5-DMP, and 3,5-DMP were not degraded. The degradation of naphthalene and acenaphthene has been shown to occur under denitrifying conditions but not under anaerobic conditions (Mihelic and Luthy, 1988). Hydroxyl substitution may render PAHs susceptible to anaerobic degradation as indicated by the degradation of naphthol under sulfate-reducing and methanogenic conditions.

Heterocyclic Compounds.   The methanogenic degradation of heterocyclic aromatic compounds that contain nitrogen (e.g., quinoline) or sulfur (e.g., benzothiophene) in their ring structures proceeds by an initial oxidation of either the homocyclic or heterocyclic ring. This is then followed by ring cleavage and a variety of reactions including oxidations and decarboxylation leading to cleavage of the remaining ring, now either a phenol or a benzoate. The initial reaction occurs rapidly and has been found to require energy (Godsy et al., 1992). The oxidized products may remain for extended periods of time before they are converted to methane and carbon dioxide. Godsy et al. (1992) have suggested that the electrons derived from this reaction are coupled with the reduction of other aromatic compounds.

Halogenated Compounds.   Reductive dehalogenation is the primary mechanism involved in the transformation of chlorinated organic compounds under anoxic or anaerobic conditions. The microbiology and biochemistry of this process have been reviewed extensively elsewhere (Mohn and Tiedje, 1992; Chaudry and Chapalamadugu, 1991; Abramowicz, 1990; Boyle, 1989; Kuhn and Suflita, 1989a; Reineke and Knackmuss, 1988; Suflita et al., 1988; Tiedje et al., 1987; Vogel et al., 1987). In this process, the halogen atoms are sequentially removed from the molecule and replaced with hydrogen (Fig. 9-2). In this type of reaction, the halogenated hydrocarbon is used as an electron acceptor, not as a source of carbon. However, this may be difficult to prove, since electron donors are added to the medium in most experiments. If the electron donor supports reductive dehalogenation and is used as a carbon substrate, it is hard to demonstrate with certainty that energy is derived from dehalogenation of chlorinated organics. However, recent studies indicate that energy may in fact be gained from dechlorination and degradation of chlorinated compounds (Freedman and Gosset, 1991; Dolfing, 1990).

Chlorinated Aliphatic Compounds.   Reductive biotransformation of chlorinated aliphatic compounds (CACs) in the absence of oxygen was

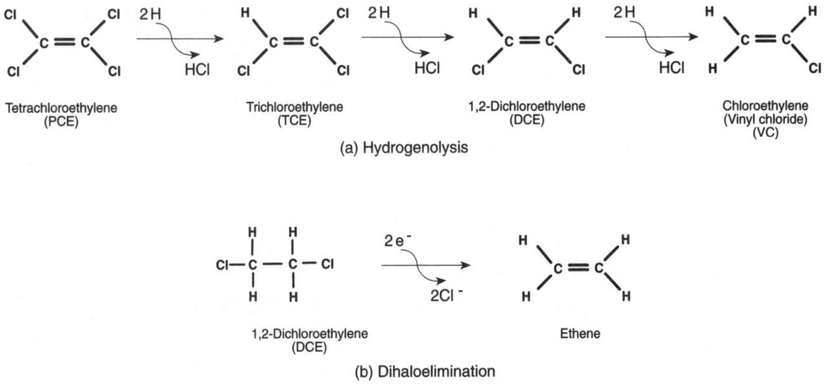

(a) Hydrogenolysis

(b) Dihaloelimination

Figure 9-2. Reductive dehalogenation. Two types of reductive dehalogenation reactions are catalyzed by anoxic/anaerobic bacteria. (*a*) Hydrogenolysis results in the direct substitution of a hydrogen atom for a halogen atom. This type of dehalogenation reaction is the most common. (*b*) Dihaloelimination, or vicinal dehalogenation, can occur when the parent compound has halogen substitutes on two adjacent carbons (vicinal carbons). The resulting compound has two fewer halogens and one additional carbon-carbon double bond.

first reported by Bouwer et al. (1981). Carbon tetrachloride (CT) transformation under denitrifying conditions was demonstrated by Bouwer and McCarty (1983). Bouwer and Wright (1988) studied the degradation of CT and other halogenated compounds under denitrifying, sulfatereducing, and methanogenic conditions. All of these compounds were transformed under methanogenic conditions. In this study, nitratereducing conditions supported the transformation of bromoform, bromodichloromethane, CT, and hexachloroethane. Rittmann et al. (1988) found CT to be completely transformed under denitrifying conditions, whereas bromoform, dibromomethane, trichloroethylene (TCE), and tetrachloroethylene (PCE) were transformed to a lesser degree. Fluorotrichloromethane (Freon 11) and 1,1,2-trichloro-1,2,2-trifluorethane (Freon 113) also have been shown to be transformed under anoxic conditions (Semprini et al., 1991).

CACs can be transformed under a variety of environmental conditions (Vogel et al., 1987). The susceptibility to degradation and the extent of degradation depend on a number of factors, as indicated by the studies described above. Compounds that are more highly substituted with halides (i.e., more oxidized) are more likely to undergo reductive dehalogenation under anoxic/anaerobic conditions. Halogenated compounds which are saturated (alkanes) undergo dehalogenation more readily than unsaturated compounds (alkenes and alkynes). Chlorine-carbon bonds have an intermediate strength. They are easier to remove than fluorine but harder than bromine or iodine. The transformation products of reductive dehalogenation are less reactive to subsequent reduction under anoxic/anaerobic conditions but subsequently become easier to degrade aerobically.

Egli et al. (1987, 1988) found pure cultures that were able to transform CT to chloroform (CF) and DCM quantitatively under sulfate-reducing and methanogenic conditions. Galli and McCarty (1989) identified a *Clostridium* species from the effluent of an anaerobic reactor that transformed trichloroethane (TCA), CF, and CT. Criddle et al. (1990a) isolated a denitrifying strain of *Pseudomonas* (strain KC) that degraded CT, 50 percent to carbon dioxide and approximately 40 percent to nonvolatile compounds. Criddle et al. (1990b) also found CT to be transformed by *E. coli* under anoxic (nitrate and fumarate respiration) and anaerobic (fermenting) conditions. The extent to which mineralization of CT to carbon dioxide takes place varies from 10 to 99 percent. Criddle (1989) proposed that several parallel pathways exist for the degradation of CT under anoxic/anaerobic conditions leading to a variety of end products, including CF, DCM, carbon dioxide, carbon monoxide, formic acid, and hexachloroethane.

TCE and PCE undergo transformation most readily under anoxic/anaerobic conditions. There is, however, enormous variability in the extent of degradation found between studies (DiStefano et al., 1992). PCE has been observed to degrade to TCE (Fathepure and Boyd, 1988), DCE (Sewell and Gibson, 1991), and vinyl chloride (Vogel and McCarty, 1985). Since vinyl chloride (VC) is carcinogenic, accumulation of this compound is problematic. It is possible that PCE can be transformed to ethene (ETH) without VC accumulation. Several field and laboratory studies have demonstrated the complete transformation of PCE to ETH (DiStefano et al., 1991; Freedman and Gossett, 1989; Major et al., 1991).

DCM has been shown to be utilized as the sole carbon and energy source under methanogenic conditions (Freedman and Gosset, 1991). Unlike other chlorinated aliphatics, which undergo reductive dechlorination, DCM is transformed under methanogenic conditions by oxidation to carbon dioxide and fermentation to acetate. Anaerobic degradation of DCM does not require the supply of an external electron donor and yields innocuous end products, unlike many of the other CACs.

Chlorinated Aromatic Compounds. Reductive aryl dehalogenation is the initial step in the degradation of chlorinated aromatic compounds in anoxic/anaerobic environments (see Fig. 9-3). Chlorophenols and

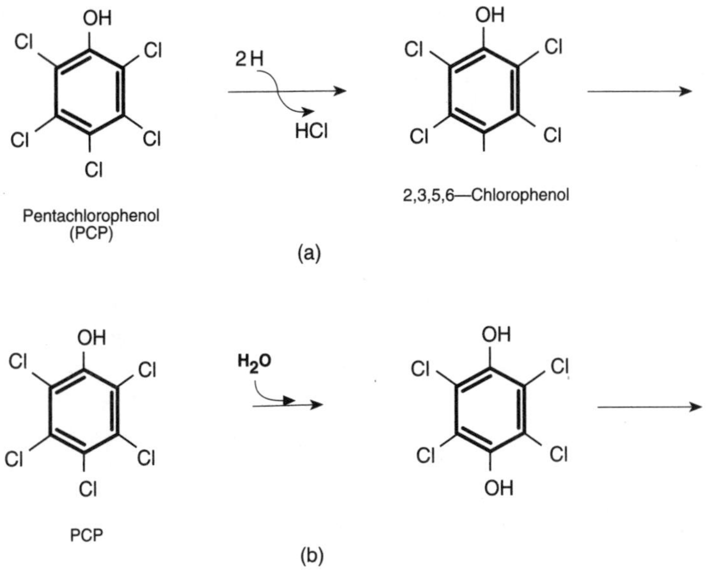

Figure 9-3. Reductive aryl dehalogenation. (*a*) Dehalogenation of aromatic compounds under anoxic/anaerobic conditions may occur via a process analogous to reductive dehalogenation of aliphatic compounds. There is a decided stereospecificity for preferential removal of halogens from compounds bearing more than one substitution. Meta-dechlorinating organisms appear to be the most widespread. (*b*) Hydrolytic dehalogenation occurs when a hydroxyl group derived from water replaces the chlorines (or other halogens) present on the parent molecule. Hydrolytic dehalogenation is frequently involved in aerobic degradation.

chlorobenzoates are dehalogenated to phenol and benzoate, respectively (Sharak-Genthner et al., 1989a). In both cases, benzoate is the final aromatic intermediate prior to ring cleavage and formation of acetate, carbon dioxide, and hydrogen (carbon dioxide and methane in the presence of methanogens). Reductive dehalogenation has been observed for several commercial PCB (Aroclor) mixtures (Quensen et al., 1990; Brown et al., 1987). Several dechlorination patterns have been observed in laboratory dechlorination studies with microbial consortia derived from various PCB-contaminated sediments which show preferences for *meta-*, *para-*, or *meta-* plus *para-*chlorine removal (Nies and Vogel, 1990; Quensen et al., 1988). These activities involve microbial consortia rather than individual strains. Meta-dechlorinating organisms appear to be more prevalent and more widely distributed among microbial groups

(Morris et al., 1992). Van Dort and Bedard (1991) have recently described the first ortho dechlorination of a PCB (2,3,5,6-CB) by an anaerobic consortium obtained from pond sediment. It is possible that this activity could be applied to the mono-(2-CB) and dichlorobiphenyls (2,2'-CB; 2,6-CB) which arise from meta and para activity in order to fully dehalogenate PCBs to biphenyl. However, the less chlorinated biphenyls are readily degraded aerobically (Abramowicz, 1990), suggesting that PCB-contaminated soils and ground water are best treated in a coupled anaerobic/aerobic process. With respect to mono and highly chlorinated aromatic compounds [e.g., pentachlorophenol (PCP)], the placement of the chlorine atoms on the molecule also has significant effects on the susceptibility of the compounds to biodegradation. Consortia have been identified which have activities where the order of the relative rates of dechlorination were ortho > meta > para, ortho > para > meta, and para > ortho > meta (Larsen et al., 1991; Bryant et al., 1991).

Reports of the anaerobic degradation of chlorophenols have mainly involved methanogenic consortia. Sulfate has been considered to be inhibitory to the anaerobic degradation of chlorophenols (Gibson and Suflita, 1986; Suflita et al., 1988). The inhibitory effect of sulfate is considered to be the result of competition for reducing equivalents between sulfate reduction and dechlorination (Colberg, 1990). Only recently has chlorophenol degradation been coupled with sulfate reduction (Häggblom and Young, 1990). Häggblom and Young (1990) reported that biomass obtained from sulfidogenic sediments and a laboratory-scale anaerobic fluidized-bed reactor for pulp mill effluent treatment could degrade 2-, 3-, and 4-chlorophenol and 2,4-dichlorophenol with sulfate as the sole electron sink. On the other hand, PCP has been shown not to be degraded under sulfate-reducing conditions (Madsen and Aamand, 1991). Madsen and Aamand (1991) have demonstrated that it may be possible that when the electron donor is present in sufficient quantities, the inhibition of dechlorination by sulfate can be alleviated.

Pure cultures of bacteria capable of reductive aryl dehalogenation have been difficult to obtain. To date, the first and best described isolate, an obligate anaerobic, *Desulfomonile tiedjei*, is capable of aryl reductive dechlorination (Shelton and Tiedje, 1984b). This organism dehalogenates a range of substrates, including benzoates, phenols, and ethenes (Mohn and Kennedy, 1992; Fathepure et al., 1987). Dehalogenation is confined to the meta position. *D. tiedjei* is able to gain energy for growth by coupling electron transport with reductive dechlorination (Mohn and Tiedje, 1992; Dolfing, 1990). More recently,

Madsen and Licht (1992) have reported the isolation of a spore-forming anaerobic chlorophenol-degrading bacterium which they have tentatively assigned to the genus *Clostridium*. This bacterium was reported to use a variety of chlorophenols, including trichlorophenols (2,4,5-TCP; 2,4,6-TCP), dichlorophenols (2,4-DCP; 3,5-DCP), and to a lesser degree pentachlorophenol. This strain differs from *D. tiedjei* in that it produces spores, does not reduce sulfate, and dehalogenates at both the ortho and meta positions.

Rhodopseudomonas palustris, a photosynthetic bacterium, is one of the few individual species isolated which has a particularly broad substrate range with respect to the anaerobic degradation of aromatic compounds (Harwood and Gibson, 1988), including phenolics and halogenated aromatic compounds (Kamal and Wyndham, 1990). An isolate of *R. palustris* was found to reductively photometabolize 3-chlorobenzoate only under anoxic/anaerobic conditions and with benzoate as a cosubstrate (Kamal and Wyndham, 1990). This strain also was able to metabolize 3-bromobenzoate. Kamal and Wyndham (1990) found other similar isolates from greenhouse pond water capable of degrading halogenated aromatics, suggesting that this activity may be prevalent in aquatic habitats. It is possible that these organisms, which may occur in illuminated surface waters and waste water treatment lagoons receiving high levels of organic carbon, may contribute to the degradation of aromatic compounds.

Determining the Feasibility of Anoxic/Anaerobic Treatment

Standardization of tests for determining the feasibility of bioremediation is highly desirable. This has not been achieved for aerobic testing methods because of the myriad of approaches, little consensus, and different conditions each site bears, requiring different testing procedures (Baker and Herson, 1990). The technical considerations for anoxic/anaerobic feasibility testing add further complications. Furthermore, with respect to the significance of anoxic/anaerobic activities in situ, it may be difficult to extrapolate from laboratory studies to field conditions (Kuhn et al., 1985; Gibson and Suflita, 1986). On the other hand, several researchers have found good agreement between microcosm studies and field observations of anoxic/anaerobic transformations of creosote components (Godsy et al., 1987), aromatic hydrocarbons (Wilson et al., 1987), and chlorinated pesticides (Lal and Saxena, 1982). The similarity between zero-order rate constants established for microcosm tests and field studies of an aromatic hydrocarbon-contaminated aquifer under anoxic conditions (Hutchins and Wilson, 1991) suggests that laboratory studies can be used to determine the feasibility of anoxic/anaerobic-based treatment.

Batch procedures employing serum bottle bioassay techniques first developed by Miller and Wolin (1974) and subsequently modified have been used for anoxic/anaerobic treatability studies (Owen et al., 1979; Shelton and Tiedje, 1984a; Cornacchio et al., 1988). The serum bottle bioassay is cost-effective when compared with continuous microcosm studies. The same assay can be used to determine the toxic effect of contaminated samples on microorganisms in a short period of time. Activity can be assessed by monitoring biogas accumulation and composition, particularly for methanogenic consortia. The aqueous fraction and solids (if present) can be sampled easily through airtight septa by syringe for subsequent analysis of compounds, intermediates, pH, etc. A commercially available respirometer (N-CON respirometer) may be modified (Cheng et al., 1992) and used to establish the biodegradation kinetics of pollutants under anoxic/anaerobic conditions. The limitations of batch-testing protocols, however, are that they are relatively short-term tests and that the testing method and results may not be indicative of in situ conditions. Since long acclimation periods, especially for anoxic/anaerobic environments (Linkfield et al., 1989), may be required to establish the potential for treatability, it may be desirable to establish simulated continuous microcosms (soil or ground water). Typically, such microcosms consist of sand-filled columns that are fed on a semicontinuous or continuous basis (Grbic-Galic, 1990). Smith and Klug (1987) have described a flow-through bioreactor that can be used to study anoxic/anaerobic biodegradation. Wilson and Noonan (1984) have reviewed many of the microcosms used in studies of ground water microorganisms.

Selection of samples from subsurface environments may influence anoxic/anaerobic treatability studies. Since many anoxic/anaerobic processes occur in regions devoid of oxygen and of lower redox potential and may involve consortia, bacteria are most likely to be found associated with solid particles. In fact, most bacteria in aquifers are considered to be associated with the aquifer solids. However, free-living ground water bacteria are capable of extensive degradation of organic contaminants (Arvin et al., 1988). Nielson et al. (1992) reported that ground water alone could be used in either aerobic or anoxic/anaerobic test systems to determine the biodegradability of aromatic hydrocarbons and chlorinated aliphatic hydrocarbons in aquifers. In this study, the extent of degradation with the ground water alone was comparable with results obtained with ground water and sediment, even though the results were not as reproducible as those obtained when sediment was used.

Many obligate anaerobic bacteria are sensitive to oxygen. Care should be taken when obtaining and transporting samples so as to minimize exposure to oxygen. Bottles or tubes completely filled to the top and sealed with tight stoppers will minimize exposure to oxygen. Any trace amounts of oxygen usually will be consumed rapidly by mixed microbial consortia. If more stringent conditions are required, the atmosphere in the container may be replaced with an oxygen-free gas such as nitrogen, or a reducing agent such as cysteine, thioglycollate, ferrous sulfide, or titanium citrate may be added (Gerhardt et al., 1981).

Although abiotic reduction of chlorinated aliphatic hydrocarbons has been observed (Vogel et al., 1987; Vogel and McCarty, 1985), the significance of this process has not been considered to be great (Mohn and Teidje, 1992). Nonetheless, abiotic processes may possibly influence the extent to which transformation of pollutants occurs when evaluating microbial activity. This may have important implications when determining the anoxic/anaerobic treatability of contaminated soils and subsurface habitats. Recently, Dunnivant and Schwarzenbach (1992) reported that natural organic matter (NOM) from a variety of sources with hydrogen sulfide as the bulk electron donor can result in significant reduction of substituted nitrobenzenes. Their results indicate that hydroquinone moieties within the NOM mediate the electron-transfer reactions. Thus the inclusion of appropriate abiotic controls is essential.

Sequential metabolism is common to anoxic/anaerobic processes, and the diversity in pathways which may be specific to individual consortia and redox conditions poses an additional level of complexity. Complete degradation of contaminants may not always take place (Vogel et al., 1987), and the various intermediates which may accumulate may pose as much of a problem as the original substrate (e.g., vinyl chloride from PCE or TCE and chloroform and possibly carbon disulfide from carbon tetrachloride). Treatability tests should include analysis for the disappearance of substrate, appearance of intermediates and end products of metabolism, and toxicity tests.

Factors Influencing Anoxic/Anaerobic Bioremediation

Adaptation/Acclimation

It is evident from studies that microbial activity in consortia obtained from contaminated habitats can be specific to congeners of compounds found only at that specific site (Bryant et al., 1991; Mikesell et al., 1991). Adaptation of microbial communities can increase the rate of conversion of contaminants and lower the long acclimation period observed for nonadapted and anoxic/anaerobic consortia. Unlike aerobic microbial communities, where the acclimation periods can be measured in hours and days, anoxic/anaerobic communities may take months to acclimate (Linkfield et al., 1989). This has important practical consequences when determining the feasibility of anoxic/anaerobic bioremediation. Linkfield et al. (1989) proposed that for sediments exposed to halogenated benzoates, the acclimation period (3 weeks to 6 months) may have been due to an induction phase with little dehalogenating activity followed by an exponential increase in activity. The acclimation period appeared to be characteristic of the chemical and the concentration. The long induction phase is possibly related to the energy that may be available to anoxic/anaerobic communities in subsurface habitats. However, other factors, which have been described for aerobic communities, also may contribute to the long acclimation periods (Wiggins et al., 1987; Lewis et al., 1986; Spain et al., 1980).

McCarty (1985) proposed an acclimation process employing a ground reactor to which substrates and nutrients could be added, a well casing reactor which operated anaerobically like a trickling filter, and the aquifer. The above-ground reactor is used to acclimate the microbial populations. The effluent from the above-ground reactor is injected into the well casing bioreactor to introduce the acclimated microbes into the aquifer or to enhance adaptation of the indigenous population to the contaminants. Once the acclimated population has developed, use of the above-ground reactors can be discontinued.

Toxic Metals and Organic Compounds

Toxic inorganic wastes, such as metals, are common contaminants in natural waters and are found in association with other pollutants.

Metals such as copper, mercury, zinc, cadmium, and chromium may be even more inhibitory to biodegradative processes than toxic organic compounds (Said and Lewis, 1991). Hydrogen sulfide and organic acids produced by anaerobic bioprocesses may precipitate with many metals. Highly reducing conditions may result in other effects. Heavy metals are more mobile under anoxic/anaerobic conditions. Metal salts can completely inhibit the transformation of chlorinated compounds, extend the acclimation period in nonadapted communities, and decrease the rate of transformation. Jones and Kong (1992) reported that sediment cultures containing higher levels of organic carbon were more resistant to the inhibitory effects of copper and cadmium salts.

High levels of aromatic compounds, when present in concentrated waste streams such as landfill leachate, may cause inhibition of anaerobic and anoxic processes. Reports of the inhibition of anaerobic bacteria, particularly methanogens, by chlorinated aromatic compounds (Patel et al., 1991) and the inhibition of denitrification by benzene and m-xylene (Bradley et al., 1992; Hutchins, 1991) emphasize the need to determine the inhibitory effects of samples from contaminated sites on microbial activity as part of a feasibility study.

Electron Donors

Electron donors provide a source of electrons required for reductive metabolism. A number of organic electron donors have been described, including simple organic acids and alcohols (Gibson and Sewell, 1992; Freedman and Gosset, 1989), hydrogen (DiStefano et al., 1992), water (Nies and Vogel, 1991), toluene (Sewell and Gibson, 1991), and dichloromethane (DCM) (DiStefano et al., 1992). Provision of an electron donor also may be important because it may serve as a primary substrate supporting microbial growth of one group of microorganisms which may satisfy nutritional requirements of a second biodegradative population (DiStefano et al., 1992). For this reason, the use of hydrogen may not be practical. Hughes and Parkin (1991) reported that for areas which may contain high concentrations of CACs, provision of adequate levels of an electron donor may be necessary to achieve sufficient microbial growth and complete biodegradation. The level of electron donors may be underestimated, since most laboratory studies involve contaminant levels of 1 mg or less.

Environmental Factors

Temperature is an important environmental determinant, and it can vary widely, particularly for anoxic/anaerobic sediments (Kohring et al., 1989). Temperature effects on microbial growth and metabolism are well known and can be described generally by the Arrhenius equation. Temperature can thus potentially influence the rate of biotransformation. The acclimation period that has been observed for the degradation of 2,4-dichlorophenol in anoxic/anaerobic sediments also has been observed to vary with temperature (Kohring et al., 1989).

Bradley et al. (1992) investigated the factors that were likely to influence microbial activity under denitrifying conditions in a JP-4 jet fuel–contaminated aquifer near North Charleston, South Carolina. Complete denitrification of amended nitrate (up to 1 mM) was observed at the site, indicating that there was minimal risk of nitrate and nitrite accumulation in the ground water. However, the rate of denitrification was influenced by pH, which ranged from approximately 4 to 8 at the site. Denitrification was 38 percent lower at pH 4 than at pH 7. Furthermore, the concentration of total organic carbon in the ground water ranged from 26 to 4487 mg/liter. A tenfold variation in sediment denitrification was observed at different levels of hydrocarbon contamination due to stimulatory and inhibitory effects. An increase in denitrification and carbon mineralization rates over the range 26 to 390 mg/liter was observed and was considered to be related to increased carbon availability. These results are consistent with the observed environmental factors which influence denitrification (Sprent, 1987; Knowles, 1982; Payne, 1981).

Field Demonstrations of Anoxic/Anaerobic Bioremediation

To date, applications of in situ anoxic/anaerobic bioremediation of contaminated soils and ground water have been limited to feasibility studies, field trials, and model simulations (USEPA, 1992; Hinchee and Olfenbuttle, 1991a, 1991b; GASReP, 1990). These have dealt largely with the degradation of aromatic hydrocarbons [benzene, toluene, ethylbenzene, and xylene (BTEX)] and chlorinated aliphatic compounds (CACs), since these are the most frequently found contaminants in ground water aquifers (MacKay and Cherry, 1989).

Petroleum: Aromatic Hydrocarbons

Anoxic/anaerobic bioremediation of hydrocarbon-contaminated soils and ground water can be promoted under denitrifying conditions. Although aromatic hydrocarbons readily degrade under aerobic conditions, the use of nitrate as an alternative electron acceptor may offer advantages over the use of oxygen (Bradley et al., 1992; Bouwer and McCarty, 1983; Kuhn et al., 1985; Zeyer et al., 1986; Major et al., 1988; Hutchins et al., 1991b).

Batterman and Werner (1984) were the first to demonstrate the field application of nitrate to clean up an oil spill in the upper Rhine catchment. Batterman (1983) attempted to stimulate denitrification in hydrocarbon-contaminated ground water by adding nitrate. The contaminated aquifer consisted of an 8- to 10-in-thick layer of sand which contained some silt and clay beds and had a ground water flow of 4 m^3/day. Phosphate was not added since it was not limiting. Removal of 1 mg of hydrocarbon required 3.3 mg of nitrate. The concentration of aliphatics declined slowly from 1.5 to about 0.7 mg/liter, whereas the concentration of aromatic compounds declined from 5.5 to about 1.5 mg/liter in approximately 1 year. Seventy-five percent (16.6 tons) of the hydrocarbons were removed. The rate of decline in the concentration of xylene was much slower than that of benzene and toluene. Water was injected during the test, which resulted in a rise in the level of the hydrocarbons as well as the water table in the unsaturated zone. There was an overall reduction of 40 percent in the concentration of hydrocarbons as a result of the treatment process. Insufficient information was available to determine whether anoxic/anaerobic processes were exclusively involved. It was possible that dissolved oxygen was delivered to the contaminated aquifer by aerated water pumped into it and that aerobic degradation of the contaminants did take place.

More recently, Kuhn et al. (1985) and Major et al. (1988) have demonstrated that benzene, toluene, and isomers of xylene (BTX) could be degraded in anoxic batch microcosms containing ground water that was obtained from contaminated aquifers. Canada Forces Base Borden, Ontario, Canada, has been the site of several field studies (Barker et al., 1987; Major et al., 1988). These studies involved the injection of ground water spiked with BTEX (0.9 mg/liter) into a shallow sand aquifer to create two separate plumes. Barker et al. (1987) found that depletion of oxygen due to initial microbial activity inhibited the degradation of BTX. In a subsequent study (Major et al., 1988), the addition of nitrate was found to stimulate the anoxic degradation of BTX. The addition of acetate, formate, or lactate was found not to significantly enhance the biodegradation of BTX. Unpublished results have demonstrated the potential for the coupling of denitrification with the in situ remediation of a JP-4 jet fuel–contaminated shallow aquifer near North Charleston, South Carolina (Bradley et al., 1992).

Hutchins and coworkers (Hutchins, 1992; Hutchins et al., 1991a, 1991b) have demonstrated the potential for remediating an aquifer contaminated with BTEX at Park City, Kansas, and at a U.S. Coast Guard facility in Traverse City, Michigan, under anoxic conditions. Microcosm tests with ground water obtained from both sites showed that toluene, ethylbenzene, xylenes (m- and p-), 1,3,5-trimethylbenzene, and 1,2,4-trimethylbenzene were degraded within 20 days when amended with nitrate (Hutchins, 1992; Hutchins and Wilson, 1991). The field-study site (Hutchins and Wilson, 1991) consisted of a 10-m^2 section of a JP-4 jet fuel spill. The area was perfused with nitrate and nutrients through an infiltration gallery, and five wells were constructed to monitor ground water quality. A lag period of 20 days was observed before nitrate removal occurred. Once denitrifying activity was established, nitrate-nitrogen removal occurred at a rate of 1 mg/liter/h. Hutchins and Wilson (1991) calculated that 1.1 kg of nitrate-nitrogen was required for each kilogram of BTEX degraded under anoxic conditions. A portion of the benzene degraded was attributed to residual oxygen present in the ground water. Nitrate enhanced xylene (m- and p-) and ethylbenzene removal; however, there was no degradation of o-xylene. The zero-order rate constants were similar between the laboratory and field studies.

By calculating the projected nitrate demand, based on reaction rates for BTEX and nitrate removal and the penetration zone of nitrate, Hutchins (1992) determined that the rate of nitrate application appears to be the primary factor that will affect remediation time. If nitrate loading rates are restricted by government regulations (nitrate is a regulated substance in ground water), this may affect the length of time required for remediation.

Bacteria capable of degrading BTEX under anoxic conditions are readily isolated from hydrocarbon-contaminated aquifers and sediments (Mikesell et al., 1991; Ball et al., 1991). Their occurrence may not be widespread in biomass obtained from uncontaminated sites or waste treatment systems (Ball et al., 1991; Leahy and Colwell, 1990). Not all degrading activities may be observed at a contaminated site either. The indigenous microbial population, hydrocarbon concentration, and extent of exposure to the contaminants may account for the differences observed from various sites. Mikesell et al. (1991) have reported that unlike aerobic hydrocarbon-degrading bacteria, anoxic BTEX-degrading bacteria can be stratified within a contaminated plume. Core samples obtained from different depths at various sites within the contaminated plume were found to contain different levels and patterns of degrading activity. Benzene degradation was the most widespread activity observed in this study. There is some controversy regarding the anoxic degradation of benzene. Observed anoxic benzene degradation may be due to dissolved oxygen introduced into microcosms or residual oxygen in ground water (Hutchins, 1992; Hutchins and Wilson, 1991). Ball et al. (1991) found toluene to be most readily degraded by biomass obtained from aquifer solids and sediment obtained from an oil refinery pond.

Chlorinated Aliphatic Compounds

The Moffett Field Naval Air Station, Mountain View, California, contains a well-characterized shallow aquifer typical of ground water contaminated with chlorinated aliphatic compounds (CACs) (Semprini et al., 1991, 1992). The site contained 1,1,1-trichloroethane (TCA), Freon 11, and Freon 113 at concentrations of 50, 6, and 3 μg/liter, respectively. Nitrate and sulfate were present at concentrations of 20 and 700 mg/liter, respectively. Dissolved oxygen levels were less than 0.2 mg/liter, and the phosphorous concentration was 0.1 mg/liter. The site did not contain appreciable concentrations of metals which may have interfered with microbial activity. Carbon tetrachloride (CT), which was not present, was added to extracted ground water (70 μg/liter) and injected back into the aquifer.

Microbial activity was promoted by the addition of acetate. Acetate and nitrate utilization was rapid, with nitrate levels being exhausted at 100 hours after acetate addition. More than 80 percent of the acetate was consumed within 1 m of transport through the injection well. A transitory buildup of nitrite was observed within the first 60 hours, which is typical of the establishment of denitrifying conditions. Sulfate reduction and methanogenesis did not develop once the nitrate was consumed. Transformation of CT to chloroform (CF) took place beyond the point where acetate and nitrate utilization was observed, indicating that a second, nondenitrifying population of bacteria was involved in the transformation. Microcosm experiments in the same study supported the hypothesis that denitrifying bacteria were not involved in the transformation of CACs and that removal of nitrate enhanced the rate of transformation of CACs at the site.

It may be possible to avoid denitrification in similar situations by providing hydrogen as the electron donor (DiStefano et al., 1992). DiStefano et al. (1992) have demonstrated that hydrogen serves as the actual electron donor for PCE degradation by a methanogenic consortium. However, by excluding organic sources of reducing equivalents, various by-products of metabolism produced by one group of bacteria may not be produced which could conceivably be required by the biotransforming population. This underscores the complexity of anoxic/anaerobic bioremediation and the need to gain a better understanding of the process requirements to effectively develop anoxic/anaerobic-based methods.

CT transformation to CF in the Moffet study was observed after 400 hours, with gradual decreases in CF taking place over the next 52 days (duration of acetate injection to the site). Between 95 and 97 percent of CT was transformed, and CF accounted for between 30 and 40 percent of transformed CT. CF is problematic and presents a limitation to the application of anoxic/anaerobic bioremediation. TCA, Freon 11, and Freon 113 were transformed at slower rates than observed for CT and to a lesser extent. The rates of transformation and the extent to which they were transformed also were enhanced by the removal of nitrate.

For anoxic/anaerobic bioremediation to be effective in the treatment of PCE- and TCE-contaminated sites, these contaminants must be fully transformed to nonchlorinated products. The rate-limiting step in the degradation of these compounds is the conversion of VC to ethene. There have been few field studies demonstrating the potential for anoxic/anaerobic bioremediation of CAC-contaminated sites. The National Priority List industrial site in St. Joseph, Michigan, which contains a TCE ground water plume, has been found to contain significant levels of ethene and methane, indicating that anaerobic transformation of TCE to nonchlorinated end products occurs under methanogenic conditions at this site (McCarty and Wilson, 1992) .

Major et al. (1991) demonstrated that dechlorination of PCE to VC and then to ethene (ETH) can take place in the field. The site, a 4.5-acre chemical transfer facility in North Toronto, was the repository for a variety of organic solvents, including methanol, methylethylketone, vinyl- and ethylacetate, butylacrylate, and PCE. The ground water was found to contain free and dissolved PCE (0.4 to 4.4 mg/liter), TCE (1.1 to 1.7 mg/liter) *cis*-dichloroethene (cDCE; 5.8 to 76.5 mg/liter), and VC (0.2 to 9.7 mg/liter). Ethene was found in the ground water at concentrations ranging from 0.02 to 0.28 mg/liter. These were found to have a good degree of distribution overlap, indicating that dechlorination of these compounds was taking place. Methanol and acetate, which were present (5 to 1770 and 430 to 715, mg/liter, respectively), were found to promote the transformation of PCE to ETH, as indicated by the strong correlation between PCE, ETH, and electron donor contours. This is consistent with laboratory studies (DiStefano et al., 1991). In ground water where ETH levels were minimal at the site, PCE was found not to be associated with methanol or acetate plumes. Although ethene may have been produced by soil microbial microorganisms, parallel microcosm studies (145-day duration) supported the biological dehalogenation of PCE to ethene. In these experiments, however, there was no clear association between methanogenesis and dehalogenation. The former was optimal with PCE plus acetate, whereas the latter activity was supported by PCE plus methanol and acetate. Subsurface sediments were found to contain low population densities of methanogens and sulfate-reducing bacteria. The results of the study indicate that other anaerobes, acetogens, may be contributing to the dechlorination of PCE to ETH. The presence of acetate in the ground water with methanol supports this hypothesis.

Other Important Applications of Anoxic/Anaerobic Processes to Environmental Problems

Chlorophenolic Compounds

Chlorophenolics are a significant component of the low-mass material often found in bleached kraft pulp mill effluent (Kingstrad and Lindstrom, 1984). These include chlorophenols, chloroguiacols, and chlorocatechols (Kingstrad and Lindstrom, 1984; Salkinoja-Salonen et al., 1981; Lindstrom et al., 1981). There have been numerous reports of the anaerobic dehalogenation of these compounds in anaerobic bioreactors which have been designed to treat pulp mill effluents. Woods et al. (1989) have shown that chlorinated vetatroles are transformed to the lesser chlorinated guiacols and catechols in an up-flow anaerobic sludge blanket (UASB) reactor. Chloroguiacols were observed to undergo demethylation and subsequent reductive dechlorination. Monochlorophenols are resistant to anaerobic degradation (Madsen and Licht, 1992; Mohn and Kennedy, 1992; Häggblom, 1990).

Fahmy et al. (1990) have demonstrated the transformation of 2,4,6-trichlorophenol to 2,4-dichlorophenol in an anaerobic fluidized bed reactor. Other research has demonstrated the ability of anaerobic systems to remove chlorophenolics (Salkinoja-Salonen et al., 1983) and the ability of biological consortia to anaerobically dechlorinate pentachlorophenol with varying levels of mineralization.

Bryant and coworkers (Amy et al., 1988; Bryant et al., 1987, 1988) have postulated that chlorinated organics in pulp mill effluent entering an aerated lagoon become sorbed to biomass and settle to the bottom, where anoxic/anaerobic processes contribute to degradation. Batch and continuous studies of organochlorine removal in simulated lagoons indicate that anoxic/anaerobic processes are important and support this hypothesis (Chernysh et al., 1993). For chlorophenols, however, aerobic processes may be more effective (Collins and Allen, 1991). In a recent study of the microbiology of a pulp mill lagoon in Northern Ontario, obligate anaerobic bacteria were found to make up 40 percent of the culturable anoxic/anaerobic population in sludge obtained from the benthic region; however, the redox levels were not typical of methanogenic or sulfate-reducing habitats (Liss and Allen, 1992). At this particular mill, the effluent was supplemented with ammonium nitrate. Although not measured directly, there was evidence that denitrification was occurring (Fulthorpe et al., 1993). To date, there have been no confirmed reports that describe the dehalogenation of chlorophenolics coupled with denitrification.

Organonitro Compounds

Hazardous energetic organonitro compounds are characterized by the presence of nitroaromatic, nitrate ester, or nitroamine groups (Kaglan, 1990). These compounds originate from the manufacture of explosives and propellants by the military and are found as contaminants in many environments. Many of these do not readily degrade under aerobic conditions.

Explosive-contaminated wastewater was once commonly disposed of in lagoons. This practice has led to an infiltration of contaminants into soil and ground water. Wastewater at two ammunition plants in the United States are now processed in treatment plants employing anaerobic microorganisms (Sisk and Keehan, 1992). Explosive-contaminated sludges and soil, on the other hand, until recently were incinerated. Recent descriptions of bacterial isolates able to use organonitro compounds (Dickel and Knachmuss, 1991; Boopathy and Kulpa, 1992) indicate the potential for using such organisms for the reclamation of trinitrotoluene (TNT)-contaminated soil and water.

Trinitrotoluene (TNT) is particularly resistant to aerobic metabolism. A *Desulfovibrio* species (B strain), a sulfate-reducing bacterium, isolated from an anaerobic digester (Boopathy and Kulpa, 1992) was able under nitrogen-limiting conditions to use TNT as a sole electron acceptor in the absence of sulfate and transform TNT via reductive deamination completely to toluene. This bacterium also was able to use dinitrophenols, dinitrotoluenes, and dinitrobenzoic acid as a source of nitrogen and as electron acceptors.

Creosote: Alkylphenols, Polycyclic Aromatic Hydrocarbons, and Heterocyclic Compounds

Alkylphenols, polynuclear aromatic hydrocarbons (PAHs), and polycyclic hydrocarbons may occur in ground water due to the infiltration by landfill leachate and the presence of creosote (Smolenski and Suflita, 1987; Goerlitz et al., 1985). Creosote-contaminated aquifers have been shown to support the degradation of cresols (*o-*, *m-*, and *p-*) under anaerobic (methanogenic and sulfate-reducing conditions) and anoxic conditions (Flyvbjerg et al., 1991; Smolenski and Suflita, 1987). PAHs, which are the most significant component of creosote (85 percent) (Nesteler, 1974), however, are not considered to be readily degraded anoxically or anaerobically.

Heterocyclic aromatic compounds, constituents of fuels and creosote, are found in contaminated subsurface environments. Alkylpyridines are particularly odorous and potentially toxic heterocyclic aromatic compounds which are known to accumulate and persist in soils and subsurface environments (Riley et al., 1981; Rogers et al., 1985). Although degradation of heterocyclic compounds proceeds most rapidly under aerobic conditions (Kaiser and Bollag, 1992; Rogers et al., 1985), microbial transformation of compounds containing nitrogen or sulfur, including pyridines, 3- and 4-picoline, nicotinic acid, furfural, quinoline, isoquinoline, and indolic acids, under anoxic/anaerobic conditions have been described (Kuhn and Suflita, 1989a; Brune et al., 1983; Bak and Widdel, 1986; Shanker et al., 1991; Kaiser et al., 1993), indicating the potential for heterocyclic compound–contaminated sites to be remediated under anoxic/anaerobic conditions. Ronen and Bollag (1992), for example, have recently reported on the ability of an *Alcaligenes* species isolated from a polluted soil to rapidly degrade pyridine under denitrifying conditions in laboratory microcosms. Transformation of heterocyclic compounds has been observed in methanogenic, denitrifying, and sulfate-reducing conditions, and heterocyclic-degrading microbial consortia have been isolated from contaminated sites (Godsy et al., 1992; Kaiser et al., 1993).

Radioactive Wastes

Bioprocessing of radioactive wastes and ores offers a cost-effective method for metal removal. Biosorption is the principal mechanism employed and may involve living or dead cells of a variety of prokaryotic and eukaryotic microorganisms. The main disadvantage with biosorption techniques is the limitation for their application for in situ remediation of ground waters. In this situation, large quantities of biomass injected into ground water may cause plugging, and possible cell lysis may result in a release of biosorbed material back into the ground water (Lovely and Philips, 1992a). Microbial reduction of soluble U(VI) to insoluble U(IV) in uranium-contaminated mine drainage waters and ground water may prove to be a valuable alternative to uranium removal. Respiratory Fe(III)-reducing bacteria have been shown to reduce U(VI) and use it as a terminal electron acceptor (Lovely et al., 1991). Lovely and Philips (1992a, 1992b) have isolated a sulfate-reducing bacterium, *Desulfovibrio desulfuricans*, which was able to reduce sulfate and U(VI) simultaneously. Their results indicate that enzymatic activity may be the major mechanism for U(VI) reduction in sulfidogenic environments. Lovely and Philips (1992a) have outlined the possible benefits to enzymatic U(VI) reduction for the remediation of uranium-contaminated wastes, ground water, and, potentially, soils. Uranyl carbonate complexes are amenable to treatment, and the resulting precipitate is a highly pure form of uranium. *D. desulfuricans* has a high capacity for uranium removal per cell. The uranium precipitate does not bind to the cells, eliminating the need to extract the metal from the biomass, as occurs in biosorption methods. U(VI)-reducing bacteria may bring about the reduction and precipitation of the radioactive contaminants and transformation of aromatic compounds which may be present in wastes containing a mixture of contaminants. Microbial enzymatic reduction of U(VI) does not appear to be affected by metals, and the process operates over a range of pH levels. In the future it may be possible to employ cell-free enzyme systems.

Chlorate Systems

Chlorate (CLO_3^-) is generated when chlorine dioxide is used for pulp bleaching. The increased use of chlorine dioxide as a substitute for elemental chlorine will require that chlorate-containing waste waters be treated to reduce chlorate and its toxic effects in receiving waters. Chlorate is a highly oxidized compound that is readily reduced under anoxic/anaerobic conditions. Malmqvist et al. (1991) have described a microbial consortium obtained from a municipal activated sludge system which utilized chlorate as a final electron acceptor when grown under anaerobic conditions in a defined medium containing acetate as the sole source of energy and carbon. The reduction of chlorate to chloride was coupled with electron-transport phosphorylation and gave a high energy yield.

Cyanide Remediation

Cyanide is a significant component of wastes generated by coal coking, precious metals mining, and nitrile polymer industries (Towill et al., 1978). Free cyanide has long been considered to be highly toxic to anoxic/anaerobic bacteria (Fedorak and Hrudey, 1987; Fedorak et al., 1986). More recently, degradation of cyanide has been reported for waste streams containing as low as 30 mg/liter (Fedorak and Hrudey, 1989) and as high as 300 mg/liter (Fallon et al., 1991). Bicarbonate and ammonia appear to be the significant end products of cyanide treatment, although the biochemisty of anoxic/anaerobic cyanide degradation is not fully understood. The presence of ammonia is undesirable, and it has been proposed by Fallon et al. (1991) that the process may be most useful as a pretreatment to lower the levels of cyanide in aqueous wastes prior to aerobic treatment.

Future Developments

As with all applications of bioremediation, the potential for anoxic/anaerobic-based technologies will be enhanced by gaining a better understanding of the microbial ecology and biogeochemistry of contaminated sites and in situ biodegradation. Advances are being made in the development of anoxic/anaerobic bioreactors, including sewage sludge digesters (Guthrie et al., 1984), anaerobic fixed-film reactors (Hendricksen et al., 1991), and UASB reactors (Hendricksen et al., 1992; Mohn and Kennedy, 1992) for treating toxic contaminants, particularly chlorophenolics, in high-strength waste streams.

Codisposal of industrial waste waters and ground water in sanitary landfills has been proposed as a method for treating hazardous waste requiring disposal (Watson-Craik, 1990). However, codisposal is either discouraged or banned in many countries, including Germany, Canada, Australia, and the United States. If designed and managed properly, codisposal may offer several advantages, including lower cost. Landfills can act as multimillion cubic meter anoxic/anaerobic bioreactors. These operate over a wide temperature range and can withstand large hydraulic and organic loads and extended down times without loss of microbial activity.

Coupling aerobic with anaerobic processes to degrade mixed wastes or recalcitrant compounds offers considerable potential. In waste water treatment, exposure of microorganisms to sequential environments has been shown to enhance biodegradation (Fig. 9-4). The

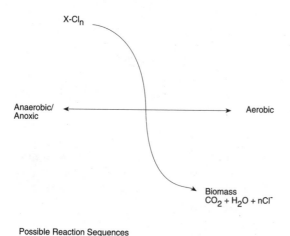

Possible Reaction Sequences

1) aerobic ⟶ anaerobic/anoxic

2) anaerobic/anoxic ⟶ aerobic

3) anaerobic/anoxic ⟶ aerobic ⟶ anaerobic/anoxic ⟶ aerobic

Figure 9-4. Conceptual design of a sequential anaerobic/aerobic treatment system.

applications and potential for sequential anaerobic/aerobic systems in waste water treatment has been reviewed recently (Zitomer and Speece, 1993).

Coupled anaerobic/aerobic processes may be necessary for complete degradation of recalcitrant pollutants, particularly highly chlorinated aromatic compounds. Brown et al. (1984) attributed the degradation of polychlorinated biphenyls (PCBs) in river sediments to the combined activities of aerobic and anaerobic consortia. Fogel et al. (1982) described the degradation conditions of the herbicide methoxychlor in soils as a sequence of anaerobic and aerobic states. The apparent requirement for aerobic bacteria to initiate reductive dehalogenation of tri- and tetrachloroethylenes (TCE and PCE, respectively) has been demonstrated (Kastner, 1991). In this study, only aerobic subcultures of the dechlorinating consortia maintained activity, whereas methanogenic subcultures did not transform the chlorinated substrates. Kastner (1991) proposed that the highly reducing conditions (-150 mV) required for transformation occurred after a transition from aerobic to anaerobic conditions following the consumption of the electron acceptor (oxygen or nitrate) and the accumulation of sulfide. It is also possible that an essential compound (e.g., nutrient) required for dehalogenation under anoxic/anaerobic conditions was produced under aerobic conditions or that the aerobic or facultative anaerobic bacteria were responsible for the reductive dehalogenation.

Sequential aerobic/anaerobic conditions for degradation may be applicable to subsurface cleanup. The potential is greatest for halogenated compounds. Aquifers could be treated like a sequencing batch reactor. PCBs, PCE, and PCP are good candidate contaminants for sequential treatment (Anid et al., 1991). Such an approach could be used to attack compounds such as PCE by first creating an anaerobic zone to promote dechlorination, followed by aerobic zones that would allow degradation of the less chlorinated products (VC) that tend to persist in anoxic/anaerobic zones but can be degraded aerobically. Methoxychlor, an insecticide, is only slightly degraded under either aerobic or anoxic/anaerobic conditions (after 3 months) (Fogel et al., 1982). When contaminated soil samples were converted from an anoxic/anaerobic to an aerobic status, mineralization of the insecticide increased significantly, 10 to 70 times that observed in soils maintained aerobically.

In the case of PCBs, the *meta-* and *para-*chlorines are removed by anaerobic reductive dehalogenation. The *ortho-*chlorines are only removed slowly under anaerobic treatment. Under aerobic conditions, the *ortho*-chlorine is removed, and complete mineralization of the compound occurs. In situ applications have inherent problems, particularly related to monitoring of the degradation process and the cost of switching from anoxic/anaerobic to aerobic conditions. Appropriate methods need to be developed. Monitoring redox levels could be a way to determine reactions rates and alter conditions. Sayles et al. (1992) have outlined a proposed methodology involving observing the disappearance of the parent contaminant and following the appearance of intermediates. The rates of degradation of parent and daughter compounds as a function of time during degradation can be used to determine the optimal time for switching conditions. Degradation rates are then correlated with redox.

Acknowledgments

We would like to thank K. A. Gilbride, Ryerson Polytechnic University, for reviewing manuscript draft. SNL's research is supported, in part, by an operating grant from the Natural Sciences and Engineering Research Council of Canada.

References

Abramowicz, D. A. 1990. Aerobic and anaerobic biodegradation of PCBs: a review. *Crit. Rev. Biotechnol.* **10**:241–251.

Aeckersberg, F., F. Bak, and F. Widdel. 1991. Anaerobic oxidation of saturated hydrocarbons to CO_2 by a new type of sulfate-reducing bacterium. *Arch. Microbiol.* **155**:5–14.

Alexander, M. 1980. Biodegradation of chemicals of environmental concern. *Science.* **211**:132.

Amy, G. L., C. W. Bryant, B. C. Allenman, and W. A. Barkley. 1988. Biosorption of organic halides in a kraft mill aerated lagoon. *J. Water Pollution Control Fed.* **60**:1445–1453.

Anid, P. J., L. Nies, and I. M. Vogel. 1991. Sequential anaerobic-aerobic biodegradation of PCBs in the river model. In R. E. Hinchee and R. F. Olfenbuttel (eds.), *On-Site Bioreclamation: Processes for Xenobiotic and Hydrocarbon Treatment.* Butterworth-Heineman, Stoneham, Mass., pp. 428–436.

Arvin, E., B. Jensen, J. Aamand, and C. Jorgenson. 1988. The potential for free-living bacteria to degrade aromatic hydrocarbons and heterocyclic compounds. *Water Sci. Technol.* **20**:109–118.

Atlas, R. M. 1991. Bioremediation of fossil fuel contaminated soils. In R. E. Hinchee and R. F. Olfenbuttel (eds.), *In Situ Bioreclamation: Applications and*

Investigations for Hydrocarbon and Contaminated Site Remediation. Butterworth-Heineman, Stoneham, Mass., pp. 14–32.

Bak, F., and F. Widdel. 1986. Anaerobic degradation of indolic compounds by sulfate-reducing enrichment cultures, and description of *Desulfobacterium indolicum* gen. nov. sp. nov. *Arch. Microbiol.* **146**:170–176.

Baker, K. H., and D. S Herson. 1990. In situ bioremediation of contaminated aquifers and subsurface soils. *Geomicrobiol. J.* **8**:133–146.

Balkwill, D. L., and W. C. Ghiorse. 1985. Characterization of subsurface bacteria associated with two shallow aquifers in Oklahoma. *Appl. Environ. Microbiol.* **50**:580–588.

Ball, H. A., M. Reinhard., and P. L. McCarty. 1991. Biotransformation of monoaromatic hydrocarbons under anoxic conditions. In R. E. Hinchee and R. F. Olfenbuttel (eds.), *In Situ Bioreclamation: Applications and Investigations for Hydrocarbon and Contaminated Site Remediation.* Butterworth-Heinemann, Stoneham, Mass., pp. 458–463.

Barker, J. F., G. C. Patrick, and D. W. Major. 1987. Natural attenuation of aromatic hydrocarbons in a shallow sand aquifer. *Ground Water Monit. Rev.* **7**:64–71.

Batterman, G. 1983. A large scale experiment on in situ biodegradation of hydrocarbon in the subsurface. In *Ground Water in Water Resources Planning,* Vol. II: *Proceedings of an International Symposium.* IASA Publication 142. International Association of Hydrological Sciences, London, p. 93.

Batterman, G., and P. Werner. 1984. Beseitigung einer untergrundkontamination mit kohlenwasserstoffen durch mikrobiellen abbau. *GWF-Wasser/Abwasser.* **125**:366.

Bechard, G., J.-G. Bisaillon, and R. Beaudet. 1990. Degradation of phenol by a bacterial consortium under methanogenic conditions. *Can. J. Microbiol.* **36**:573–578.

Bisaillon, J.-G., F. Lepine, R. Beaudet, and M. Sylvestre. 1991. Carboxylation of *o*-cresol by an anaerobic consortium under methanogenic conditions. *Appl. Environ. Microbiol.* **57**:2131–2134.

Blake, C. L., and G. D. Hegeman. 1987. Plasmid pCB1 carries genes for anaerobic benzoate catabolism in *Alcaligenes xylosoxidana* subs. *denitrificans* PN-1. *J. Bacteriol.* **169**:4878–4883.

Boopathy, R., and C. F. Kulpa. 1992. Trinitrotoluene (TNT) as a sole nitrogen source for a sulfate-reducing bacterium *Desulfovibrio* sp. (B strain) isolated from an anaerobic digester. *Cur. Microbiol.* **25**:235–241.

Boulton, C. A., and C. Ratledge. 1984. The physiology of hydrocarbon-utilizing microorganisms. *Top. Enzyme Ferment. Biotechnol.* **9**:11–77.

Bouwer, H. 1984. Elements of soil science and ground water hydrology. In G. Bitton and C. P. Gerba. (eds.), *Ground Water Pollution Microbiology.* John Wiley and Sons, Inc., New York, p. 9.

Bouwer, E. J., and P. L. McCarty. 1983. Transformations of halogenated organic compounds under denitrification conditions. *Appl. Environ. Microbiol.* **45**:1295–1299.

Bouwer, E. J., and P. L. McCarty. 1984. Modeling of trace organics biotransformed in the subsurface. *Ground Water* **22**:433–440.

Bouwer, E. J., and J. P. Wright. 1988. Transformations of trace halogenated aliphatics in anoxic biofilm columns. *J. Contam. Hydrol.* **2**:155–169.

Bouwer, E. J., B. E. Rittmann, and P. L. McCarty. 1981. Anaerobic degradation of halogenated 1- and 2-carbon organic compounds. *Environ. Sci. Technol.* **15**:155–169.

Boyle, M. 1989. The environmental microbiology of chlorinated aromatic decomposition. *J. Environ. Qual.* **18**:395–402.

Bradley, P. M., C. M. Aelion, and D. A. Vrobleskey. 1992. Influence of environmental factors on denitrification in sediment contaminated with JP-4 jet fuel. *Ground Water* **30**:843–848.

Brown, J. F., R. E. Wagner, D. L. Bedard, M. J. Brennan, J. C. Carnahan, and R. J. Man. 1984. PCB transformations in Upper Hudson sediments. *Northeast Environ. Sci.* **3**:167.

Brown, J. F., D. L. Bedard, M. J. Brennan, J. C. Carnahan, H. Feng, and R. E. Wagner. 1987. Polychlorinated biphenyl dechlorination in aquatic sediments. *Science.* **236**:709–712.

Brune, G., S. M. Schobert, and H. Sahm. 1983. Growth of a strict anaerobic bacterium on furfural (2 furaldehyde). *Appl. Environ. Microbiol.* **46**:1187–1192.

Bryant, C. W., G. L. Amy, and B. C. Allman. 1987. Organic halide and organic carbon distribution and removal in a pulp and paper wastewater lagoon. *J. Water Pollution Control Fed.* **59**:890–896.

Bryant, C. W., G. L. Amy, R. Neill, and S. Ahmad. 1988. Partitioning of organic chlorine between bulk water and benthal interstitial water through a kraft pulp mill aerated lagoon. *Wat. Sci. Technol.* **20**:73–79.

Bryant, F. O., D. D. Hale, and J. E. Rogers. 1991. Regiospecific dechlorination of pentachlorophenol by dichlorophenol-adapted microorganisms in freshwater anaerobic sediment slurries. *Appl. Environ. Microbiol.* **57**:2293–2301.

Chaudhry, G. R., and S. Chapalamadugu. 1991. Biodegradation of halogenated organic compounds. *Microbiol. Rev.* **55**:59–79.

Cheng, J., M. T. Suidan, and A. D. Venosa. 1992. Evaluation of anaerobic respirometry to quantify intrinsic anaerobic biodegradation kinetics of recalcitrant organic compounds. In U. S. Environmental Protection Agency, (ed.), *Bioremediation of Hazardous Wastes.* EPA/600/R-92/126. U.S. EPA, Washington.

Chernysh, A., S. N. Liss, and D. G. Allen. 1993. A batch study of the aerobic and anaerobic removal of chlorinated organic compounds in an aerated lagoon. *Water Pollution Res. J. Can.* (in press).

Colberg, P. J. S. 1990. Role of sulfate in microbial transformations of environmental contaminants: Chlorinated aromatic compounds. *Geomicrobiol. J.* **8**:147–165.

Collins, T., and D. G. Allen. 1991. A bench-scale aerated lagoon for studying the removal of chlorinated organics. *Tappi* **74**:231–234.

Cornacchio, L., E. R. Hall, and J. T. Trevore. 1988. Modified anaerobic serum bottle testing procedures for industrial wastewaters. *Water Pollution Res. J. Can.* 23:450–459.

Criddle, C. S. 1989. Reductive Dehalogenation in Microbial and Electrolic Model Systems. Ph.D. Thesis, Civil Engineering, Stanford University, Stanford, Calif.

Criddle, C. S., J. T. DeWitt, D. Grbic-Galic, and P. L. McCarty. 1990a. Transformation of carbon tetrachloride by *Pseudomonas* sp. strain KC under denitrifying conditions. *Appl. Environ. Microbiol.* 56:3240–3246.

Criddle, C. S., J. T. DeWitt, and P. L. McCarty. 1990b. Reductive dehalogenation of carbon tetrachloride by *Escherichia coli* K12. *Appl. Environ. Microbiol.* 56:3247–3254.

Dagley, S. 1984. Microbial degradation of aromatic compounds. *Dev. Ind. Microbiol.* 25:53–65.

Dickel, O., and H.-J. Knackmuss. 1991. Catabolism of 1,3-dinitrobenzene by *Rhodococcus* sp. QT-1. *Arch. Microbiol.* 157:76–79.

DiStefano, T. D., J. M. Gossett, and S. H. Zinder. 1991. Reductive dechlorination of high concentrations of tetrachloroethene to ethene by an anaerobic enrichment culture in the absence of methanogenesis. *Appl. Environ. Microbiol.* 57:2287–2292.

DiStefano, T. D., J. M. Gossett, and S. H. Zinder. 1992. Hydrogen as an electron donor for dechlorination of tetrachloroethene by an anaerobic mixed culture. *Appl. Environ. Microbiol.* 58:3622–3629.

Dolfing, J. 1990. Reductive dechlorination of 3-chlorobenzoate is coupled to ATP production and growth in an anaerobic bacterium, strain DCB-1. *Arch. Microbiol.* 153:264–266.

Dolfing, J., and B. K. Harrison. 1992. Gibbs free energy of formation of halogenated aromatic compounds and their potential role as electron acceptors in anaerobic environments. *Environ. Sci. Technol.* 26:2213–2218.

Dunnivant, F. M., and R. P. Schwarzenbach. 1992. Reduction of substituted nitrobenzenes in aqueous solutions containing natural organic matter. *Environ. Sci. Technol.* 26:2133–2144.

Egli, C., R. Scholtz, A. M. Cook, and T. Leisinger. 1987. Anaerobic dechlorination of tetrachloromethane and 1,2-dichloroethane to degradable products by pure cultures of *Desulfobacterium* sp. and *Methanobacterium* sp. *FEMS Microbiol. Lett.* 43:257–261.

Egli, C., T. Tschan, R. Scholtz, A. M. Cook, and T. Leisinger. 1988. Transformation of tetrachloromethane to dichloromethane and carbon dioxide by *Acetobacterium woodii*. *Appl. Environ. Microbiol.* 54:2819–2824.

Ehrlich, G. G., E. M. Godsy, D. F. Gorelitz, and M. F. Hult. 1983. Microbial ecology of a creosote-contaminated aquifer at St. Louis Park, Minn. *Dev. Ind. Microbiol.* 24:233–245.

Ehrlich, H. L. 1985. Bacteria and their products in food webs. In E. R. Leadbetter and J. S. Poindexter (eds.), *Bacteria in Nature*, Vol. 1: *Bacterial Activities in Perspective.* Plenum Press, New York, pp. 199–219.

Evans, W. C., and G. Fuchs. 1988. Anaerobic degradation of aromatic compounds. *Annu. Rev. Microbiol.* **42:**289–317.

Fahmy, M., E. Heinzle, and O. M. Kut. 1990. Treatment of bleaching effluents in aerobic/anaerobic fluidized biofilm systems. In *Proceedings of the 3rd IAWAPCR Symposium on Forest Industry Wastewaters, June 5–8, Tempere, Finland.*

Fallon, R. D., D. A. Cooper, R. Speece, and M. Henson. 1991. Anaerobic biodegradation of cyanide under methanogenic conditions. *Appl. Environ. Microbiol.* **57:**1656–1662.

Fathepure, B. Z., and S. A. Boyd. 1988. Reductive dechlorination of perchloroethylene and the role of methanogens. *FEMS Microbiol. Lett.* **49:**149–156.

Fathepure, B. Z., J. P. Nengu, and S. A. Boyd. 1987. Anaerobic bacteria that dechlorinate perchloroethene. *Appl. Environ. Microbiol.* **53:**2671–2674.

Federle, T. W., D. C. Dobbins, J. R. Thornton-Manning, and J. D. Jones. 1986. Microbial biomass, activity, and community structure in subsurface soils. *Ground Water* **24:**365.

Fedorak, P. M., and S. E. Hrudey. 1987. Inhibition of anaerobic degradation of phenolics and methanogenesis by coal-coking wastewater. *Water Sci. Technol.* **19:**219–228.

Fedorak, P. M., D. J. Roberts, and S. E. Hrudey. 1986. The effects of cyanide on the methanogenic degradation of phenolic compounds. *Water Res.* **10:**1315–1320.

Fedorak, P. M., and S. E. Hrudey. 1989. Cyanide transformation in anaerobic phenol-degrading methanogenic cultures. *Water Sci. Technol.* **21:**1315–1320.

Flyvbjerg, J., E. Arvin, B. K. Jensen, and S. K. Olsen. 1991. Biodegradation of oil- and creosote-related aromatic compounds under nitrate-reducing conditions. In R. E. Hinchee and R. F. Olfenbuttel (eds.), *In Situ Bioreclamation: Applications and Investigations for Hydrocarbon and Contaminated Site Remediation.* Butterworth-Heinemann. Stoneham, Mass., pp. 471–479.

Focht, D. D. 1992. Diffusional constraints on microbial processes in soil. *Soil Sci.* **154:**300–307.

Fogel, S., R. L. Lancione, and A. E. Sewall. 1982. Enhanced biodegradation of methoxychlor in soil under sequential environmental conditions. *Appl. Environ. Microbiol.* **44:**113–119.

Freedman, D. L., and J. M. Gossett. 1989. Biological reductive dechlorination of tetrachloroethylene and trichloroethylene to ethylene under methanogenic conditions. *Appl. Environ. Microbiol.* **55:**2144–2151.

Freedman, D. L., and J. M. Gossett. 1991. Biodegradation of dichloromethane and its utilization as a growth substrate under methanogenic conditions. *Appl. Environ. Microbiol.* **57:**2847–2857.

Fulthorpe, R. R., S. N. Liss, and D. G. Allen. 1993. Characterization of bacteria isolated from a bleached kraft pulp mill wastewater treatment system. *Can. J. Microbiol.* **39:**13–24.

Gallert. C., G. Koll, and J. Winter. 1991. Anaerobic carboxylation of phenol to benzoate. Use of deuterated phenols revealed carboxylation exclusively in the C_4 position. *Appl. Microbiol. Biotechnol.* **36**:124–129.

Galli, R., and P. L. McCarty. 1989. Biotransformation of 1,1,1-trichloroethane, trichloromethane, and tetrachloromethane by *Clostridium* sp. *Appl. Environ. Microbiol.* **55**:837–844.

GASReP, National Ground Water and Soil Remediation Program. 1990. In Situ Bioremediation: Considerations, Limitations, Potential and Future Direction. Unpublished report no. 87-90-01, Environment, Canada.

Gerhardt, P., R. G. E. Murray, R. N. Castilow, E. W. Nester, W. A. Wood, N. R. Krieg, and C. B. Phillips (eds.). 1981. *Manual of Methods for General Bacteriology.* American Society for Microbiology, Washington.

Ghiorse, W. C., and D. L. Balkwill. 1983. Enumeration and morphological characterization of bacteria indigenous to subsurface environments. *Dev. Ind. Microbiol.* **24**:213–224.

Gibson, S. A., and G. W. Sewell. 1992. Stimulation of reductive dechlorination of tetrachloroethene in anaerobic aquifer microcosms by addition of short-chain organic acids or alcohols. *Appl. Environ. Microbiol.* **58**:1392–1393.

Gibson, S. A., and J. M. Suflita. 1986. Extrapolation of biodegradation results to ground water aquifers: Reductive dehalogenation of aromatic compounds. *Appl. Environ. Microbiol.* **52**:681–688.

Godsy, E. M., D. R. Gorelitz, and D. Grbic-Galic. 1992. Methanogenic degradation of heterocyclic aromatic compounds by aquifer-derived microcosms. In U.S. Environmental Protection Agency (ed.), *Bioremediation of Hazardous Wastes.* EPA/600/R 92/126. U.S. EPA, Washington, pp. 89–92.

Godsy, E. M., D. F. Goerlitz, and D. Grbic-Galic. 1987. Anaerobic biodegradation of creosote contaminants in natural and simulated ground water ecosystems. In B. J. Franks (ed.), *U.S. Geological Survey Program on Toxic Waste-Ground Water Contamination.* U.S. Geological Survey, Tallahassee, Fla., pp. A-17–A-19.

Goerlitz, D. F., D. E. Troutman, E. M. Godsy, and B. J. Franks. 1985. Migration of wood-preserving chemicals in contaminated ground water in a sand aquifer at Pensacola, Florida. *Environ. Sci. Technol.* **19**:955–961.

Grbic-Galic, D. 1990. Anaerobic microbial transformation of nonoxygenated aromatic and alicyclic compounds in soil, subsurface, and freshwater sediments. In J.-M. Bollag and G. Stotzky (eds.), *Soil Biochemistry,* Vol. 6. Marcel Dekker, New York, pp. 117–189.

Guthrie, M. A., E. J. Kirsch, R. F. Wukasch, and C. P. L. Grady. 1984. Pentachlorophenol biodegradation: II. Anaerobic. *Water Res.* **18**:451–461.

Häggblom, M. M. 1990. Mechanisms of bacterial degradation and transformation of chlorinated monoaromatic compounds. *J. Basic Microbiol.* **30**:115–141.

Häggblom, M. M., and L. Y. Young. 1990. Chlorophenol degradation coupled to sulfate reduction. *Appl. Environ. Microbiol.* **56**:3255–3260.

Harvey, R. W., R. L. Smith, and L. George. 1984. Effect of organic contaminants upon microbial distribution and heterotrophic uptake in a Cape Cod, Massachusetts aquifer. *Appl. Environ. Microbiol.* **48**:1197–1202.

Harwood, C. S., and J. Gibson. 1988. Anaerobic and aerobic metabolism of diverse aromatic compounds by the photosynthetic bacterium *Rhodopseudomonas palustris. Appl. Environ. Microbiol.* **54:**712–717.

Hendricksen, H. V., S. Larsen, and B. K. Ahring. 1991. Anaerobic degradation of PCP and phenol in fixed-film reactors: The influence of an additional substrate. *Water Sci. Technol.* **24:**431–436.

Hendriksen, H. V., S. Larsen, and B. K. Ahring. 1992. Influence of supplemental carbon source on anaerobic dechlorination of pentachlorophenol in granular sludge. *Appl. Environ. Microbiol.* **58:**365–370.

Hinchee, R. E., and R. F. Olfenbuttel (eds.), 1991a. *In Situ Bioreclamation: Applications and Investigations for Hydrocarbon and Contaminated Site Remediation.* Butterworth-Henemann, Stoneham, Mass.

Hinchee, R. E., and R. F. Olfenbuttel (eds.). 1991b. *On Site Bioreclamation: Processes for Xenobiotic and Hydrocarbon Treatment.* Butterworth-Henemann, Stoneham, Mass.

Horowitz, A., J. M. Suflita, and J. M. Tiedje. 1983. Reductive dehalogenations of halobenzoates by anaerobic lake sediment microorganisms. *Appl. Environ. Microbiol.* **45:**1459–1465.

Horowitz, A., D. R. Shelton, C. P. Corwell, and J. M. Tiedje. 1982. Anaerobic degradation of aromatic compounds in sediments and digested sludge. *Dev. Ind. Microbiol.* **23:**435–444.

Hughes, J. B., and G. F. Parkin. 1991. The effect of electron donor concentration on the biotransformation of chlorinated aliphatics. In R. E. Hinchee and R. F. Olfenbuttel (eds.), *On Site Bioreclamation: Processes for Xenobiotic and Hydrocarbon Treatment.* Butterworth-Heinemann, Stoneham, Mass., pp. 59–76.

Hutchins, S. R. 1992. Use of nitrate to bioremediate a pipeline spill at Park City, Kansas: Projecting from a treatability study to full-scale remediation. In U. S. Environmental Protection Agency (ed.), *Bioremediation of Hazardous Wastes.* EPA/600/R = 92/126. EPA, Washington, pp. 41–42.

Hutchins, S. R. 1991. Optimizing BTEX biodegradation under denitrifying conditions. *Environ. Toxicol. Chem.* **10:**1437–1448.

Hutchins, S. R., and J. T. Wilson. 1991. Laboratory and field studies on BTEX biodegradation in a fuel-contaminated aquifer under denitrifying conditions. In R. E. Hinchee and R. F. Olfenbuttel (eds.), *In Situ Bioreclamation: Applications and Investigations for Hydrocarbon and Contaminated Site Remediation.* Butterworth-Heinemann, Stoneham, Mass., pp. 157–172.

Hutchins, S. R., G. W. Sewell, D. A. Kovacs, and G. A. Smith. 1991a. Biodegradation of aromatic hydrocarbons by aquifer microorganisms under denitrifying conditions. *Environ. Sci. Technol.* **25:**68–76.

Hutchins, S. R., W. C. Downs, J. T. Wilson. G. B. Smith, D. A. Kovacs, D. D. Fine, R. H, Douglas, and D. J. Hendrix. 1991b. Effect of nitrate addition on biorestoration of fuel-contaminated aquifer: Field demonstration. *Ground Water* **29:**571–580.

Jones, J., and I. C. Kong. 1992. Effects of metals on the reductive dechlorination of chlorophenols. In U.S. Environmental Protection Agency (ed.), *Bioremediation of Hazardous Wastes.* EPA/600/R-92/126. U.S. EPA, Washington, pp. 55–56.

Kaglan, D. L. 1990. Biotransformation pathways of hazardous energetic organonitro compounds. In D. Kamely (ed.), *Biotechnology and Biodegradation.* Gulf Publishing Company, Houston, Tex., pp. 155–180.

Kaiser, J. P., and J.-M. Bollag. 1992. Influence of soil inoculum and redox potential on the degradation of several pyridine derivatives. *Soil Biol. Biochem.* **24**:351–357.

Kaiser, J. P., and K. W. Hanselmann. 1982. Fermentative metabolism of substituted monoaromatic compounds by a bacterial community from anaerobic sediments. *Arch. Microbiol.* **133**:185–194.

Kaiser, J. P., R. D. Minard, and J.-M. Bollag. 1993. Transformation of 3- and 4-picoline under sulfate-reducing conditions. *Appl. Environ. Microbiol.* **59**:701–705.

Kamal, V. S. and R. C. Wyndham. 1990. Anaerobic phototrophic metabolism of 3-chlorobenzoate by *Rhodopseudomonas palustris* WS17. *Appl. Environ. Microbiol.* **56**:712–717.

Kastner, M. 1991. Reductive dechlorination of tri- and tetrachloroethylenes depends on transition from aerobic to anaerobic conditions. *Appl. Environ. Microbiol.* **57**:2039–2046.

Kingstrad, K. P., and K. Lindstrom 1984. Spent liquors from pulp bleaching. *Environ. Sci. Technol.* **18**:236A–248A.

Knowles, R. 1982. Denitrification. *Microbiol. Rev.* **46**:43–70.

Kobayashi, T., T. Hashinoga, E. Mikami, and T. Suzuki. 1989. Methanogenic degradation of phenol and benzoate in acclimated sludges. *Water Sci. Technol.* **21**:55–65.

Kohring, G.-W., J. E. Rogers, and J. Wiegel. 1989. Anaerobic biodegradation of 2,4-dichlorophenol in freshwater lake sediments at different temperatures. *Appl. Environ. Microbiol.* **55**:348–353.

Kuhn, E. P., P. J. Colberg, J. L. Schnoor, O. Warner, A. B. J. Zehnder, and R. P. Schwarzenbach. 1985. Microbial transformation of substituted benzenes during infiltration of river water to ground water: Laboratory column studies. *Environ. Sci. Technol.* **19**:961.

Kuhn, E. P., and J. M. Suflita. 1989a. Microbial degradation of nitrogen, oxygen, and sulfur heterocyclic compounds under anaerobic conditions. Studies with aquifer samples. *Environ. Toxicol. Chem.* **8**:1149–1158.

Kuhn, E. P., and J. M. Suflita. 1989b. Dehalogenation of pesticides by anaerobic microorganisms in soils and ground water: A review. In B. L. Sawhney and K. Brown (eds.), *Reactions and Movements of Organic Chemicals in Soils.* Soil Science Society of America and American Society of Agronomy, Madison, Wis., pp. 111–180.

Kuhn, E. P., J. Zeyer, P. Eicher, and R. P. Schwarzenbach. 1988. Anaerobic degradation of alkylated benzenes in denitrifying laboratory aquifer columns. *Appl. Environ. Microbiol.* **54**:490–496.

Lal, R., and D. M. Saxena. 1982. Accumulation, metabolism, and effects of organochlorine insecticides on microorganisms. *Microbiol. Rev.* **46**:95–127.

Larsen, S., H. V. Hendriksen and B. K. Ahring. 1991. Potential for thermophilic (50°C) anaerobic dechlorination of pentachlorophenol in different ecosystems. *Appl. Environ. Microbiol.* **57**:2085–2090.

Leahy, L. G., and R. R. Colwell. 1990. Microbial degradation of hydrocarbons in the environment. *Microbiol. Rev.* **54**:305–315.

Lee, M. D., and C. H. Ward. 1985. Microbial ecology of a hazardous waste disposal site: Enhancement of biodegradation. In N. N. Durham and A. E. Redelfs (eds.), *Proceedings of the 2nd International Conference on Ground Water Resources Quality Research.* Oklahoma State University Printing Services, Stillwater, Okla., p. 25.

Lee, M. D., J. M. Thomas, R. C. Borden, P. B. Bedient, C. H. Ward, and J. T. Wilson. 1988. Biorestoration of aquifers contaminated with organic compounds. *CRC Crit. Rev. Environ. Control* **18**:29–89.

Lewis, D. L., H. P. Kollig, and R. E. Hodson. 1986. Nutrient limitation and adaptation of microbial populations of chemical transformations. *Appl. Environ. Microbiol.* **51**:598–603.

Lindstrom, K., J. Nordin, and F. Osterberg. 1981. Chlorinated organics of low and high relative molecular mass in pulp mill bleaching effluents. In L. H. Keith (ed.), *Advances in the Identification and Analysis of Organic Pollutants,* Vol. 2. Ann Arbor Science Publishers, Ann Arbor, Mich.

Linkfield, T. G., J. M. Suflita, and J. M. Tiedje. 1989. Characterization of the acclimation period before anaerobic dehalogenation of halobenzoates. *Appl. Environ. Microbiol.* **55**:2773–2778.

Liss, S. N., and D. G. Allen. 1992. Microbiological study of a bleached kraft pulp mill aerated lagoon. *J. Pulp Paper Sci.* **18**:J216–J221.

Lovely, D. R., and E. J. P. Philips. 1992a. Bioremediation of uranium contamination with enzymatic uranium reduction. *Environ. Sci. Technol.* **26**:2228–2234.

Lovely, D. R., and E. J. P. Philips. 1992b. Reduction of uranium by *Desulfovibrio desulfuricans. Appl. Environ. Microbiol.* **58**:850–856.

Lovely, D. R., E. J. Phillips, Y. A. Gorby, and E. R. Landa. 1991. Microbial reduction of uranium. *Nature* **350**:413–416.

Lyngkilde, J., T. H. Christensen, B. Skov, and A. Foverskov. 1991. Redox zones downgradient of a landfill and implications for biodegradation of organic compounds. In Hinchee, R. E. and R. F. Olfenbuttel (eds.), *In Situ Bioreclamation: Applications and Investigations for Hydrocarbon and Contaminated Site Remediation.* Butterworth-Heinemann, Stoneham, Mass., pp. 363–376.

McCarty, P. L. 1985. Application of biological transformation in ground water. In N. N. Durham and A. E. Redelfs (eds.), *Proceedings of the 2nd International Conference on Ground Water Quality Research.* Oklahoma State University Printing Services, Stillwater, Okla., p. 6.

McCarty, P. L., and J. T. Wilson. 1992. Natural anaerobic treatment of a TCE plume St. Joseph, Michigan, NPL site. In U.S. Environmental Protection Agency (ed.), *Bioremediation of Hazardous Wastes*. EPA/600/R-92/126. U.S. EPA, Washington, pp. 47–50.

MacKay, D. M., and J. A. Cherry. 1989. Ground water contamination: Pump-and-treat remediation. *Environ. Sci. Technol.* **23**:630–636.

Madsen, T., and J. Aamand 1991. Effects of sulfuroxy anions on degradation of pentachlorophenol by a methanogenic enrichment culture. *Appl. Environ. Microbiol.* **57**:2453–2458.

Madsen, T., and D. Licht. 1992. Isolation and characterization of an anaerobic chlorophenol-transforming bacterium. *Appl. Environ. Microbiol.* **58**:2874–2878.

Major, D. W., W. W. Hodgins, and B. J. Butler. 1991. Field and laboratory evidence of in situ biotransformation of tetrachloroethene to ethene and ethane at a chemical transfer facility in North Toronto. In R. E. Hinchee and R. F. Olfenbuttel (eds.), *On Site Bioreclamation: Processes for Xenobiotic and Hydrocarbon Treatment*. Butterworth-Heinemann, Stoneham, Mass., pp. 141–171.

Major, D. W., C. I. Mayfield, and J. F. Barker. 1988. Biotransformation of benzene by denitrification in aquifer sand. *Ground Water* **26**:8–14.

Malmqvist, A., T. Welander, and L. Gunnarsson. 1991. Anaerobic growth of microorganisms with chlorate as an electron acceptor. *Appl. Environ. Microbiol.* **57**:2229–2232.

Mihelic, J. R., and R. G. Luthy. 1988. Microbial degradation of acetanaohthalene and naphthalene under denitrification conditions in soil-water systems. *Appl. Environ. Microbiol.* **54**:1188–1198.

Mikesell, M. D., R. H. Olson, and J. J. Kukor. 1991. Stratification of anoxic BTEX-degrading bacteria at three petroleum-contaminated sites. In R. E. Hinchee and R. F. Olfenbuttel (eds.), *In Situ Bioreclamation: Applications and Investigations for Hydrocarbon and Contaminated Site Remediation*. Butterworth-Heinemann, Stoneham, Mass., pp. 351–362

Miller, T. L., and M. J. Wolin. 1974. A serum bottle modification of the Hungate technique for cultivating obligate anaerobes. *Appl. Environ. Microbiol.* **27**:985–987.

Moat, A. G., and J. W. Foster. 1988. *Microbial Physiology*, 2d ed. John Wiley & Sons, Inc., New York.

Mohn, W. W., and K. J. Kennedy. 1992. Limited degradation of chlorophenols by anaerobic sludge granules. *Appl. Environ. Microbiol.* **58**:2131–2136.

Mohn, W. W., and J. M. Tiedje. 1992. Microbial reductive dehalogenation. *Microbiol. Rev.* **56**:483–507.

Morris, P. J., W. M. Mohn, J. F. Quensen III, J. M. Tiedje, and S. A. Boyd. 1992. Establishment of a polychlorinated biphenyl-degrading enrichment culture with predominantly meta dechlorination. *Appl. Environ. Microbiol.* **58**:3088–3094.

Murray, R. E., L. L. Parsons, and M. Scott Smith. 1990. Aerobic and anaerobic growth of rifampin-resistant denitrifying bacteria in soil. *Appl. Environ. Microbiol.* **56**:323–328.

Nesteler, F. H. M. 1974. Characterization of wood-preserving coal-tar creosote by gas-liquid chromatography. *Anal. Chem.* **46:**46–53.

Nielson, P. H., P. E. Holm, and T. H. Christenson. 1992. A field method for determination of ground water and ground water sediment associated potentials for degradation of xenobiotic compounds. *Chemosphere* **25:**449–462.

Nies, L., and T. M. Vogel. 1990. Effects of organic substrates on dechlorination of Aroclor 1242 in anaerobic sediments. *Appl. Environ. Microbiol.* **56:**2612–2617.

Nies, L., and T. M. Vogel. 1991. Identification of the proton source for the microbial reductive dechlorination of 2,3,4,5,6-pentachlorobiphenyl. *Appl. Environ. Microbiol.* **57:**2771–2774.

Owen, W. F., D. C. Stuckey, J. B. Healy, Jr., L. Y. Young, and P. L. McCarty. 1979. Bioassay for monitoring biochemical methane potential and anaerobic toxicity. *Water Res.* **13:**485–492.

Patel, G. B., B. J. Agnew, and C. J. Dicaire. 1991. Inhibition of pure cultures of methanogens by benzene ring compounds. *Appl. Environ. Microbiol.* **57:**2969–2974.

Payne, W. J. 1981. *Denitrification.* John Wiley & Sons, Inc., New York, pp. 118–133.

Piet, G. J., and J. G. M. M. Smeenk. 1985. Behavior of organic pollutants in pretreated Rhine water during dune infiltration. In C. H. Ward and P. L. McCarty (eds.), *Ground Water Quality.* John Wiley & Sons, Inc., New York, pp. 122–144.

Quensen, J. F., III, S. A. Boyd, and J. M. Tiedje. 1990. Dechlorination of four commercial polychlorinated biphenyl mixtures (Aroclors) by anaerobic microorganisms from sediments. *Appl. Environ. Microbiol.* **56:**2360–2369.

Quensen, J. F., III., J. M. Tiedje, and S. A. Boyd. 1988. Reductive dechlorination of polychlorinated biphenyls by anaerobic organisms form sediments. *Science.* **242:**752–754.

Ratledge, C. 1978. Degradation of aliphatic hydrocarbons. In R. J. Watkinson (ed.), *Developments in Biodegradation of Hydrocarbons,* Vol. 1. Applied Science, Ltd., London, pp. 1–46.

Rehm, H. J., and T. Reiff. 1981. Mechanism and occurrence of microbial oxidation of long chain alkanes. *Adv. Biochem. Eng.* **19:**175–215.

Reinhard, M., N. L. Goodman, and J. F. Barker. 1984. Occurrence and distribution of organic chemicals in two landfill leachate plumes. *Environ. Sci. Technol.* **18:**953–961.

Reineke, W., and H.-J. Knackmuss. 1988. Microbial degradation of haloaromatics. *Annu. Rev. Microbiol.* **42:**263–287.

Riley, R. G., T.G. Garland, K. Shiosaki, D. C. Mann, and R. E. Wildung. 1981. Alkylpyridines in surface waters, ground waters, and subsoils of a drainage located adjacent to an oil shale facility. *Environ. Sci. Technol.* **15:**697–701.

Rittmann, B. E., A. J. Valocchi, J. E. Odencrantz, and W. Bae. 1988. *In Situ Bioreclamation of Contaminated Ground water.* Illinois Hazardous Waste Research and Information Center, HWRIC RR 031.121.

Roberts, D. J., P. M. Fedorak, and S. E. Hrudey. 1990. CO_2 incorporation and 4-hydroxy-2-methylbenzoic acid formation during anaerobic metabolism of m-cresol by a methanogenic consortium. *Appl. Environ. Microbiol.* **56**:472–478.

Rogers, J. E., R. G. Riley, S. W. Li., M. L. O'Malley, and B. L. Thomas. 1985. Microbial transformation of alkylpyridines in ground water. *Water Air Soil Pollution* **24**:443–454.

Ronen, Z., and J.-M. Bollag. 1992. Rapid anaerobic mineralization of pyridine in a subsurface sediment inoculated with a pyridine-degrading *Alcaligenes* sp. *Appl. Microbiol. Biotechnol.* **37**:264–269.

Said, W. A., and D. L. Lewis. 1991. Quantitative assessment of the effects of metals on microbial degradation of organic chemicals. *Appl. Environ. Microbiol.* **57**:1498–1503.

Salkinoja-Salonen, M. S., M.-L. Saxelin, T. Jaakkola, J. Saarikoski, R. Hakulinen, and O. Koistinen. 1981. Analysis of toxicity and biodegradability of organochlorine compounds released into the environment in bleaching effluents of kraft pulping. In L. H. Keith (ed.), *Advances in the Identification and Analysis of Organic Pollutants in Water.* Ann Arbor Science Publishers, Ann Arbor, Mich., pp. 1131–1161.

Salkinoja-Salonen, M. S., R. Valo, J. Apajalahti, R. Hakulinen, L. Silakoski, and T. Jaakkola. 1983. Biodegradation of chlorophenolic compounds in wastes from wood-processing industry. In M. J. Klug and C. A. Reddy (eds.), *Current Perspectives in Microbial Ecology.* American Society for Microbiology, Washington, pp. 668–673.

Sayles, G. D., G. M. Lopez, D. F. Bishop, I. S. Kim, G. Young, M. J. Kupperie, and D. S. Lipton. 1992. Sequential anaerobic-aerobic treatment of contaminated soils and sediments. In U.S. Environmental Protection Agency (ed.), *Bioremediation of Hazardous Wastes.* EPA/600/R-92/126. U.S. EPA, Washington, pp. 103–104.

Schennen, U., K. Braun, and H.-J. Knackmuss. 1985. Anaerobic degradation of 2-fluorobenzoate by benzoate degrading, denitrifying bacteria. *J. Bacteriol.* **161**:373–380.

Schink, B. 1985. Degradation of unsaturated hydrocarbons by methanogenic enrichment cultures. *FEMS Microbiol. Ecol.* **31**:69–77.

Schink, B., and A. Tschech. 1988. Fermentative degradation of aromatic compounds. In S. R. Hagedorn, R. S. Hanson, and D. A. Kunz (eds.), *Microbial Metabolism and the Carbon Cycle.* Harwood Academic Publishing, London, pp. 213–226.

Schwarzenbach, R. P., W. Giger, E. Hoehn, and J. K. Schneider. 1983. Behavior of organic compounds during infiltration of river water to ground water: Field studies. *Environ. Sci. Technol.* **17**:472–479.

Semprini, L., G. D. Hopkins, P. V. Roberts, and P. L. McCarty. 1991. "In situ biotransformation of carbon tetrachloride, Freon 113, Freon 11 and 1,1,1-T CA under anoxic conditions. In R. E. Hinchee and R. F. Olfenbuttel (eds.), *On Site Bioreclamation: Processes for Xenobiotic and Hydrocarbon Treatment.* Butterworth-Heinemann, Stoneham, Mass., pp. 41–59.

Semprini, L., G. D. Hopkins, P. L. McCarty, and P. V. Roberts. 1992. In situ biotransformation of carbon tetrachloride and other halogenated compounds resulting from biostimulation under anoxic conditions. *Environ. Sci. Technol.* **26:**2454–2461.

Sewell, G. W., and S. A. Gibson. 1991. Effects of alternate electron acceptors and specific metabolic inhibitors on the reductive dechlorination of tetrachloroethene (PCE) in aquifer microcosms (Abstract Q-38). 91st General Meeting of the American Society for Microbiology. American Society for Microbiology, Washington.

Shanker, R., J.-P. Kaiser, and J.-M. Bollag. 1991. Microbial transformation of heterocyclic molecules in deep subsurface sediments. *Microb. Ecol.* **22:**305–316.

Sharak Genthner, B. R., W. A. Price II, and P. H. Pritchard. 1989a. Anaerobic degradation of chloroaromatic compounds in aquatic sediments under a variety of enrichment conditions. *Appl. Environ. Microbiol.* **55:**1466–1471.

Sharak Genthner, B. R., W. A. Price II, and P. H. Pritchard. 1989b. Characterization of anaerobic dechlorinating consortia derived from aquatic sediments. *Appl. Environ. Microbiol.* **55:**1472–1476.

Shelton, D. R., and J. M. Tiedje. 1984a. General method for determining anaerobic biodegradation potential. *Appl. Environ. Microbiol.* **47:**850–857.

Shelton, D. R., and J. M. Tiedje. 1984b. Isolation and partial characterization of bacteria in an anaerobic consortium that mineralizes 3-chlorobenzoic acid. *Appl. Environ. Microbiol.* **48:**840–845.

Sisk, W., and K. Keehan. 1992. Biotreatment of hazardous waste in the United States Army. *Microbial Clean-Up.* **1:**5.

Smith, G. A., J. S. Nickels, B. D. Kerger, J. D. Davis, S. P. Collins, J. T. Wilson, J. F. McNabb, and D. C. White. 1986. Quantitative characterization of microbial biomass and community structure in subsurface material: A prokaryotic consortium responsive to organic contamination. *Can. J. Microbiol.* **32:**104–109.

Smith, R. L., and M. J. Klug. 1987. Flowthrough reactor flasks for study of microbial metabolism in sediments. *Appl. Environ. Microbiol.* **53:**371–374.

Smolenski, W., and J. M. Suflita. 1987. The microbial metabolism of cresols in anoxic aquifers. *Appl. Environ. Microbiol.* **53:**710–716.

Spain, J. C., P. H. Pritchard, and A. W. Bourquin. 1980. Effects of adaptation on biodegradation rates in sediment/water cores from estuarine and freshwater environments. *Appl. Environ. Microbiol.* **40:**726–734.

Sprent, J. I. 1987. *The Ecology of the Nitrogen Cycle.* Cambridge University Press, New York, pp. 49–66.

Steiert, J. G., and R. L. Crawford. 1986. Catabolism of pentachlorophenol by a *Flavobacterium* sp. *Biochem. Biophys. Res. Commun.* **141:**825–830.

Stetzenbach, L. D., L. M. Kelly, and N. Sinclair. 1986. Isolation, identification and growth of well water bacteria. *Ground Water* **24:**8.

Suflita, J. M., S. A. Gibson, and R. E. Beeman. 1988. Anaerobic biotransformation of pollutant chemicals in aquifers. *J. Indust. Microbiol.* **3:**179–194.

Suflita, J. M., A. Horowitz, D. R. Shelton, and J. M. Tiedje. 1982. Dehalogenation: A novel pathway for the anaerobic biodegradation of haloaromatic compounds. *Science.* **218:**1115–1117.

Tarvin, D., and A. M. Buswell. 1934. The methane fermentation of organic acids and carbohydrates. *J. Am. Chem. Soc.* **56:**1751–1755.

Taylor, B. F., W. L. Hearn, and S. Pincus. 1979. Metabolism of monofluoro- and monochlorobenzoates by denitrifying bacteria. *Arch. Microbiol.* **122:**301–306.

Thauer, R. K., K. Jungermann, and K. Decker. 1977. Energy conservation in chemotrophic anaerobic bacteria. *Bacteriol. Rev.* **41:**100–180.

Tiedje, J. M., S. A. Boyd, and B. Z. Fathepure. 1987. Anaerobic degradation of chlorinated aromatic hydrocarbons. *Dev. Indust. Microbiol.* **27:**117–127.

Towill, L. E., J. S. Drury, B. L. Whitfield, E. B. Lewis, E. L. Galyan, and A. S. Hammons. 1978. *Review of the Environmental Effects of Pollutants,* Vol. 5: *Cyanide.* U.S. Environmental Protection Agency, Cincinnati, Ohio.

Tschech, A. T., and G. Fuchs. 1987. Anaerobic degradation of phenol by pure cultures of newly isolated denitrifying pseudomonads. *Arch. Microbiol.* **148:**213–217.

Urbanek, M., T. Styck, C. Wyndham, and M. Goldner. 1989. Use of bromobenzoate for cross-adaptation of anaerobic bacteria in Lake Ontario sediments for degradation of chlorinated aromatics. *Lett. Appl. Microbiol.* **9:**191–194.

U.S. Environmental Protection Agency. 1992. *Bioremediation of Hazardous Wastes.* EPA/600/R-92/126. U.S. EPA, Washington.

Van den Tweel, W. J. J., J. B. Kok, and J. A. M. DeBont. 1987. Reductive dechlorination of 2,4-dichlorobenzoate to 4-chlorobenzoate and hydrolytic dehalogenation of 4-chloro, 4-bromo, and 4-iodobenzoate by *Alcaligenes denitrificans* NTB-1. *Appl. Environ. Microbiol.* **53:**810–815.

Van Dort, H. M., and D. L. Bedard. 1991. Reductive *ortho* and *meta* dechlorination of a polychlorinated biphenyl congener by anaerobic microorganisms. *Appl. Environ. Microbiol.* **57:**1576–1578.

Vogel, T. M., and D. Grbic-Galic. 1986. Incorporation of oxygen from water into toluene and benzene during anaerobic fermentative transformation. *Appl. Environ. Microbiol.* **52:**200–202.

Vogel, T. M., and P. L. McCarty. 1985. Biotransformation of tetrachloroethylene to trichloroethylene, dichloroethylene, vinyl chloride, and carbon dioxide under methanogenic conditions. *Appl. Environ. Microbiol.* **49:**1080–1083.

Vogel, T. M., C. S. Criddle, and P. L. McCarty. 1987. Transformations of halogenated aliphatic compounds. *Environ. Sci. Technol.* **21:**722–736.

Ward, T. E. 1985. Characterizing the aerobic and anaerobic microbial activities in surface and subsurface soils. *Environ. Toxicol. Chem.* **4:**727–737.

Watson-Craik, I. A. 1990. Landfill co-disposal. In E. Senior (ed.), *Microbiology of Landfill Sites.* CRC Press, Inc., Boca Raton, Fla. pp. 159–214.

White, D. C., H.F. Frederickson, M. J. Gerhon, G. A. Smith, and R. F. Marck.

1983. The ground water aquifer microbiota: Biomass, community structure, and nutritional status. *Dev. Ind. Microbiol.* **24**:189–199.

Wiggins, B. A., S. H, Jones, and M. A. Alexander. 1987. Explanations of the acclimation period preceding the mineralization of organic chemicals in aquatic environments. *Appl. Environ. Microbiol.* **53**:791–796.

Williams, R. J., and W. C. Evans. 1975. The metabolism of benzoate by *Moraxella* species through anaerobic nitrate respiration. *Biochem. J.* **148**:1–10.

Wilson, B. H., B. Bledsoe, and D. Kampbell. 1987. Biological processes occurring at an aviation gasoline site. In R. C. Averett and D. M. McKnight (eds.), *Chemical Quality of Water and the Hydrologic Cycle*. Lewis Publishers, Chelsea, Mich., pp. 125–137.

Wilson, J., and M. J. Noonan. 1984. Microbial activity in model aquifer systems. In G. Bitton and C.P. Gerba (eds.), *Ground Water Pollution Microbiology*. John Wiley and Sons, Inc., New York, pp. 117–133.

Wilson, J. T., and J. F. McNabb. 1983. Biological transformation of organic pollutants in ground water. *EOS* (*Am. Geophys. Union Trans.*) **64**:505–506.

Wilson, J. T., and C. H. Ward. 1987. Opportunities for bioreclamation of aquifers contaminated with petroleum hydrocarbons. *Dev. Ind. Microbiol.* **27**:109–116.

Wilson, J. T., J. F. McNabb, D. L. Balkwill, and W. C. Ghiorse. 1983. Enumeration and characterization of bacteria indigenous to a shallow water-table aquifer. *Ground Water* **21**:134–142.

Woods, S. L., J. F. Ferguson, and M. M. Benjamin. 1989. Characterization of chlorophenol and chloromethoxybenzene biodegradation during anaerobic treatment. *Environ. Sci. Technol.* **23**:62–68.

Zeyer, J., E. P. Kuhn, and R. P. Schwarzenbach. 1986. Rapid microbial mineralization of toluene and 1,3-dimethylbenzene in the absence of molecular oxygen. *Appl. Environ. Microbiol.* **52**:944–947.

Zhang, X., and J. Wiegel. 1990. Sequential anaerobic degradation of 2,4-dichlorophenol in freshwater sediments. *Appl. Environ. Microbiol.* **56**:1119–1127.

Zhang, X., T. V. Morgan, and J. Wiegel. 1990. Conversion of ^{13}C-1 phenol to ^{13}C-4 benzoate, an intermediate step in the anaerobic degradation of chlorophenols. *FEMS Microbiol. Lett.* **67**:63–66.

Zitomer, D. H., and R. E. Speece. 1993. Sequential environments for enhanced biotransformation of aqueous contaminants. *Environ. Sci. Technol.* **27**:226–244.

10
Bioremediation and Government Oversight

Katherine Devine
DEVO Enterprises, Inc., Washington, D.C.

Definition of Terms
Legislation, Regulation, and Policy

While sometimes used interchangeably by those not in the government, each of the terms *legislation, regulation,* and *policy* has a unique meaning. *Legislation* refers to law, which is enacted by Congress and which gives mandates or general direction to the particular government entity(ies) having jurisdiction over the activity(ies) at which the legislation is directed.

Government organizations enact the legislation guidance or directives through the promulgation of *regulations*. Regulations are first released as "proposed" to the public for comment. The purpose of a proposed regulation is to allow the public, both the target regulated community and public interest groups and other interested parties, the opportunity to comment. Comments are taken into account, with the final regulation reflecting such comments or defending the government's stance for not incorporating certain comments.

Policies are interpretations of existing regulations or legislation issued by a government body, and as such, they do not carry strict legal authority. The extent to which the target community may respond and adhere to such guidance can vary, in many cases heeding the written direction as if it were a regulation. Some policies provide the states with the ultimate interpretive authority. Policy statements can be released as a proposed work for public comment, with a final statement issued after comments have been reviewed by the government issuing the proposal.

The U.S. Environmental Protection Agency

The federal government organization that has produced the most regulations and policies that affect the bioremediation industry is the U.S. Environmental Protection Agency (EPA). The EPA is a relatively young agency, created in 1970 under the Nixon administration. As an "agency," the EPA does not have presidential cabinet status, as do federal departments, such as the Department of Energy or the Department of Defense. However, legislation was introduced in Congress in May of 1991 to have the EPA made a cabinet-level position (HMCRI, 1991).

The U.S. EPA is organized with its headquarters in Washington and with 10 regional offices throughout the United States, each representing several different states. EPA headquarters is responsible primarily for producing the regulations and programs or establishing policies that the regions carry forward in their respective locales.

EPA headquarters is divided into subunits termed *offices*. Various offices have been created to work under the mandate of particular environmental legislation (e.g., the Office of Toxic Substances was formed to carry out the mandate of the Toxic Substances Control Act and the Office of Solid Waste was created to carry out the Resource Conservation and Recovery Act). Offices are further subdivided into sister offices.

The following flowchart (Fig. 10-1) shows the major offices and the relationship of the offices to each other:

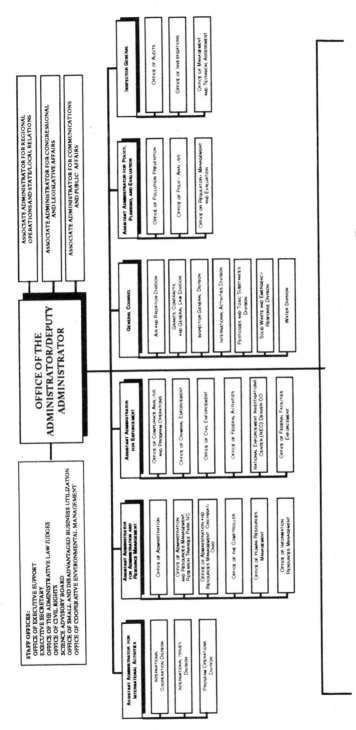

Figure 10-1. U.S. Environmental Protection Agency. Flowchart showing the major offices.

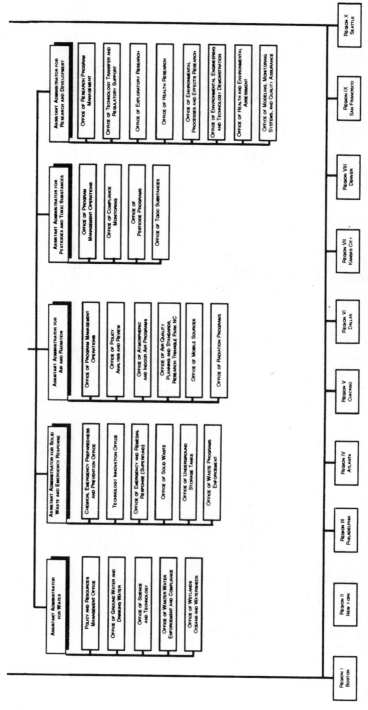

Figure 10-1. (*Continued*) U.S. Environmental Protection Agency. Flowchart showing the major offices.

Overview of Current Government Regulations and Programs of Interest to the Bioremediation Community

The ramifications of legislation relevant to the bioremediation industry written 10 to 15 years ago are being felt today through the proposal, promulgation, and enactment of select environmental regulations and policies. The purpose of this chapter is to highlight the number and scope of government actions that affect those involved in bioremediation. Government actions, organized according to select EPA offices that have issued various regulations and policies, then other federal department actions, and the Canadian government's activities, will be discussed. This is intended to familiarize the reader with some of the federal actions of importance to the bioremediation community and is not meant to be all-encompassing or to substitute for legal counsel.

U.S. Environmental Protection Agency

Office of Solid Waste

The Land Disposal Restrictions. The Resource Conservation and Recovery Act (RCRA) was passed in 1976, with the Hazardous and Solid Waste Amendments (HWSA) to RCRA being passed in 1984. The purpose of the amendments was to expand the scope of the original RCRA concerning primarily land disposal of hazardous waste. According to EPA, a waste is hazardous if it exhibits any of the four characteristics of ignitability, corrosivity, reactivity, or toxicity or the waste is listed by the EPA as hazardous. Currently, there are over 450 process waste streams and commercial chemical and off-specification products considered hazardous according to these specifications.

RCRA created a tracking and permitting system for hazardous waste from generation through treatment, storage, and disposal. Among other things, HWSA required EPA to look at all hazardous waste to determine if any should be banned from land disposal. The amendments prohibit land disposal of untreated hazardous waste unless a petitioner can demonstrate that there will be "no migration of hazardous constituents...for as long as the wastes remain hazardous." Any waste banned from land disposal must be treated and rendered less hazardous before it can be disposed of on the land. HSWA also directed EPA to develop treatment standards for all hazardous wastes. Wastes meeting these standards are not subject to this restriction of land disposal of wastes, or what is known as the *land disposal restrictions* (LDRs) or *land ban* (EPA, 1989a, 1990a).

Congress mandated that LDRs be promulgated on a phased-in basis. Hence a total of five rulemakings have taken place. The five rules and their mandated date of promulgation are

1. Solvents and dioxins, November 1986

2. The California list wastes (so called because this is the listing of hazardous wastes developed by the state of California as the subject of regulations to restrict land disposal of hazardous waste containing these constituents), and the remainder of the listed wastes were divided into thirds, based on their volume and toxicity, July 1987

3. The first third, August 1988

4. The second third, June 1988

5. The third third, May 1990 (EPA, 1989a, 1990a)

EPA is required to promulgate regulations specifying levels or methods of treatment which substantially reduce the toxicity or mobility of RCRA hazardous wastes and their hazardous constituents. Treatment standards are based on the concept of "best demonstrated available technology" (BDAT):

- A treatment technology is *demonstrated* for a given waste if it is currently used to treat the waste; bench- and pilot-scale technologies are not demonstrated.

- A technology is *available* if it can be purchased or leased. In addition, the technology must substantially diminish the toxicity of the waste or substantially reduce the likelihood of migration of hazardous constituents from the waste.

- After determining the technologies that are demonstrated and available, a statistical analysis is used which incorporates a variability factor to account for normal process variations in well-designed and well-operated treatment systems to determine the *best* technology (Jonesi, 1989).

Because the regulations list, by hazardous waste, either the cleanup standard or the technology(ies) to be used *prior* to land placement, biotreatment cannot be used in instances where a technology has been specified nor in instances where the cleanup level is too low for bioremediation to meet. Consequently, there are instances where bioremediation, which could be used effectively, is prohibited. These rules have far-reaching effects in that other agency programs, such as Superfund and the RCRA corrective action program (discussed below), are subject to the LDRs.

While currently hindering the use of biotreatment in the above-described situations, these same regulations also create new or expand existing market opportunities. For example, biotreatment of wastes in place, or in situ, is a means of avoiding the LDRs. Additionally, the LDRs will motivate hazardous waste generators to explore more thoroughly source-reduction options. Such actions will expand existing biotreatment markets or create new applications or microbial product usage.

Toxicity Characteristic Leaching Procedure. As stated above, RCRA specifies four characteristics to identify a solid waste as a hazardous waste. One of the characteristics used for identification is *toxicity.* In June of 1986, The EPA proposed to revise its existing toxicity characteristics identification system with a new test, the *toxicity characteristic leaching procedure* (TCLP). The intent of the proposed rule was to broaden the scope of the hazardous waste program. On March 29, 1990, the final rule, Hazardous Waste Management System; Identification and Listing of Hazardous Waste; Toxicity Characteristics Revisions; Final Rule, was promulgated (EPA, 1990b).

This final rule implements two major changes to the hazardous waste identification regulations: 25 chemicals have been added to the original list of 8 metals and 6 pesticides that are to be tested to determine if a waste is hazardous. The 25 additional chemicals and maximum allowable concentrations are

Benzene, 0.5 ppm

Carbon tetrachloride, 0.5 ppm

Chlordane, 0.03 ppm

Chlorobenzene, 100.0 ppm

Chloroform, 6.0 ppm

o-Cresol, 200.0

m-Cresol, 200.0

p-Cresol, 200.0

1,4-Dichlorobenzene, 7.5 ppm

1,2-Dichloroethane, 0.5 ppm

1,1-Dichlorethylene, 0.7 ppm

2,4-Dinitrotoluene, 0.13 ppm

Heptachlor, 0.008 ppm

Hexachlorobenzene, 0.13 ppm

Hexachloro-1,3-butadiene, 0.5 ppm

Hexachloroethane, 3.0 ppm

Methylethyl ketone, 200.0 ppm

Nitrobenzene, 2.0 ppm

Pentachlorophenol, 100.0 ppm

Pyridine, 5.0 ppm

Tetrachloroethylene, 0.7 ppm

Trichloroethylene, 0.5 ppm

2,4,5-Trichlorophenol, 400.0

2,4,6-Tichlorophenol, 2.0

Vinylchloride, 0.2 ppm

Any of these wastes exceeding these levels, when analyzed with the TCLP test, will be regulated as a hazardous waste (EPA, 1990c).

RCRA has certain requirements for those who are defined as hazardous waste generators and treatment/storage/disposal (TSD) facilities. By bringing previously unregulated wastes into regulation, previously unregulated parties may now fall under the definition of generator or TSD and become subject to the requirements. Previously nonhazardous wastes such as petroleum industry–related chemical substances (e.g., gasoline) may now be regulated (API, 1990).

RCRA defines waste generators on the basis of amount of hazardous waste produced per month. Small-quantity generators (SQGs) are defined as producing between 220 and 2200 lb of hazardous waste per month (EPA, 1990b). SQGs may now become large-quantity generators (LQGs), those producing 2200 lb or more a month of hazardous waste, subject to LQG stipulations of RCRA.

Additionally, existing generators may have to apply for TSD permits and existing TSDs may have to modify their permits or permit applications.

Corrective Action Program Rule. *Corrective action* is defined by the EPA as activity undertaken to clean up environmental contamination resulting from the mismanagement of soils and hazardous waste. The purpose of the corrective action program is to require the remediation of historical contamination at facilities that currently manage hazardous wastes. Since 1982, the RCRA program has been implementing corrective action requirements for releases to ground water from regulated units through permits. In 1984, the corrective action provisions of HSWA significantly expanded the scope of the RCRA program. Every facility seeking a permit to run a TSD facility after November 8, 1984 is required to undertake corrective action of all releases of hazardous waste from solid waste management units (SWMUs) at the facility, regardless of when the wastes were placed in the unit involved. An SWMU, according to EPA, is any discernible unit at which solid wastes have been placed at any time, irrespective of whether the unit was intended for management of solid or hazardous waste (EPA, 1990d).

On July 27, 1990, the EPA issued Corrective Action for Solid Waste Management Units at Hazardous Waste Management Facilities, Proposed Rule in order to establish a regulatory framework for corrective action. Comments received on this proposal are in the process of being reviewed, and a final rule is not expected for another year (EPA, 1991b).

There are several procedural components to the corrective action plan:

RCRA facility assessment (RFA)—determines if there is a sufficient evidence of a release to require the owner/operator to undertake additional steps to characterize the release

RCRA facility investigation (RFI)—characterizes the nature, extent, and rate of migration of releases identified in the RFA

Corrective measures study (CMS)—identifies the appropriate corrective measures

Corrective measures implementation (CMI)—design, construction, operation and maintenance of the selected response action

Interim measures (IM)—conducts short-term measures to respond to immediate threats (EPA, 1990d)

Where contamination is identified during the facility investigation, EPA or an authorized state will have to make a decision on whether further analysis, including analysis of potential remedies, is appropriate or whether the contamination is at an insignificant level (EPA, 1990d).

A primary goal of corrective action is to achieve cleanup consistent with existing media-specific cleanup standards or, when such standards do not exist, to achieve protection against risks to human health such that the excess lifetime risk from exposure to a carcinogenic hazardous waste constituent in soil, air, ground water, or surface water does not exceed 10^{-6} (EPA, 1990d).

One of the major drawbacks of this proposed plan, from the bioremediation community's perspective, is that the LDRs (see above) pertain to corrective action sites. Therefore, the same ramifications as experienced under the land ban will be encountered in the corrective action program (see above).

If a biotreatment company is either a treatment, storage, or disposal facility, then it would be subject to the corrective action program.

Contaminated Soil and Debris Advanced Notice of Proposed Rulemaking. Soil and debris [*debris* defined by the EPA, in part, as solid material that has been originally manufactured or processed, or plant or animal matter, or natural geologic material, such as gravel, cobbles, and boulders (EPA, 1991c)] contaminated with hazardous wastes are considered to be hazardous wastes subject to the land disposal restrictions. Certain types of contaminated soil and debris, such as those resulting from response actions taken under the Comprehensive Environmental Response, Compensation and Liability Act of 1980 (CERCLA) or corrective actions under RCRA, pose a special problem for the EPA because they are, in general, highly variable in nature and may not be treatable to the levels established for the contaminating wastes. The BDAT levels established for wastes are usually based on data from treatment of pure industrial process wastes, which are typically less difficult to treat than those that are highly variable (EPA, 1990e).

To resolve this problem where highly variable wastes may not be treatable to established levels, EPA intends to institute treatment standards specifically for these types of CS&D using data from actual treatment tests on these CS&D wastes. The EPA issued an Advanced Notice of Proposed Rulemaking (ANPRM) in May of 1991, with a call for data to assist in making the determinations. A similar call for data concerning contaminated soil is scheduled for the fall of 1991 (EPA, 1991d).

Currently, the only relief is to petition the EPA for a treatability variance if the waste is a CERCLA or RCRA corrective action cleanup. If the variance is granted, an exemption from the BDAT levels is achieved. It takes a significant amount of time for EPA review. In fact, such review authority has recently been granted to the regional offices in an attempt to speed up the petition process.

The EPA expects that this action will bring some relief to the current situation in which any soil with a hazardous waste is considered a hazardous waste itself and therefore subject to the BDAT stipulations. Therefore, the use of bioremediation in certain instances is excluded, as discussed in the section on the LDRs.

Office of Emergency and Remedial Response

The Comprehensive Environmental Response, Compensation and Liability Act. The Comprehensive Environmental Response, Compensation and Liability Act (CERCLA) was enacted in 1980 in response to the public's concern over health and environmental threats posed by uncontrolled, abandoned hazardous waste sites and the perceived inadequacy of prior environmental laws to address this problem. It was estimated that at that time, more than 90 percent of the hazardous waste produced was disposed of in an environmentally unsafe manner. There also were inadequate funds at the time for cleaning up hazardous waste sites (Cooke, 1990).

CERCLA established the basic federal program for addressing the problem of hazardous substances that have been released into the environment. CERCLA contained two basic types of provisions: those relating to the investigation and cleanup of waste sites and those relating to the imposition of liability for these activities. CERCLA was funded by a 5-year $1.8 billion "superfund." Superfund is financed by a series of special taxes, federal revenues, and cleanup cost recoveries. Superfund authorized EPA to perform long- and short-term cleanups on hazardous substances, recover costs from responsible parties, and advance the EPA's scientific and technical compatibilities with respect to alternative technologies. Traditional cleanup technologies have been source controls and residual contamination management (capping or excavation and landfilling) (Cooke, 1990).

In October of 1986, the Superfund Amendments and Reauthorization Act (SARA) of 1986 was enacted. SARA limits EPA's ability to use source controls. SARA established a preference for on-site treatment and significantly expanded research capabilities to develop cost-effective innovative treatment technologies. The fund amount was expanded to $8.5 billion in 1986 through 1991 (EPA, 1989b). At the end of 1990, SARA was extended to September of 1994, with the taxing authority extended to 1996. A total of $5.1 billion also was approved (EPA, 1991e).

The process of determining the technology to be used for cleanup includes a remedial investigation/feasibility study (RI/FS). The RI is a detailed investigation/characterization of the site. The FS is an evaluation, based on the data generated from the RI, of the alternative treatment options. A record of decision (ROD) is then prepared that lists the treatment(s) chosen for cleanup (Bakst, 1991).

Bioremediation companies can bid for contracts to perform remedial cleanups. From 1982 through 1989, EPA selected bioremediation for use at Superfund sites 22 times. Based on ROD data, bioremediation was chosen 5 times for Superfund site cleanup in 1990 (EPA, 1991f). The number of RODs signed in which bioremediation is a treatment of choice is expected to continue to grow.

Site Program. As stated above, SARA required EPA to establish an alternative or innovative treatment technology research and demonstration program where the technology could be used in response actions to achieve more permanent protection of human health and the environment. In response, EPA established the SITE program.

The SITE program, which was jointly established and is jointly run by OSWER and the Office of Research and Development (ORD), supports the development of technology for assessing and treating wastes for Superfund sites. The program provides the opportunity for technology developers to demonstrate their technologies' capabilities to successfully process and remediate Superfund waste. EPA evaluates the technology and provides an assessment of potential for future use for Superfund cleanups. The SITE program includes the demonstration program (for field-level work), the emerging-technologies program (for bench- and pilot-scale work), and the monitoring and measurement technology program (technologies to assess contamination) (EPA, 1991g).

The SITE program can provide financial incentives for bioremediation companies to develop their innovative technology. As of late August 1991, there were 16 biotreatment (primarily bioremediation plus more historical microbial applications) projects out of a total of 75 projects in the SITE demonstration program (Martin, 1991).

Office of Water

The National Contingency Plan. The first national contingency plan (NCP) of the United States was published in 1968 after an accident involving 37 million gallons of crude oil being spilled off the coast of England in March of 1967. The plan's scope has been broadened over the years. For oil spills, the national system operates under the Federal

Water Pollution Control Act as amended by the Clean Water Act (CWA) of 1977. For hazardous substances, operational authority is under the CERCLA (the National Response Team, 1990). The Oil Pollution Act of 1990 (OPA) amended the CWA to call for revisions to parts of the NCP (OTA, 1991).

The OPA directed EPA to prepare a list of dispersants, other chemicals, and other spill-mitigating devices and substances that may be used to execute the plan. Subpart J of the NCP governs the use of additives for marine oil spills. It provides for preauthorization for the use of regulated agents through an advance planning procedure (OTA, 1991).

On-scene coordinators are authorized to use biological additives that have been preapproved by regional response teams. The EPA emphasizes preplanning and has been working to develop a listing of microbial agents. A company selling microbial products for use in aquatic oil spill environments should be registered for its product to be considered for oil spill cleanup.

Office of Toxic Substances

Proposed Biotechnology Rule. In 1976, Congress passed the Toxic Substances Control Act (TSCA), designed to be a "gap-filling" measure as pertained to existing legislation at that time. This legislation included provisions for a notification system for manufacturers of new chemical substances. In a 1984 Proposal for a Coordinated Framework for Regulation of Biotechnology; Notice, the EPA, FDA, and USDA made statements regarding their respective federal oversight of biotechnology. In its statement, the EPA defined microbes as chemical substances and thus subject to regulations under TSCA. EPA stated that microorganisms produced by rDNA, rRNA, and cell fusion should be subject to EPA review. Much public comment was received criticizing the process-based orientation of the document.

Based on comment received, in 1986, the Coordinated Framework of Regulation of Biotechnology, Announcement of Policy and Notice for Public Comment was issued. In addition to the EPA, the FDA, USDA, OSHA, and NIH contributed to this policy notice.

The EPA's revised policy, a product-based approach, stated that intergeneric microorganisms manufactured or imported for commercial purposes are subject to premanufacture notification (a PMN is currently required for every chemical imported or manufactured that is not on an EPA inventory of chemicals); intergeneric microorganisms manufactured or imported for research and development for release are subject to reporting; pathogens or microbes with pathogenic material are subject to reporting requirements similar to those of the PMN program; and all other microorganisms are subject to general reporting requirements (OSTP, 1986). This modified perspective also has been the cause of much criticism and debate since its issuance. The EPA continues to work under this policy while a proposed rule is being written for public comment.

Currently, the EPA should be alerted prior to environmental release of intergeneric and pathogenic microorganisms, including microorganisms with pathogenic material for TSCA uses, in order for the EPA to determine the level of review deemed necessary prior to release. To date, a few field experiments have been reviewed that involve microbes for TSCA uses that have been created by movement of genetic material. These experiments included enhancement of nitrogen-fixing capabilities of *Rhizobium* and *Bradyrhizobium* and addition of tracer capability to a *Pseudomonas* (Zeph, 1991).

As of the fall of 1991, the EPA was drafting a proposed rule with a somewhat different focus from the 1986 policy statement. The primary factor would be to extend its oversight to the environmental release of organisms with deliberately modified hereditary traits created by scientifically advanced techniques.

PCB Permits. Congress gave the EPA explicit regulatory oversight of PCBs in the Toxic Substances Control Act (TSCA). Organizations or persons wishing to dispose of PCBs are required to use approved methods and to obtain a permit from the EPA. In 1986, the Office of Toxic Substances issued guidelines for persons applying to the EPA for approval of PCB disposal by methods other than incineration (EPA, 1986).

Alternative methods of PCB destruction include, but are not limited to, catalytic dehydrochlorination, chlorolysis, plasma arc, ozonation, catalyzed oxidation, and microbiological and sodium-catalyzed decomposition of the PCB molecules. Methods for decontamination of PCB-contaminated materials by removal and concentration of the PCBs also are considered alternative methods of PCB destruction (EPA, 1986).

For a company who is dealing with a cleanup that involves PCBs, an application for a permit must be filed for review and granting of the permit prior to any cleanup work. Activities such as treatability studies are also subject to this stipulation. As of the late summer of 1991, there have been seven permits issued for bioremediation of PCBs (Blake, 1991).

Department of Transportation, Public Health Service, and U.S. Postal Service: Shipping Regulations Affecting Microbial Products

Over the past 2 years, the Department of Transportation (DOT), the Centers for Disease Control's Public Health Service (PHS), and the U.S. Postal Service (USPS) have proposed or finalized regulations pertaining to the transportation of microbiological products. This action has been the result of public concern for adequate worker protection from exposure to select microbial products, such as the AIDS virus.

As written, these rules have a broad, all-encompassing effect that will potentially bring a significant number of microbial product–related shipping activities under regulation. This effect is due to the extremely broad definitions found in these rules pertaining to microbial products.

The U.S. Postal Service (USPS) published its final regulation, The Final Rule on Mailability of Etiologic Agents, in August of 1989. With guidance from the PHS and DOT, the USPS issued a rule that stipulates four categories of biological materials to be regulated: (1) etiologic agents, (2) etiologic agent preparations, (3) clinical specimens, and (4) biological products. The definitions relevant to the biotreatment industry are *etiologic agents* and *etiologic agent preparations*. *Etiologic agent* is defined as "a microbiological agent or its toxin that causes, or can cause, human or animal disease"; *etiologic agent preparation* means "a culture or suspension of an etiologic agent and includes purified or partially purified spores or toxins" (Wick, 1990).

This rule requires that (1) etiologic agent preparations must be packaged in both a primary (e.g., vial) and secondary (outer) container, with the ability to absorb the primary contents in case of breakage, and (2) the total volume of the primary container must not exceed 50 ml (Wick, 1990).

On December 21, 1990, the Department of Transportation (DOT) published a final rule in which an *etiologic agent* or *infectious substance* is defined as "a viable microorganism or its toxin, which causes or may cause disease in humans or animals, and includes those agents listed in 42 CFR 72.3 of the regulations of the Department of Health and Human Services or any other agent that has the potential to cause severe, disabling or fatal disease" (DOT, 1990).

Such agents are to be (1) shipped in inner packaging, comprising a water-tight primary receptacle, and secondary packaging if necessary, absorbent material between the primary and secondary packaging that is capable of absorbing the contents of the primary package. Packaging must be capable of passing specified tests delineated in the rule, and (2) the current 50-ml volume exemption has been abolished in this new rule.

In response to public comment concerning the etiologic agent definition, DOT issued a notice in February which delays the effective date of the etiologic definition to September 30, 1991.

The Public Health Service's (PHS) proposed rule of March of 1990, Interstate Shipment of Etiologic Agents, states that an *etiologic agent* is a "microbiological agent or its toxin that causes, or may cause, human disease." An *etiologic agent preparation* is defined as "a culture or suspension of an etiologic agent and includes purified spores or toxins that are themselves etiologic agents." This rule proposed to abolish PHS current written guidance to the regulated community in the form of an etiologic agent listing mechanism.

Under this rule, (1) a primary container and secondary container are necessary, with each set of primary and secondary containers enclosed in an outer shipping container constructed of specified materials, and (2) the total volume of etiologic agent shipped in any package must not exceed 50 ml.

The etiologic definition has been suspended by DOT until September 30, 1991. The definitions of etiologic agent and etiologic agent preparation can be all-inclusive of industries dealing with the transportation of microorganisms. Companies in the bioremediation industry who routinely ferment or re-formulate and ship microorganisms to clients for on-site waste cleanup are potentially subject to these rules.

The abolition of the 50-ml exemption in the DOT rule, coupled with the etiologic agent definition, could mean that routine bioremediation activities, such as taking samples of indigenous populations from the remediation site to the laboratory, would be subject to regulatory oversight.

The party who prepares the microorganisms for transport should be able to provide evidence that the microorganisms are non-health-threatening, for example, by knowing the genus and species of all of the microorganisms being handled and by being knowledgeable as to whether the various genuses and species are considered by health experts, such as the PHS, to be health-threatening.

The U.S. Department of Agriculture (USDA)

The USDA, through the Federal Plant Pest Act (FPPA), regulates the importation and interstate transport of plant pests. The FPPA defines a plant pest as that which causes disease directly or indirectly to plants or plant parts or any processed, manufactured, or other products of plants (Bakst, 1991).

If a company imports a plant pest or transports a plant pest across state lines for bioremediation purposes, a permit must be obtained from USDA.

Animal and Plant Health Inspection Service (APHIS) issued regulations in 1987 that require a permit be obtained for certain genetically engineered microorganisms that are also plant pests and which are either imported, transported across state lines, or released into the environment. If a microorganism is the product of genetic engineering and is included in USDA's list of genera and taxa for organisms classified as plant pests and is either imported, moved across state lines, or released into the environment, a permit must be obtained (Bakst, 1991).

A company transporting a microorganism across state lines for bioremediation purposes, such as treatability work, could be subject to USDA requirements if the microbe is considered a plant pest.

Environment Canada: Proposed Canadian Biotechnology Regulations

In September of 1990, Environment Canada issued Proposed Notification Regulations for Biotechnology Products Under the Canadian Environmental Protection Act (CEPA). The Canadian Environmental Protection Act (CEPA), passed in June of 1988, contains provisions for the assessment of environmental and health effects of substances new to Canada. Environment Canada and Health and Welfare Canada have developed notification requirements designed to assess the potential environmental and health effects of the manufacture and use of biotechnology products. These regulations are slated to be promulgated under the new substances provisions of CEPA (Environment Canada, 1990).

The regulations will specify information that must be supplied prior to the manufacture or importation into Canada of new biotechnology products. The proposed rules define a series of notifications based on the stage of development of the product toward commercialization. The five developmental stages have been defined as R&D for a contained facility; not for R&D in a contained facility; small-scale field trial; commercial use, sale, distribution, or test marketing, or testing by potential commercial customers or users and other than that covered by the previous two stages. Each stage has a prescribed notification and assessment period, as well as a corresponding schedule, which sets out the information to be submitted (Environment Canada, 1990).

CEPA defines a new substance as one not on the domestic substances list (DSL). The DSL that was to be published at the end of last year did not contain any biotechnology products. Consequently, all biotechnology products will be considered "new substances" and notifiable under the proposed new substances notification regulations. Should manufacturers or importers provide satisfactory evidence to prove that specific products were in commerce in Canada between January 1, 1984 and December 31, 1986, consideration will be given to include those on the DSL. The listing mechanism also will be used for certain biotechnology products that may have an established safety record when used for specific purposes (Environment Canada, 1990).

Naturally occurring microorganisms would be subject to this rule. If enacted as it now stands, companies in the United States shipping microbial cultures for decontamination of Canadian waste sites would be subject to reporting requirements. This proposed rule also could affect those who take indigenous samples from a Canadian site to a U.S. laboratory and return microorganisms of choice to the site. The ambiguity associated with the current proposed listing scheme also could mean that a listing for a particular microorganism would be necessary every time that microbe is shipped from the United States to Canada for bioremediation purposes. Public comments are being taken into account and a revised rule written.

References

API, American Petroleum, Institute. 1990. *Applying the Revised Toxicity Characteristic to the Petroleum Industry.* API, Washington, D.C.

Bakst, J. 1991. Impact of Present and Future Regulations on Biotechnology and Toxic Waste Degradation. Society of Industrial Microbiology, 4th Annual Meeting, Orlando, Fla., July 29–August 3.

Blake, J. 1991. Office of Toxic Substances, EPA. Personal communication, August 26.

Cooke, S. M. (ed.). 1990. *The Law of Hazardous Waste: Management, Cleanup, Liability and Litigation.* Matthew Bender and Co., Inc.

DOT, Department of Transportation. 1990. Performance-Oriented Packaging Standards; Changes to Classification, Hazard Communication, Packaging and Handling Requirements Based on UN Standards and Agency Initiative; Final Rule (49 CFR Part 107, et al.). DOT, Washington, December 21.

Environment Canada. 1990. Proposed Notification Regulations for Biotechnology Products Under the Canadian Environmental Protection Act. Commercial Chemicals Branch, September.

EPA. 1986. Draft Guidelines for Permit Applications and Demonstration Test Plans for PCB Disposal by Non-Thermal Alternative Methods. Office of Toxic Substances, Chemical Regulation Branch, Washington, August 21.

EPA. 1989a. Overview of the Land Disposal Restrictions. Presented by the Office of Solid Waste, Washington.

EPA. 1989b. Office of Solid Waste and Emergency Response. Superfund: Getting into the Act—Contracting and Subcontracting Opportunities in the Superfund Program. EPA/540/G-89/003a. EPA, Washington.

EPA. 1990a. Environmental Fact Sheet: Final Rule for Third Scheduled Wastes Completes Statutory Requirements for Land Disposal Restrictions. Office of Solid Waste and Emergency Response, Office of Solid Waste (OS-305). EPA/530-SW-90-046. EPA, Washington.

EPA. 1990b. Hazardous Waste Management System; Identification and Listing of Hazardous Waste; Toxicity Characteristics Revisions; Final Rule (40 CFR Part 261 et al.). EPA, Washington, March 29.

EPA. 1990c. Environmental Fact Sheet Toxicity Characteristic Rule Finalized. EPA/530-SW-89-045. Office of Solid Waste, EPA, Washington.

EPA. 1990d. Corrective Action for Solid Waste Management Units at Hazardous Waste Management Facilities; Proposed Rule (40 CFR Parts 264, 265, 270 and 271). EPA, Washington, July 27.

EPA. 1990e. Quality Assurance Project Plan for Characterization Sampling and Treatment Tests Conducted for the Contaminated Soil and Debris (CS&D) Program. Office of Solid Waste, EPA, Washington, April 30.

EPA. 1991a. Headquarters Telephone Directory. EPA, Washington.

EPA. 1991b. Office of Solid Waste. Personal communication with staff.

EPA. 1991c. Land Disposal Restrictions; Potential Treatment Standards for Newly Identified and Listed Wastes and Contaminated Debris. (40 CFR Part 268). EPA, Washington, May 30.

EPA. 1991d. Office of Solid Waste. Personal communication with staff.

EPA. 1991e. Office of Emergency and Remedial Response. Personal communication with staff.

EPA. 1991f. Status Report—Innovative Treatment Technologies. Technology Innovation Office, Office of Solid Waste and Emergency Response, EPA, Washington.

EPA. 1991g. Superfund Innovative Technology Evaluation (SITE) Program. EPA/540/8-91/005. EPA, Washington.

Jonesi, G. 1989. Impact of RCRA Land Disposal Restrictions on Bioremediation. Enforcement and Compliance Monitoring, EPA. Presented at HMCRI Second National Conference on Biotreatment, Washington, November 27–29.

HMCRI. 1991. Hazardous Materials Control Research Institute. *Focus,* Vol. 7, No. 7, August.

Martin, J. 1991. SITE Program, Office of Research and Development. EPA, personal communication, August 27.

OTA, Office of Technology Assessment. 1991. Bioremediation for Marine Oil Spills. Congress of the United States, Washington.

OSTP, Office of Science and Technology Policy. 1984. Proposal for a Coordinated Framework for Regulation of Biotechnology, Notice. *Federal Register,* Vol. 49, No. 252, December 31.

OSTP, Office of Science and Technology Policy. 1986. Coordinated Framework for Regulation of Biotechnology; Announcement of Policy and Notice for Public Comment. *Federal Register,* Vol. 51, No. 123, June 26.

Wick, C. B. 1990. Governmental agencies plan to restrict bioproduct shipment. *Genetic Engineering News,* Vol. 10, No. 6, June.

Zeph, L. 1991. Office of Pesticides and Toxic Substances. Personal communication, August 27.

Index

A (*see* Cross-sectional area)
Abiotic transformations, 45, 204, 313
Above-ground reactor, 97, 314
Above-ground treatment, 4, 262
Acclimation (*see* Adaptation)
Acenaphthene, 41, 43, 268, 306
Acenaphthylene, 41
Acetate, 45, 227, 289, 318, 319, 321, 325
Acetogen(s), 299, 321
Acetone, 5, 270, 289
Acidity, 214, 244
Activated carbon, 232, 268, 293
Activated sludge, 2, 111, 112, 113, 136, 142, 270, 293, 325
Adaptation, 2, 20, 113, 130, 135, 141, 143, 226, 236, 238, 285, 312, 314, 316
ADMTMs (*see* Advective-dispersive microbial transport models)
Adsorption (*see* Sorption)
Advection, 71, 72, 73, 82, 207, 242, 243
Advective-dispersive microbial transport models (ADMTMs), 81, 82
Aeration, 184, 185, 218, 222, 230–232, 235–238, 243, 244, 275, 276, 284, 290, 291, 317
Aerators, 284, 291
Aerobic, 3, 5, 16, 21, 22–25, 36, 37, 38, 41, 43, 45, 132, 138, 143, 211, 221, 245, 293, 297, 299, 300, 301, 303, 325, 327, 328
Aerobic biodegradation, 33, 36, 37, 38, 41, 43, 45, 47, 48, 50, 183, 184, 237, 317, 318, 323
Aerobic microorganisms, 41, 42, 183, 221, 222, 229, 231, 235, 283, 284, 293, 319, 327
Aerobic-anaerobic, 293
Aerobic-anaerobic cycling, 284, 290
Air, 1, 3, 6, 33, 68, 175, 184, 186, 212, 222, 230, 240, 241, 272, 273, 352
Air emissions, 185, 206, 218, 222, 231, 237, 244, 266, 275, 291, 292
Air injectors, 291
Air sparging, 222

Air stripping, 188, 236, 268, 269, 270, 273, 275, 276
Airflows, 240–242
Alcaligenes, 324
  denitrificans, 303
Alcohols, 304
Alfalfa, 246
Alicyclic hydrocarbons, 28
Aliphatic hydrocarbons, 26–27, 30–32, 37, 229, 269, 307, 308, 317
  (*See also* Specific compounds)
Alkanes (*see* Aliphatic hydrocarbons)
Alkenes (*see* Aliphatic hydrocarbons)
Alkylbenzenes, 26
  (*See also* Aromatic hydrocarbons)
Alkylkbenzene sulfonate, 209
Alkyl halide, 298
Alkylphenols, 323
Alkylpyridines, 323
Alkynes (*see* Aliphatic hydrocarbons)
Alternative Treatment Technology Information Center (ATTIC), 6, 135, 179
Ames test, 158
Aminoaromatic compounds, 301
Ammonia, 13, 17, 48, 100, 127–128, 218–221, 243, 273, 325
Ammonium, 181, 219
Ammonium nitrate, 294
Anaerobic, 5, 16, 17, 21, 43, 48, 138, 143, 144, 160, 181, 275, 297, 299, 300, 301, 303, 308, 316, 317, 320, 323, 324, 325, 326, 327, 328
Anaerobic biodegradation, 32, 37, 38, 43, 45, 50
Anaerobic bioreactors, 321
Anaerobic digestor(s), 323
Anaerobic fixed-film reactors, 326
Anaerobic microorganisms, 284, 289, 292, 298, 313, 325, 327
Anaerobic transformation, 320
Anaerobic-aerobic cycling (*see* Aerobic-anaerobic cycling)

ABOUT THE EDITORS

KATHERINE H. BAKER is president of Environmental Microbiology Associates, Inc., an environmental consulting firm in Harrisburg, Pa. She is a visiting scholar at the University of Delaware, where she teaches a course in Bioremediation. She was formerly Assistant Professor of Biology at Millersville University of Pennsylvania.

Dr. Baker received her Ph.D. in microbial ecology from the University of Delaware, and was a postdoctoral fellow at the University of Virginia. She is a frequent contributor to major scientific and engineering journals.

DIANE S. HERSON is vice president of Environmental Microbiology Associates, Inc. and Associate Professor in the School of Life and Health Sciences, University of Delaware, where she teaches courses in Microbiology, Microbial Physiology, Microbial Ecology, and Bioremediation. She received her Ph.D. from Rutgers Univesity, and she is a frequent contributor to major scientific journals.